Stimulus–Response Compatibility Principles

Data, Theory, and Application

Stimulus–Response Compatibility Principles

Data, Theory, and Application

Robert W. Proctor
Kim-Phuong L. Vu

Taylor & Francis
Taylor & Francis Group
Boca Raton   London   New York

CRC is an imprint of the Taylor & Francis Group,
an informa business

Published in 2006 by
CRC Press
Taylor & Francis Group
6000 Broken Sound Parkway NW, Suite 300
Boca Raton, FL 33487-2742

© 2006 by Robert W. Proctor and Kim-Phuong L. Vu
CRC Press is an imprint of Taylor & Francis Group

No claim to original U.S. Government works
Printed in the United States of America on acid-free paper
10 9 8 7 6 5 4 3 2 1

International Standard Book Number-10: 0-415-31536-0 (Hardcover)
International Standard Book Number-13: 978-0-415-31536-4 (Hardcover)

Library of Congress Cataloging-in-Publication Data

Catalog record is available from the Library of Congress

Taylor & Francis Group
is the Academic Division of Informa plc.

Visit the Taylor & Francis Web site at
http://www.taylorandfrancis.com

and the CRC Press Web site at
http://www.crcpress.com

Contents

Preface

The link between perception and action and the role played by attention and intention have been of concern to researchers in human performance since the 1800s. Foremost among the phenomena involved in this link are stimulus–response compatibility effects, which are differences in performance that occur as a function of the relations between stimulus and response sets, and of the mappings of the individual set members to one another. Compatibility effects have a long history of both basic and applied interest within psychology and human factors/ergonomics. This interest is because they provide insight into the basic processes involved in selection and control of action and into the applied problems that arise from those processes.

F. C. Donders took note of compatibility effects in his seminal use of reaction time to study information-processing stages in the 1860s. Nearly a century later, Paul M. Fitts formalized the investigation of compatibility effects. In the 53 years since publication of Fitts's first article on the topic (Fitts & Seeger, 1953), compatibility effects have been a continual topic of research. The 1990s saw an upsurge of research on the topic that has continued to the present. This research has yielded not only a large and varied set of empirical results but also significant theoretical advances. The purpose of this book is to organize and summarize the current knowledge regarding stimulus–response compatibility effects broadly defined, and to describe implications of this knowledge for the design of human–machine interfaces.

Two books on the topic of stimulus–response compatibility have been published to date: *Stimulus–Response Compatibility: An Integrated Perspective* (1990), edited by R. W. Proctor and T. G. Reeve, and *Theoretical Issues in Stimulus–Response Compatibility* (1997), edited by B. Hommel and W. Prinz. Both of the books are edited volumes that include chapters by authors who describe their unique lines of research and theoretical perspectives in considerable detail. The books provide a wealth of information concerning compatibility effects and have had considerable influence on research. However, neither book provides the systematic and organized presentation of research topics that an authored book like the present one does nor a detailed treatment of the applied implications and aspects of compatibility effects. Although, the Proctor and Reeve book contains a section of four chapters on applied research, those chapters are not linked closely to the chapters on more basic research, and the Hommel and Prinz book has a basic theoretical orientation.

Many of the chapters in our book place an emphasis on basic research devoted to theoretical issues concerning stimulus–response compatibility in particular and the relation between perception and action more generally. Knowledge of this research is important for researchers interested in a range of theoretical issues in a variety of areas of psychology. We also think that it is valuable for the human factors practitioner to have such knowledge. Understanding of the factors that influence compatibility and determine when and how compatibility effects will arise is a necessary foundation for appropriately applying compatibility principles in design and for evaluating the relative compatibility of alternative designs. We hope that the thorough, up-to-date coverage of basic and applied research on compatibility effects provided in our book, coupled with the emphasis on relating fundamental theory to applications, will be of interest and value to researchers, practitioners, and students in cognitive science and human factors/ergonomics.

We thank Tony Moore of Taylor & Francis for encouraging us to write this book. We extend special thanks to Addie Johnson, professor of human performance and ergonomics at the University of Groningen, the Netherlands, for reading and commenting on drafts of each chapter.

The Authors

Robert Proctor is a professor of psychology at Purdue University. He is a fellow of the Association for Psychological Science and American Psychological Society, and an honorary fellow of the Human Factors and Ergonomics Society. Dr. Proctor has been conducting research on stimulus–response compatibility for more than 20 years. He co-edited the first book on compatibility effects, *Stimulus–Response Compatibility: An Integrated Perspective*, in 1990.

Kim Vu is an assistant professor in human factors at California State University Long Beach. While in graduate school, she was a recipient of the U.S. Department of Education Jacob K. Javits fellowship and was awarded the designation of Student Member with Honors by the Human Factors and Ergonomics Society. Dr. Vu has research interests in human performance, human factors, and human–computer interaction. She is co-editor of the *Handbook of Human Factors in Web Design*.

The authors have been working together on basic and applied research on stimulus–response compatibility for the past seven years. One of their goals has been to relate knowledge gained from basic research to applied problems.

1 Stimulus–Response Compatibility and Selection of Action: Basic Concepts

By a great idea, I mean a simple concept of great reach, an acorn of an idea that ramifies into a great oak tree of application, a spider of an idea that can spin a great web and draw in a feast of explanation and elucidation.

Peter Atkins, 2003

1.1 INTRODUCTION

In his book, *Galileo's Finger: The Ten Great Ideas of Science,* Atkins (2003) emphasizes that great ideas in science are simple concepts with far-reaching consequences. Within the study of human performance and, more generally, psychology, the concept of stimulus–response (S-R) compatibility introduced by Paul Fitts and colleagues a little over 50 years ago qualifies as a great idea. It has served to define a large body of basic and applied research of a variety of types since its introduction, with the concept taking central stage in recent years as researchers have focused increasingly on understanding perception–action relations. Our goal in this book is to convey the broad range of theoretical and applied issues that have been investigated from the compatibility perspective and to acquaint readers with the progress that has been made and many of the issues that remain to be resolved.

The concept of S-R compatibility can be understood by considering relations between displays and controls, some of which are much more natural and intuitive than others. When the relation between displays and controls, or stimuli and responses, is direct and natural, it is described as being compatible. In contrast, when the relation is indirect and unnatural, it is described as incompatible. For example, a stovetop typically has four burners, arranged in a 2×2 layout. If the controls are also arranged in a 2×2 pattern, it is obvious which control operates the corresponding burner, and errors of selecting the wrong control to turn will seldom occur. However, if the controls are arranged linearly, as is often the case, there is no obvious spatial relation between the controls and burners (Chapanis & Lindenbaum, 1959). Consequently, the wrong control is more likely to be selected, and failure to detect this error may cause food to be burned, cookware to be damaged, or, even worse, a fire to be started. Another example of good and bad interfaces concerns multiple

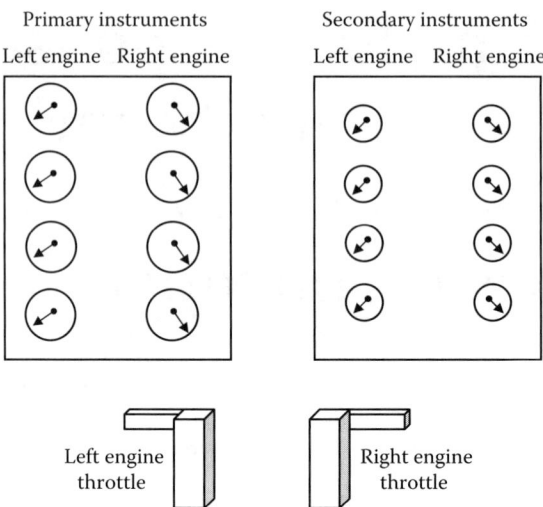

FIGURE 1.1 Schematic depiction of the primary and secondary instrument displays and engine throttles for the left and right engines in the British Midland Airways Boeing 737-400 aircraft that crashed on January 8, 1989.

overhead lights in a room operated by switches on the same panel. When one light is located to the left of the other in the room, the most natural pairing of switches with lights is the left switch to the left light and the right switch to the right light. If the switches that operate the lights have the opposite mapping, as is the case for the dining and cooking areas of the kitchen in the first author's house, you will often unintentionally turn on or off the wrong light.

The burner and light examples involve simple situations that rarely have consequences beyond annoyance, but for more complex interfaces such as those used in modern computers, airplanes, and automobiles, both the complexity of the relations that must be considered for a good interface design and the consequences of poor design increase substantially. As an example, spatial incompatibility apparently was a factor in the crash of a British Midland Airways Boeing 737-400 aircraft on January 8, 1989, in which the functioning right engine was shut down instead of the failing left engine (Learmount & Norris, 1990). The engine displays were arranged in the center of the cockpit instrument panel, grouped into left and right rectangular blocks, as depicted in Figure 1.1. The left block contained the primary instruments for both the left and right engines and the right block contained the secondary instruments for both engines. Within each block, the instruments for the left engine were in the left column and those for the right engine in the right column. The control throttles for the left and right engines were aligned with the primary and secondary instrument panels, respectively. Thus, although the left–right display–control correspondence was maintained relative to one reference frame (the columns within each block), it was not maintained relative to another (the left and right blocks themselves). The cockpit recordings indicate that the pilots were confused about which engine to shut down due to the conflicting location cues, before inadvertently turning off the functioning engine instead of the malfunctioning one.

The speed and accuracy with which we can select what actions to take in response to events occurring in the environment vary as a function of many factors, with spatial correspondence being only one. For example, if a single response is prepared and held in readiness until a signal to execute it occurs, responding will be fast. In contrast, if there are many alternative stimulus events that can occur, each with a unique response assigned to it, responding will be relatively slow, and errors are likely to occur. Factors of the type described in the previous paragraph that affect response time and accuracy are collectively called S-R compatibility effects, after the terminology of Fitts and colleagues. Loosely speaking, S-R compatibility effects are differences in reaction time (RT) and accuracy of responding between situations for which the relation between stimuli and responses is compatible and ones for which it is not. A mapping of stimuli and responses is considered to be highly compatible when the responses that are to be made to displayed information are the most natural ones and incompatible when they are not.

As noted, S-R compatibility effects and related phenomena have been the subject of considerable basic and applied research since the early 1950s. The basic research focuses on explaining the nature of the representations and processes involved in relating perception to action (see, e.g., Hommel & Prinz, 1997; Proctor & Reeve, 1990). In contrast, the applied research is concerned primarily with how to design interfaces for increasingly complex systems in such a manner that the display and control arrangements and mappings are highly compatible (e.g., Andre & Wickens, 1990; Kantowitz, Triggs, & Barnes, 1990). Because of their different concerns, the basic research tends to emphasize tightly controlled experiments in artificial settings, whereas the applied research tends to emphasize less well-controlled experiments in more complex situations that are closer approximations to real-world environments. A complete understanding of compatibility effects requires integrating the results from both the basic and applied research. However, because the volume of research on basic aspects of S-R compatibility is much larger than that on applied aspects, much of our coverage is by necessity devoted to fundamental analyses of compatibility effects and their determinants. One of our goals is to highlight the implications and principles of these analyses that are relevant to applied design issues.

1.2 CLASSIC STUDIES

1.2.1 DONDERS'S EARLY INVESTIGATIONS

Compatibility effects were first noted in the 1860s by the Dutch physiologist Franz C. Donders (1868/1969). Donders's influential work on the use of RT to study the duration of information-processing stages included experiments that employed compatibility manipulations. Donders had subjects respond with the left or right hand to left or right stimuli (an electrical impulse presented to the left or right foot) or to the red or white color of a light stimulus. He discovered that response selection took less time for the former than the latter (66 ms vs. 122 ms). Donders also found that the response–selection time when repeating aloud one of two possible vowel sounds was 56 ms. From these effects, he concluded that choice among two S-R alternatives is faster when the stimuli are paired with their natural responses than when they are not. Donders also briefly described the first instance of spatial mapping effects. He noted that, in comparison to the condition in which the assigned response was on the same side as the stimulus, "when movement of the right hand was required with stimulation of the left side or the other way round, then the time lapse was longer and errors common" (p. 421).

1.2.2   FITTS'S SEMINAL STUDIES

It was nearly a century later before Fitts and colleagues formalized the concept of S-R compatibility in two classic studies, those of Fitts and Seeger (1953) and Fitts and Deininger (1954). We describe those studies in detail because they established terminology, major findings, and theoretical views concerning compatibility effects that provided the basis for much of the contemporary research on compatibility effects. Fitts and Seeger's article used the term "S-R compatibility" in the title, and they are given credit for coining the term, although in a footnote they acknowledge that A. M. Small first used the term "compatibility" in a paper presented at the 1951 meeting of the Ergonomics Research Society (see Small, 1990).

Fitts and Seeger's (1953) study investigated performance of an eight-choice–reaction task in which one or two styluses were moved from a home key along a pathway to a response location. In their Experiment 1, three stimulus panels and three response panels were used, and the response panels were assigned to the stimulus panels in all combinations (see Figure 1.2). For stimulus panel A, eight stimulus lights were arranged in a circle, and for response panel A, the eight response locations were configured in a similar manner. For stimulus panel B, only four stimulus lights were used (located at 0°, 90°, 180°, and 270°), and the eight stimuli consisted of the four individual lights and the pairs of adjacent lights. For example, the upper and right lights together signaled "upper right." Response panel B was arranged so that the initial movement was straight along the horizontal or vertical axis, with a subsequent movement needed if the response location was at a corner. For stimulus panel C, there were two subpanels, with left–right locations on one and top–bottom locations on the other. Again, the eight stimuli were the single lights and appropriate pairs. In this case, the upper light of the vertical pair and the right light of the horizontal pair designated "upper right." Response panel C likewise consisted of two subpanels for which the left hand moved to designate left–right location and the right hand to designate top–bottom location. For all combinations of stimulus and response panels, the most natural mapping of the stimulus and response elements was used. For example, for stimulus panel A, the correct response to the upper right light was the upper right response position for response panels A and B and the upper right combination for response panel C.

The major finding from Fitts and Seeger's (1953) Experiment 1 was that responses were faster and more accurate when the configurations of the stimulus and response panels corresponded than when they did not. This can be seen in Table 1.1, which shows that RT was 138 ms shorter and error rate 7.7% lower when the stimulus and response sets corresponded (the cells along the diagonal) than when they did not (the remaining cells). In their Experiment 2, subjects responded with

Stimulus panels   Response panels

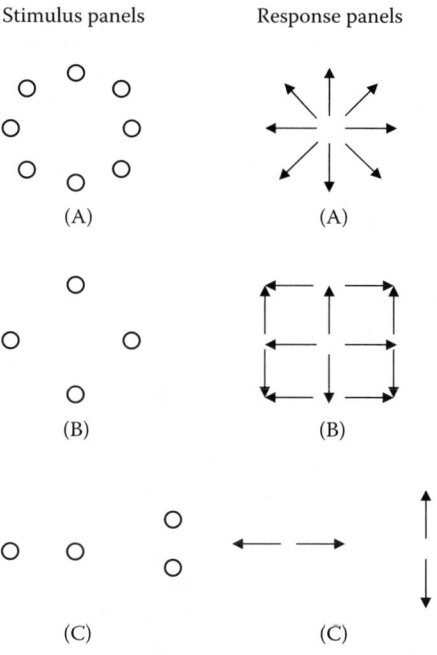

FIGURE 1.2 Illustration of the three stimulus panels and three response panels used by Fitts and Seeger (1953).

TABLE 1.1
Reaction Time (in ms) and Percentage Error (in Parentheses) from Fitts and Seeger's (1953) Experiment 1 for the Stimulus and Response Panels Shown in Figure 1.2

Response Panel	Stimulus Panel		
	A	**B**	**C**
A	390 (4.4)	430 (7.5)	580 (11.6)
B	450 (6.6)	410 (3.4)	580 (17.8)
C	770 (16.3)	580 (18.8)	480 (8.4)

response panel A to the three different stimulus panels for 32 sessions spread over 2½ months. In sessions 1 through 26, the task was performed alone with all three stimulus panels in each session, and in session 27, a concurrent mental arithmetic task was introduced. Mean RT decreased across the initial 26 sessions for all three stimulus panels, with stimulus panel C showing the largest benefit of practice over the first 10 sessions. However, even in sessions 17 through 26, significant differences existed between the three display panels (RT = 272, 286, and 355 ms for panels A, B, and C, respectively, with error rates of 6.9%, 10.0%, and 11.9%). The addition of the secondary task in session 27 increased RT for all stimulus panels and reduced, but did not eliminate, the compatibility effect. Thus, Fitts and Seeger demonstrated not only the importance of S-R compatibility, but also that "some S-R compatibility effects are relatively unaffected by extended practice" (p. 208).

In Fitts and Seeger's (1953) study, the most natural mappings of individual stimuli and responses were used for all pairings of display and response panels, and compatibility involved the degree of correspondence between the configurations of the two panels. Fitts and Deininger (1954) demonstrated compatibility effects as a function of the mappings of individual S-R pairs within the same S-R sets. In their Experiment 1, subjects responded by moving a stylus to one of eight locations of a circular response configuration, response panel A of Fitts and Seeger's study. Different groups of subjects responded to stimuli from one of three stimulus displays with one of three mappings (see Figure 1.3). The first display was a circular stimulus configuration in which the eight stimulus locations corresponded spatially with the eight response locations (display panel A used in Fitts and Seeger's study). The second was a symbolic, circular display, in which three to four digits of a digital clock display, representing eight equidistant points along the face of a clock, were mapped to the eight response locations. The third was a spatial, linear display, in which the eight stimuli were arranged in a row and mapped to the eight response locations. With the maximum mapping (i.e., the one that was maximally compatible), each stimulus was mapped to a corresponding or ordered response location. With the mirrored mapping, each stimulus was mapped to the mirror opposite response location compared to the maximum mapping. With the random mapping, there was no systematic relation between the stimuli and their assigned responses.

As in Fitts and Seeger's (1953) study, performance with the maximum mapping was best when the display panel corresponded spatially with the response panel (see Table 1.2). Furthermore, within each S-R set, performance was best with the maximal mapping, intermediate with the mirrored mapping, and worst with the random mapping. The difficulty with the random mapping (relative to the maximal mapping) was largest for the circular display and least for the symbolic display. These results indicate that the cost of an incompatible mapping is greatest when the stimulus and response sets also correspond. In addition, they illustrate that performance can be quite good with a mapping that provides a systematic relation between stimuli and responses that do not correspond (i.e., the mirrored mappings). Fitts and Deininger (1954) concluded that their results

> support the assertion that the degree of S-R compatibility characterizing a perceptual-motor task depends not so much upon the particular set of stimuli nor upon the particular set of responses involved in the task as upon (a) the selection of congruent stimulus and response sets, and (b) the generation of congruent pairings of these stimulus and response elements in the formation of an S-R ensemble. (p. 490)

1.2.3 SIMON'S EFFECT OF IRRELEVANT LOCATION INFORMATION

J. R. Simon conducted a series of studies in the late 1960s investigating compatibility effects that occur when stimulus location is irrelevant to tasks but the responses are defined along a spatial dimension (see Simon, 1990, for a review). For example,

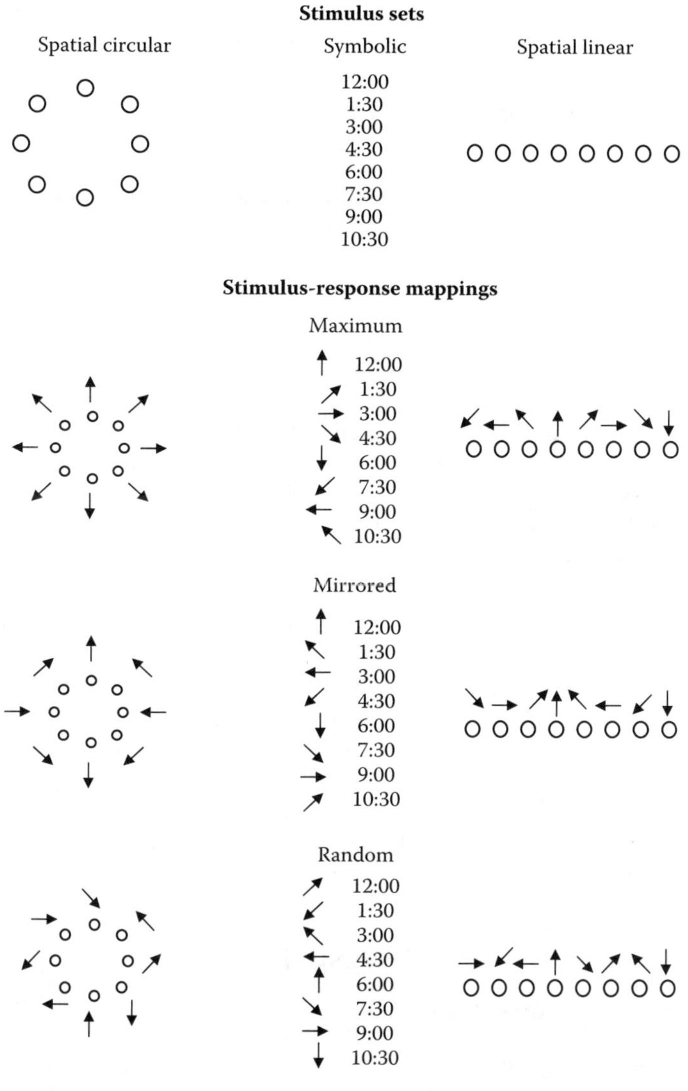

Stimulus sets

Spatial circular Symbolic Spatial linear

12:00
1:30
3:00
4:30
6:00
7:30
9:00
10:30

Stimulus-response mappings

Maximum

12:00
1:30
3:00
4:30
6:00
7:30
9:00
10:30

Mirrored

12:00
1:30
3:00
4:30
6:00
7:30
9:00
10:30

Random

12:00
1:30
3:00
4:30
6:00
7:30
9:00
10:30

FIGURE 1.3 Illustration of the three stimulus sets and three S-R mappings used by Fitts and Deininger (1954). The arrows indicate the direction in which the stylus was to be moved in response to the onset of the stimulus at that location.

Simon and Rudell (1967) had subjects respond to the spoken words *left* and *right* with a left or right keypress, respectively. The words were presented over headphones to the left or right ear, but location was irrelevant to the task. The left response was 39 ms faster when the word *left* occurred in the left ear than when it occurred in the right ear, and the right response was 44 ms faster when the word *right* occurred in the right ear. Thus, even though stimulus location was irrelevant, responses averaged 42 ms faster when the stimulus and response locations corresponded than

TABLE 1.2
Reaction Time (in ms) and Percentage Error from
Fitts and Deininger's (1954) Experiment 1 for the
Stimulus Sets and S-R Mappings Shown in Figure 1.3

	Stimulus Set					
	Spatial Circular		Symbolic		Spatial Linear	
S-R Mapping	RT	PE	RT	PE	RT	PE
Maximum	387	1.9	675	5.0	793	12.4
Mirrored	541	4.4	777	7.2	838	15.5
Random	1111	15.1	885	10.0	1232	13.8

when they did not. Simon (1968) obtained a similar correspondence effect when the responses were unimanual movements of a control handle to the left or right.

With the procedure used by Simon and Rudell (1967) and Simon (1968), the relevant stimulus dimension (the word *left* or *right*) also referred to location. Simon and Small (1969) showed that this was not a critical factor producing the correspondence effect. In their experiment, the relevant stimulus dimension was the pitch of a tone (high or low) presented to one ear or the other. The high-pitch tone was assigned to the left response key and low-pitch tone to the right response key, or vice versa. Responses were again approximately 50 ms faster when the location of the stimulus corresponded with the location of the response assigned to it. Hedge and Marsh (1975) called the effect of irrelevant location correspondence the "Simon effect," and this is the name by which it has continued to be referred up to the present. The Simon effect and its variants, to which Chapter 4 is devoted, have been the subject of considerable research in the recent past to try to understand why a correspondence effect for stimulus location occurs when it is irrelevant to the task.

1.3 TERMINOLOGY AND DISTINCTIONS

1.3.1 SET- AND ELEMENT-LEVEL COMPATIBILITY

Kornblum, Hasbroucq, and Osman (1990) emphasized the distinction originally made by Fitts and Deininger (1954) between two determinants of S-R compatibility: "(*a*) the selection of congruent stimulus and response sets, and (*b*) the generation of congruent pairings of these stimulus and response elements" (p. 490). They coined the terms *set-level* and *element-level* compatibility, respectively, to refer to these two aspects of compatibility effects. Set-level compatibility manipulations involve different pairings of stimulus and response sets, typically with the most compatible mapping of individual stimuli and responses as in the combinations of display and response arrangements examined by Fitts and Seeger (1953). Another example of a set-level compatibility effect is that the location words *left* and *right* are relatively more compatible with their spoken names than with left and right keypresses, but left and right physical stimulus locations are more compatible with the keypresses than with the vocal *left–right* responses (e.g., Proctor & Wang, 1997a; Wang & Proctor, 1996).

Element-level compatibility manipulations involve differences in performance within the same stimulus and response sets as a function of the mapping of the individual stimuli and responses. The mapping effects studied by Fitts and Deininger (1954) can be categorized as ones of element-level compatibility. As another example, the mapping of the words *left* and *right* to left and right keypresses, respectively, is more compatible than the opposite mapping. Set-level and element-level compatibility are related, in that element-level compatibility effects are typically larger for sets with high set-level compatibility than for sets with low set-level compatibility. Continuing the example, the difference in performance with the compatible and incompatible mappings is larger when location words are assigned to vocal responses than to keypresses and when stimulus locations are assigned to keypresses rather than vocal responses (Proctor & Wang, 1997a).

Kornblum et al. (1990) noted that S-R compatibility effects occur when there is similarity, or what they called dimensional overlap, between the stimulus and response sets. Kornblum (1991, p. 5) stated, "We define dimensional overlap as the degree to which two sets of items are physically or conceptually similar." Left–right physical locations have conceptual similarity with the spoken responses *left* and *right* and both conceptual and physical, or perceptual, similarity with left and right keypress responses. Kornblum and Lee (1995) subsequently expanded the concept to include structural similarity to account for S-R compatibility effects that occur for stimulus and response sets that are related in terms of structure. For example, the ordered set of stimuli A, B, and C has structural similarity to the ordered set of responses "one," "two," and "three."

Kornblum et al. (1990) proposed a task taxonomy based on dimensional overlap, distinguishing four categories of tasks according to the dimensions on which the tasks overlap (see Table 1.3). In the taxonomy, ensemble type 1 includes tasks for which there is no dimensional overlap among relevant and irrelevant stimulus dimensions or the response dimension. For example, the colors red and green have no overlap with left and right keypresses. Such ensembles yield no element-level mapping effects. For ensemble type 2, the tasks have dimensional overlap between the relevant stimulus dimension and the response dimension. This category includes the standard spatial S-R compatibility tasks for which differences in performance are obtained as a function of the S-R mapping. Ensemble type 3 tasks are those for which the irrelevant stimulus dimension overlaps with the response dimension, as in the Simon task. Type 4 ensembles have dimensional overlap between the relevant and irrelevant stimulus dimensions, but no overlap of either stimulus dimension with the response dimension. Tasks in this category include the Eriksen flanker task (Eriksen & Eriksen, 1974), for which a left or right response is made to a target letter that is flanked by instances of an irrelevant potential target letter. Responding in such tasks is slower when the irrelevant stimulus (e.g., the letter A) is assigned to a response other than that to which the target letter (e.g., the letter S) is assigned.

Kornblum and Lee (1995) added four more categories to the taxonomy (ensemble types 5 through 8) to include tasks with dimensional overlap on two or more pairs of dimensions. These additional ensemble types are also summarized in Table 1.3.

TABLE 1.3
Dimensional Overlap Taxonomy for Classifying Tasks into Eight Ensemble Types

| Ensemble Type | Overlapping Dimensions | | | Example | | |
| | S-R Dimensions | | S-S Dimensions | Example Stimulus Sets | | Example Response Sets |
	Relevant	Irrelevant		Relevant	Irrelevant	
1	No	No	No	Colors	Letters	Keypresses
2	Yes	No	No	Locations	Letters	Keypresses
3	No	Yes	No	Colors	Locations	Keypresses
4	No	No	Yes	Colors	Color words	Keypresses
5	Yes	Yes	No	Colors	Locations	Keypresses on colored keys
6	Yes	No	Yes	Locations	Colors and color words	Keypresses
7	No	Yes	Yes	Colors	Color words/ locations	Keypresses
8	Yes	Yes	Yes	Locations	Location words	Keypresses

Note: Based on Kornblum (1994).

Kornblum and colleagues (1990) presented the task taxonomy in the context of their dimensional overlap model, described later, in which the effects of the three different sources of dimensional overlap are attributed to different processes. Moreover, at least initially, the idea was that the different sources producing the compatibility effects in ensembles 2 through 4 would combine additively to yield the effects in ensembles 5 through 8. However, there has been considerable debate about these processing assumptions (see Chapter 11), and the value of the taxonomy for our purposes is primarily as a means for distinguishing the different types of tasks.

1.3.2 RESPONSE SELECTION

It is customary in models of human information processing to distinguish a minimum of three stages (Proctor & Vu, 2003): stimulus identification (perception), response selection (cognition), and motor execution (action). The duration of stimulus-iden-tification processes is presumed to be primarily a function of stimulus properties, and the duration of motor-execution processes primarily a function of response properties. S-R compatibility effects are attributed mainly to response–selection processes, which are those that mediate between perceptual and motor responses, or perception and action. As expected if S-R compatibility effects are mainly due to response selection, studies often show little or no interaction of compatibility manipulations with those that affect perceptual and motoric processes. For example, Frowein and Sanders (1978) had subjects perform a four-choice task in which the preferred hand was placed on a start button and moved to one of four response buttons, arranged at the corners of a rectangle. The stimuli were pairs of horizontal and diagonal lines that converged at (i.e., "pointed to") one of the four corners. With a compatible S-R mapping, the response was to move to the button to which the lines pointed; with an incompatible mapping, the response was to move to the next button in a clockwise direction. Stimulus degradation (undegraded or degraded by superimposing a visual noise pattern of nonsense shapes) and foreperiod duration (1.5-s and 10.5-s intervals between a 500-ms flash of a figure composed of all possible line elements) were also varied between trial blocks. Although RT was slowed by stimulus degradation and a long foreperiod, the effects of these variables did not interact with each other or with the compatibility effect, which was approximately 50 ms.

For reasons described in Chapter 3, response selection is presumed to be based on stimulus and response codes that represent the stimulus information and the potential actions that can be taken in response to them, respectively. There is little dispute that S-R compatibility proper (i.e., effects of relevant S-R mapping) and the Simon effect are attributable primarily to response–selection processes. Most authors also favor a response–selection account of the flanker effect, but Kornblum and colleagues attribute the effect primarily to perceptual processes because the dimensional overlap is between the relevant and irrelevant stimulus dimensions (Zhang & Kornblum, 1998).

The most widely accepted models of compatibility effects at present are dual-process models that distinguish between two response–selection routes, one that is automatic in the sense of producing response activation that is largely independent from intended actions and one that is intentional in the sense of involving translation of the relevant stimulus information to arrive at the response that is correct according to the task instructions. Various terms are used to refer to these processing routes, with the automatic route sometimes called direct or unconditional, and said to rely on activation produced by long-term S-R associations that are innate or overlearned. The Simon effect is typically attributed to automatic activation of the spatially corresponding response. The intentional route is also called indirect, conditional, or response identification and is said to rely on short-term associations defined for the specific task. Some models attribute the effects of S-R compatibility proper entirely to this intentional route, whereas others attribute it in part to this route and in part to the automatic route. Several recent findings described later in the book show that the relation between automatic and intentional processing is much more complex than implied by a simple dual-process model.

1.4 TECHNIQUES FOR STUDYING S-R COMPATIBILITY

1.4.1 BEHAVIORAL MEASURES

As should be apparent from the studies described so far in this chapter, mean RT and error rate are the main measures used in studies of S-R compatibility. Consequently, our emphasis in this book will be primarily on those behavioral measures. Because the two measures are usually positively correlated for compatibility manipulations, we will often report only RT data. Many issues concerning S-R compatibility effects can be evaluated by examining the influences on mean RT of different stimulus and response sets, S-R mappings within sets, and properties of the stimuli, responses, and their configurations.

In recent years, there has been increased use of analyses that provide a more detailed picture of performance on different trial types. One such analysis examines sequential effects, for example, obtaining mean RT for the different conditions as a function of whether certain features of the stimulus or response or both repeat or change from the previous trial (e.g., Hommel, Proctor, & Vu, 2004). The examination of sequential effects can give a more dynamic depiction of how processing of one event affects the processing of a subsequent one. One finding that has attracted considerable interest is that the Simon effect is large immediately following a trial on which stimulus and response locations corresponded but small or absent immediately following a trial on which they did not (e.g., Stürmer, Leuthold, Soetens, Schröter, & Sommer, 2002). Some authors have attributed this finding to gating of the automatic, or direct, response–selection route, with automatic activation of the corresponding response suppressed following a trial on which the incorrect response was activated and released following a trial on which it was correct (e.g., Stürmer et al., 2002; Wühr, 2005; see Chapter 6 for discussion of this issue).

Another type of analysis is to examine RT distributions (e.g., De Jong, Liang, & Lauber, 1994). For a distribution analysis, the RTs for each condition are ranked from fastest to slowest and partitioned into bins, usually 5 or 10, with an equal percentage of responses in each bin. Mean RT for each bin is computed, and the S-R compatibility effect of interest is calculated for each bin. RT bin analyses have been used most often in studies of the Simon effect, with the Simon effect frequently being largest at the shortest RT bin and decreasing at the longer bins. For example, De Jong et al. found a Simon effect of slightly more than 30 ms at the shortest of five RT bins when the task was to press a blue or red key in response to a stimulus of the same color, and this effect decreased monotonically to 0 ms at the longest RT bin. De Jong et al. and others have interpreted this result pattern as suggesting that the activation of the spatially corresponding response diminishes over time. However, alternative interpretations of the distribution functions can be provided that do not attribute the changes to temporal properties of activation (e.g., Zhang & Kornblum, 1997).

1.4.2 Electrophysiological and Brain Imaging Techniques

Over the past 25 years, considerable technological advances have been made that allow measurement of brain activity as tasks are being performed. These measures help determine which neural mechanisms underlie particular aspects of behavior and provide evidence relevant to some issues that are difficult to resolve on the basis of behavioral measures alone. For example, examining whether there is activation of areas of the brain that are associated with motor responses can provide evidence regarding whether response activation occurs in certain conditions.

For purposes of evaluating models of choice–reaction tasks, it is often useful to record event-related potentials (ERPs) from various sites by placing electrodes on the skull. ERPs are changes in brain activity that are elicited by an event (e.g., stimulus presentation or response initiation) and are usually obtained by averaging across many trials of a task to remove baseline brain activity. Different components of the ERP are linked to different aspects of processing, thus providing information about what processes are affected by particular variables. The components are distinguished by whether their peaks are positive (P) or negative (N) and the order in which they occur after the event. Components that occur soon after stimulus onset (e.g., N1) are associated with early perceptual processes. Later components (e.g., P3) reflect cognitive processes such as attention and response selection. By comparing the effects of task manipulations on the magnitudes of various ERP components, their onset latencies, and their scalp distributions, relatively detailed inferences about cognitive processes can be made.

An ERP component that has been used widely in research on S-R compatibility is the lateralized readiness potential (LRP; Eimer, 1998). The LRP is a measure of differential activation of the lateral motor areas of the cortex that occurs shortly before and during execution of a response by the left or right hand. The asymmetric activation favors the motor area contralateral to the hand making the response, because this is the area that controls the hand. Of importance, the LRP has been obtained in situations in which no overt response is ever executed, allowing it to be used as an index of covert, partial response activation. The LRP is thus a measure of the difference in activity from the two hemispheres of the brain that can be used as an indicator of covert response tendencies, to determine whether a response has been prepared even when it is not actually executed.

Masaki, Wild-Wall, Sangals, and Sommer (2004) reported an experiment that provides an illustration of the use of ERPs in evaluating S-R compatibility effects. Subjects performed a two-choice task, for which the stimuli were the letters R (for right) and L (for left), presented at the center of a computer monitor, and the responses were isometric extensions (lifts) of the left or right middle finger. S-R compatibility (compatible and incompatible S-R mapping) was manipulated between trial blocks to selectively influence response selection, as was movement velocity (short versus long time to peak force) to selectively influence motor programming. RT was approximately 100 ms longer with the incompatible S-R mapping than with the compatible mapping, and the movement velocity manipulation had no significant effect on RT. The most important finding was that S-R compatibility selectively influenced the interval between stimulus and LRP onsets, whereas movement velocity selectively influenced the interval between LRP onset and response. The authors interpreted this pattern of results as indicating that response selection is completed prior to onset of the LRP, with the onset coinciding with the beginning of motor programming. Results from several other studies (e.g., Smulders, Kok, Kenemans, & Bashore, 1995) are generally in agreement with this conclusion.

Brain imaging allows examination of activity in different parts of the brain while a task is being performed. The two most common techniques are event-related positron emission tomography (PET) and functional magnetic resonance imaging (fMRI), which measure brain activity by monitoring blood flow. For both techniques, subjects perform a task (e.g., a choice–reaction task) while in a scanner, and an image of brain activity is recorded. The brain activity during this task is subtracted from the activity present during baseline testing (i.e., without the specific cognitive aspect of interest) in order to isolate the activity that is unique to the process of interest. In this way, cerebral activity associated with specific task components can be isolated.

A study by Schumacher and D'Esposito (2002) illustrates use of fMRI to localize effects of S-R compatibility on brain activation. They had subjects perform a four-choice task in which the stimuli were four circles arranged in a row and the responses were four buttons, also arranged in a row, pressed with the index and middle fingers of each hand. On each trial, outlines of all four circles appeared, following which, after a brief delay, they were filled in. Three of the filled circles were dark gray distractors and the fourth was lighter in appearance than the rest, to designate it as the target stimulus. In different blocks of eight or nine trials, the target stimulus was white (high contrast) or light gray (low contrast), and the stimuli were mapped to responses at the corresponding response locations (compatible mapping) or in an unordered manner (incompatible mapping).

Each subject performed all trial blocks during each of 12 fMRI runs, and brain activation in several regions of interest, identified from prior studies, was examined. Stimulus contrast selectively influenced activation in the extrastriate cortex (a region of the occipital lobe involved in vision that is located next to the primary visual cortex), with low-contrast stimuli producing more activation than high-contrast ones. S-R compatibility selectively influenced the dorsal prefrontal cortex (an area of the frontal lobes that has been implicated in planning) and superior parietal cortex (the upper part of the parietal lobe involved in movement planning). In addition, both stimulus contrast and S-R compatibility had effects on activation in anterior cingulate and lateral premotor cortices, which may reflect increased response activation under more difficult conditions. This study demonstrates that it is possible to identify specific regions of the brain that are associated with response selection and other processes.

In summary, ERPs can provide precise temporal data about the time course of information processing, but they generally do not supply detailed localization information about which brain areas are involved in particular activities. In contrast, the imaging techniques allow localization of brain functions, but they yield lower temporal resolution (this may change in the future as the imaging technology improves). Thus, both techniques can provide distinct, useful information regarding the neural correlates of response selection and other processes (e.g., Eimer & Schlaghecken, 2003).

1.4.3 NEUROPSYCHOLOGICAL PATIENTS

One technique for analyzing physiological mechanisms underlying behavior in non-human species is to lesion particular areas of the brain and examine the effect on the behavior of interest. For ethical reasons, research of this type cannot be conducted on human subjects. Instead, researchers must rely on individuals who have received damage to a part of the brain through natural causes (e.g., a stroke or head injury from an accident) or through a surgical procedure performed for other medical purposes (e.g., callosotomy patients whose corpus callosum has been severed). These studies can provide useful insights, but they are limited in that the damage is often not well localized and typically few patients with a given disorder are available.

As an example, researchers have examined issues regarding dual-task interference and cross-talk between two tasks as a function of their S-R mappings using callosotomy patients (Ivry & Hazeltine, 2000). Ivry, Franz, Kingstone, and Johnston

(1998) had a callosotomy patient and control subjects perform two 2-choice spatial compatibility tasks concurrently, one with the left hand (Task 1) and the other with the right hand (Task 2). The stimulus for Task 1, for which the S-R mapping was always spatially compatible, appeared 50 to 1,000 ms prior to that for Task 2, for which the mapping was compatible in one condition and incompatible in another. For the control subjects, RT for Task 1 was more than 100 ms shorter when the mapping for Task 2 was consistent with its mapping (i.e., also compatible) than when it was not (i.e., an incompatible Task 2 mapping), which is a commonly obtained finding (see Chapter 10). In contrast, for the callosotomy patient, inconsistency of mappings slowed RT on the compatibly mapped Task 1 only slightly (by 20 ms). This outcome and others suggested to the authors that the callosotomy patient was able to maintain the mappings for each task separately, whereas the control subjects were not able to do so.

1.5 MODELS FOR S-R COMPATIBILITY EFFECTS

Numerous attempts have been made to develop models for various aspects of S-R compatibility. In many cases, conceptual models have been used that are stated qualitatively. Predictions about patterns of results can be made from models of this type, and much research has focused on evaluating such predictions. In recent years, there has been a move toward developing quantitative and computational models that also specify magnitudes of effects. In this section, we provide brief descriptions of several models of various types to illustrate how S-R compatibility effects can be modeled and introduce several of the theoretical issues that are discussed in later chapters. We return to these models and others in Chapter 11, giving more detailed consideration to their strengths and weaknesses.

1.5.1 AN EARLY CONCEPTUAL MODEL OF S-R COMPATIBILITY

Deininger and Fitts (1955) developed a conceptual model to explain the results from their 1954 study, described earlier. They stated,

> The fundamental assumption for the model used in developing the S-R compatibility concept is that all perceptual-motor performance involves one or more steps of information transformation, since the response clearly must always be encoded in a manner different from the stimulus. (p. 318)

This model also assumes that "the efficiency of performance ... varies depending on the way in which the information must be transformed or recoded as stimulus information is converted to response information" (Deininger and Fitts, 1955, p. 318). Performance is most efficient when a minimum number of operations are required to make this transformation. Thus, the performance is directly related to the number of transformation operations that are required between perception and the execution of the response.

Fitts, Bahrick, Briggs, and Noble (1959) indicated that transformation efficiency varies "as a function of unlearned and/or highly overlearned behavior patterns, which are also determinants of population stereotypes, including stereotypes that involve the utilization of concepts and preliminary adjustments (sets, presets, or expectancies)" (pp. 9.13–9.14). This quote indicates that Fitts et al. saw a close relationship between the S-R compatibility effects obtained with speeded responses and the degree of stereotypic behavior observed in nonspeeded, free-choice situations. Furthermore, the quote conveys their opinion that preparatory set must be taken into consideration in the development of any theory of response selection and that certain acquired concepts "can be translated almost immediately into appropriate expectancies and utilized in responding quickly to stimuli that occur in a probabilistic situation" (p. 9.21). Fitts et al. also characterized the difference in response–selection efficiency for compatible and incompatible tasks as being analogous to computer programs of different complexity:

> We might think of compatible tasks as being ones for which efficient "programs" for processing information are available and can quickly be put to use, whereas less compatible tasks are ones for which no efficient program is available, and less efficient programs involving excess operations must be employed. Concepts permit the substitution of efficient general rules for less efficient specific ones and place less load on immediate memory or "buffer storage." (pp. 9.21–9.22)

1.5.2 THE ALGORITHMIC MODEL OF S-R COMPATIBILITY

Rosenbloom (1986) developed an algorithmic model of S-R compatibility (see also Rosenbloom & Newell (1987)). This model hypothesizes only a single, intentional response–selection route and attributes S-R compatibility effects to the number of transformational operations that have to be performed in this route, as in the conceptual model proposed by Deininger and Fitts (1955) and Fitts et al. (1959). The algorithmic model of S-R compatibility "is based on the supposition that people perform reaction-time tasks by executing *algorithms* (or programs) developed by them for the task" (Rosenbloom, 1986, p. 156). There are two steps to providing a compatibility metric within the algorithmic framework:

1. A task analysis is performed to determine the algorithms that can be used to perform the task.
2. A complexity analysis for each algorithm is performed by estimating the time for the basic operations and determining the number of operations of each type that would be required to perform on a given trial.

The result of such an algorithmic analysis of different task conditions is a quantitative prediction of the RT effects.

Rosenbloom (1986) developed algorithms for the conditions of several studies of S-R compatibility and showed that the algorithms predicted the RT differences relatively well. As an example, consider their analysis for stimulus set A and response set A of Fitts and Seeger's (1953) study, described earlier, compared with stimulus set B and response set A (see Figure 1.2). For stimulus set A and response set A, both arrangements are circular and the subject is to respond at the location corresponding to the stimulus. The algorithm for this task is

```
Algorithm Fitts & Seeger S_A-R_A:

    BEGIN

    Stimulus ← Get-Stimulus ("On-Light")

    Angle ← Get-Value (Stimulus, "Angle")

    Make-Response ("Push-Lever", Angle)

    END
```

The number of operations to execute this algorithm is three.

For stimulus set B, either one or two adjacent lights are presented. When using response set A, if a single light comes on, the subject is to respond by moving to the corresponding location. If two lights appear, the subject is to respond by moving to the response location that is intermediate to the two stimulus locations. A task algorithm for this condition is

```
Algorithm Fitts & Seeger S_B-R_A:

BEGIN

Stimulus ← Get-Stimulus ("On-Light")

Angle ← Get-Value (Stimulus, "Angle")

IF SUCCEEDED One-Light-On? () THEN

    Make-Response ("Push-Lever", Angle)

ELSE

    IF SUCCEEDED Many-Lights-On? () THEN

    BEGIN

    Stimulus2 ← Get-Second-Stimulus ("On-Light",
       Angle)

    Angle2 ← Get-Value (Stimulus, "Angle")

    Angle ← Average (Angle, Angle2)

    Make-Response ("Push-Lever", Angle)

    END

END
```

For the trials in which more than one light is presented, the selection of the first light is arbitrary, as is the order of executing the **One Light On?** or **Many Lights On?** steps. Consequently, Rosenbloom (1986) averaged the number of operations to execute each of the possible algorithms, which results in a value of 6.

Comparing the number of steps required to select a response when response set A is paired with stimulus set A with that required when it is paired with stimulus set B, it is easy to see that RT is predicted to be shorter with the former stimulus set than with the latter. Rank orderings of the times for all conditions in Fitts and Seeger's (1953) study can be obtained by estimating the number of operations for each of the nine combinations of displays and response sets. Exact values for the differences can be predicted as well by providing time estimates for the different types of operations, as mentioned above. Rosenbloom (1986) implemented the algorithms for the different conditions of Fitts and Seeger's study, as well as those for several other studies, within a production-system architecture and showed that computer simulations provided good fits to the data.

1.5.3 DIMENSIONAL OVERLAP MODEL

Whereas Deininger and Fitts's (1955) and Rosenbloom and Newell's (1987) models emphasized intentional translation between stimuli and responses, most recent models have provided two-factor accounts of S-R compatibility in which an automatic activation component is included in addition to the translational one. Kornblum et al.'s (1990) dimensional overlap model is the best-known model of this type. The original model included only a response-production stage, consisting of the right half of Figure 1.4. Kornblum and Lee (1995) later augmented this model with a stimulus identification stage, shown in the left half of the figure. The response-production stage has two response–selection routes. The first is an automatic activation route through which, when dimensional overlap is present, the response corresponding to the stimulus is identified and programmed, regardless of whether it is the response assigned for the task. The second is an intentional response-identification route by way of which the response assigned to the stimulus for the task is identified. This intentional response identification can proceed by search of different possible responses. When dimensional overlap between the relevant stimulus dimension and the response dimension is present, intentional response identification may be accomplished through use of a rule. The fastest rule is considered to be the identity rule, which is applicable for tasks in which the instructions are to make the response that corresponds to the stimulus. When the intentional route completes identification of the assigned response, a verification process compares the identified and programmed responses. If they are the same, the programmed response is executed. If they are different, the programmed response is aborted and a motor program for the identified response is retrieved and executed.

According to the dimensional overlap model, dimensional overlap affects performance in two ways. First, when overlap is present, the automatically activated corresponding response is correct for the compatible mapping and facilitates responding, but it is incorrect for the incompatible mapping and interferes with responding. Second, dimensional overlap allows the assigned response to be identified through the intentional route more quickly when the mapping is compatible than when it is incompatible because the identity rule (respond at the corresponding location) can be applied.

The stimulus-identification module was added by Kornblum (1994) to account for situations in which relevant and irrelevant stimulus codes overlap, as in the Eriksen flanker task. When the stimulus is bidimensional (labeled "2 Tags" in Figure 1.4), the relevant and irrelevant stimulus codes are compared. If they are the same,

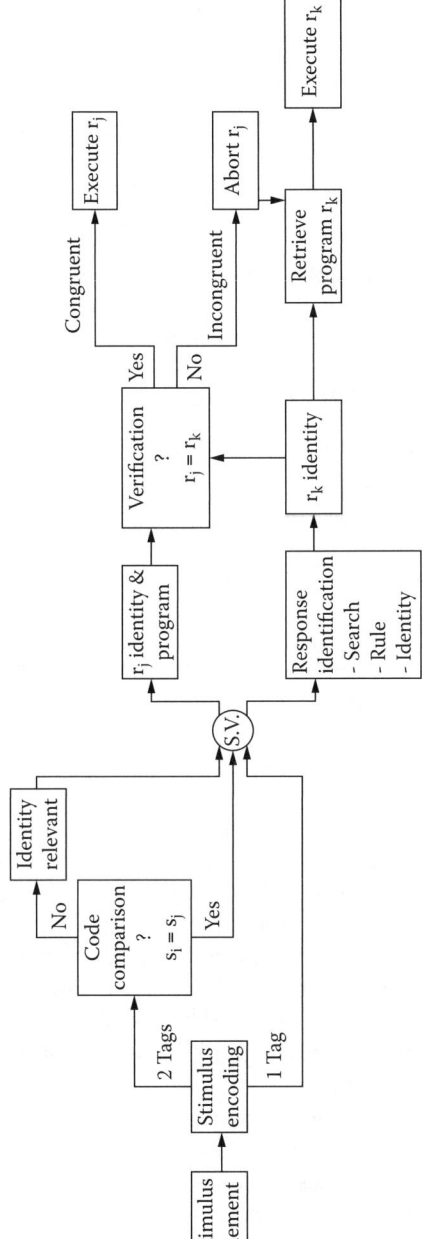

FIGURE 1.4 Illustration of the dimensional overlap model. The left side depicts stimulus-identification processes and the right side shows response-identification and execution processes. S.V. stands for stimulus vector; s_i = a stimulus attribute that overlaps with another stimulus attribute; s_j = a stimulus attribute that overlaps with a response attribute; r_j = an automatically activated response; r_k = the correct response. From "Stimulus–response compatibility with relevant and irrelevant stimulus dimensions that do and do not overlap with the response," by S. Kornblum and J.-W. Lee, 1995, *Journal of Experimental Psychology: Human Perception and Performance, 21,* p. 858. Copyright 1995 by the American Psychological Association. Reprinted with permission.

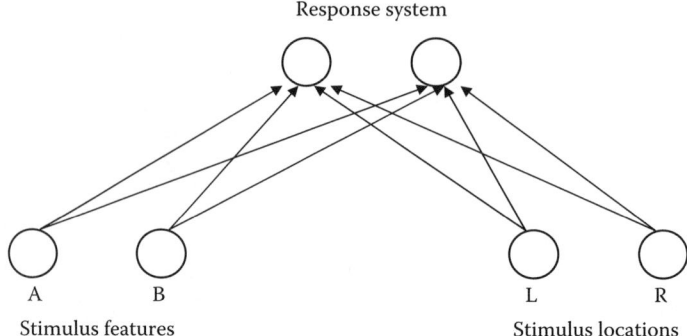

FIGURE 1.5 Illustration of Zorzi and Umiltà's (1995) connectionist model for the Simon effect. The relevant stimulus-feature nodes are designated as A and B, whereas the irrelevant stimulus-location nodes are designated as L (left) and R (right).

then the stimulus information is passed on to the response-production stage. If they are different, then extra processing must be performed to identify which code is relevant before passing this information on to the response-production stage. Thus, according to this model, the Eriksen flanker effect is based entirely in this stimulus-identification stage, in contrast to the S-R compatibility and Simon effects, which arise in the response-production stage.

1.5.4 Connectionist Models

The role of automatic activation and intentional translation can also be discussed in terms of associations between stimuli and responses, and several models of S-R compatibility effects have emphasized contributions of short-term and long-term associations rather than response–selection via dual routes (e.g., Barber & O'Leary, 1997; Stoffer & Umiltà, 1997). According to these models, the experimental instructions relate short-term, task-defined associations of the relevant stimulus dimension to the response, and the long-term, pre-existing associations are those between stimulus locations and response locations that are overlearned. The S-R compatibility effect is attributed to both automatic activation and the directness of the short-term S-R associations, whereas the Simon effect is attributed entirely to automatic activation of the corresponding response via the long-term associations.

Zorzi and Umiltà (1995) developed a connectionist model for the Simon effect that distinguishes between the contributions of short-term and long-term associations (see Figure 1.5). Their model consists of a network of nodes that represent stimulus locations, features, and responses. Each node has an activation level, and the nodes are interconnected by weighted links that can be excitatory or inhibitory. The location nodes are connected to the response nodes by long-term links that relate stimulus locations and their corresponding response locations through prior experience. The feature nodes are connected to the response nodes by short-term memory links that are established based on the task instructions. At stimulus onset, the irrelevant feature and location nodes are activated to the maximal extent, and this activation decreases over time. Activation of the feature nodes is delayed, with the length of the delay being a function of the difficulty of the required stimulus discrimination. Activation propagates from the input nodes to the response nodes, and a response occurs when the activation of a response node exceeds a threshold. Zorzi and Umiltà's model generates the Simon effect through the long-term links between position nodes and responses, and accounts for the fact that the Simon effect tends to decrease across time through the decay of the initial activation of the position node across time after stimulus onset.

Zhang, Zhang, and Kornblum (1999) developed a connectionist model based on the dimensional overlap model to explain a broader range of S-R compatibility effects. Their connectionist model retains the assumption that overlap of relevant stimulus dimensions, irrelevant stimulus dimensions, and responses is crucial for determining the type of correspondence effect likely to occur, but unlike the dimensional overlap model, processing occurs in parallel at multiple layers (input, intermediate, and output), with partial activation from earlier layers sent to later layers. This model maintains most of the distinctions emphasized by the dimensional overlap model, but allows quantitative predictions to be derived, rather than just qualitative predictions of the type made by the original model.

1.5.5 THE THEORY OF EVENT CODING

Hommel, Müsseler, Aschersleben, and Prinz (2001) proposed the theory of event coding as a metatheory of perception and action planning. The essence of the model is based on the notion of common coding. That is, perception and action are represented in the same coding system. The second assumption is that action plans, as well as stimulus representations, are composed of feature codes. The third assumption is that the cognitive codes, those for both perception and action, refer to external events rather than to the physical motor output.

Of central importance to the theory is the concept of an event code. An event code consists of the integrated feature codes that represent an event. These codes receive input from the sensory systems and modulate the activities of motor systems. Figure 1.6 illustrates two abstract feature codes. One implication of this theory is that when a stimulus possessing particular stimulus features is perceived, it will prime those actions that produce those same features. The theory accounts for S-R compatibility effects primarily through such priming of response features by stimulus features. Note that, although it shares many characteristics with Kornblum et al.'s (1990) dimensional overlap model, which is also intended to provide a comprehensive account of response selection, the theory emphasizes distal action effects (e.g., turning on a light) rather than effectors (e.g., moving the finger downward to press a button). The theory also implies that action should affect perception.

1.5.6 SUMMARY

The models described in this section are intended only to illustrate the range of types of models that will be encountered throughout the book. Models that characterize the qualitative aspects of response selection have been, and continue to be, widely used, as illustrated by Deininger and Fitts's (1955) early conceptual model and the recent theoretical frameworks provided by Kornblum et al.'s (1990) dimensional

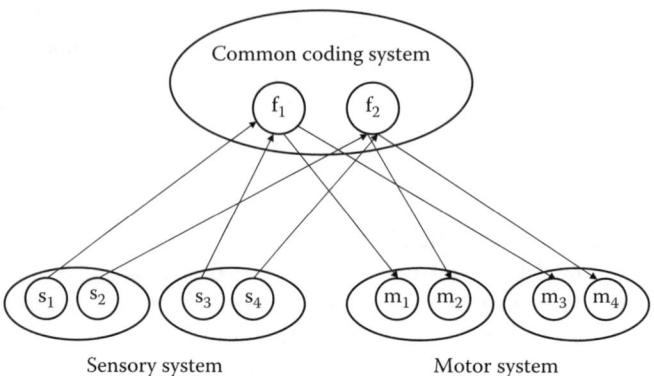

FIGURE 1.6 Illustration of common coding of abstract features in the theory of event coding. Stimulus codes are designated by s, motor codes by m, and feature codes by f. From "The theory of event-coding (TEC): A framework for perception and action planning," by B. Hommel, J. Müsseler, G. Aschersleben, and W. Prinz, 2001, *Behavioral and Brain Sciences, 24,* p. 862. Copyright 2001 by Cambridge University Press. Reprinted with permission.

overlap model and Hommel et al.'s (2001) theory of event coding. Models that make quantitative predictions, as illustrated by Rosenbloom and Newell's (1987) algorithmic model and Zorzi and Umiltà's (1995) connectionist model, are being used increasingly. Some models emphasize effects of S-R compatibility proper (e.g., Deininger & Fitts, 1955; Rosenbloom & Newell, 1987), others focus on the Simon effect (e.g., Zorzi & Umiltà, 1995), and still others attempt to provide broad accounts of response–selection and perception–action relations in general (e.g., Hommel et al., 2001; Kornblum et al., 1990). Models concerned with S-R compatibility proper tend to emphasize the nature of intentional S-R translation, whereas models concerned with the Simon effect tend to emphasize the role of automatic activation. We will return to the models described in this section later in the book and will introduce many others.

1.6 CHAPTER SUMMARY

The concept of S-R compatibility is a "great idea" that has many basic and applied implications for human information processing and performance. Since the first systematic demonstrations of compatibility effects in the 1950s by Fitts and colleagues, they have continually been the subject of numerous investigations. Compatibility effects have been found to influence performance in a variety of situations that range quite far from those studied in the initial investigations. In this chapter, we have introduced several classic studies and basic findings, terminology that is used in the literature, a variety of techniques used to investigate the effects, and types of models that have been developed to explain compatibility effects. This chapter provides the foundation from which we build in subsequent chapters as we explore the full range of basic and applied issues that have been addressed through studies of S-R compatibility effects and closely related phenomena.

2 Factors in Addition to S-R Compatibility that Affect Response– Selection Efficiency

2.1 INTRODUCTION

Fast and accurate actions in response to perceptual events are required in many situations. For example, if a person is driving in the middle lane of a freeway that has three or more lanes in each direction, and a vehicle in the right adjacent lane starts to move into the middle lane, the driver must make a rapid controlling action upon detection of this event in order to avoid a collision. Depending on the context in which this scenario occurs, the required action may be to step on the brake, to accelerate quickly, or to steer the vehicle into the lane immediately to the left. There are many factors that influence the speed and accuracy with which an appropriate action can be taken. Although some of these do not involve response selection (e.g., perceiving the stimulus event), many of them do.

When the stimulus and response environment promotes natural or easy response selection, performance can be dramatically better than when it does not. Fitts and Deininger's (1954) experiment, described in Chapter 1, which used an eight-choice spatial task with different S-R mappings, provides a vivid illustration of this point. In their experiment, RT was approximately 3 times longer and error rate 7 times higher with a random mapping of stimuli to responses (1,111 ms and 15.1%) than with a spatially corresponding mapping (387 ms and 1.9%). Although performance is influenced by a variety of factors, response–selection efficiency must be high if performance with any display–control configuration is to approach its maximum.

S-R compatibility effects make up only one of several categories of phenomena that are known to affect the efficiency of response–selection processes (see, e.g., Sanders, 1998). These other categories include (a) the tradeoff between response speed and accuracy; (b) increases in RT as the number of possible S-R alternatives, that is, uncertainty, increases; (c) decreases in RT when this uncertainty is reduced by precuing subsets of S-R alternatives; (d) performance benefits for trials on which the S-R mapping or pair from the previous trial is repeated; and (e) reductions in RT that occur as a function of practice. In this chapter, we describe each of these phenomena in turn, summarizing the major findings and indicating how the phenomena are influenced by S-R compatibility.

2.2 THE SPEED–ACCURACY TRADEOFF

Most people are aware that, in everyday life, the speed of responding to a situation can be traded off with accuracy. For example, a driver may choose to drive through an obstacle course at a high speed but, as a consequence, may not avoid all the obstacles. That same person could elect to drive more slowly and make sure that no obstacles were struck. In sporting events where points are given for speed and deductions are taken for "mistakes," the speed–accuracy criterion adopted by athletes can determine whether they will win the event. It is also important to realize that the speed–accuracy criterion can be changed during the course of an event. For example, in an equestrian event, the rider may start off riding very quickly through the course. However, if the horse misses a jump or knocks over a barrier, then the rider may slow the horse down in the approach to the next obstacle to ensure that the horse will be able to clear that obstacle.

In agreement with the introspection that speed and accuracy can be traded off, considerable evidence indicates that people can intentionally trade speed for accuracy when performing choice–reaction tasks in the laboratory (Pachella, 1974). Given the same task, but with instructions or payoffs differing in the relative emphasis placed on speed versus accuracy, a range of RTs and error rates can be obtained. One way to conceive of this speed–accuracy tradeoff function is in terms of the criterion amount of information favoring one response or another that is required before a choice is made. When speed of responding is stressed, the criterion is set low, whereas when accuracy is stressed, the criterion is set high. Because the process of analyzing and evaluating the stimulus is noisy, the response will be incorrect more often when the speed–accuracy criterion is low than when it is high.

Given that response selection occurs more rapidly with a compatible S-R mapping than with an incompatible one, the tradeoff of speed for accuracy should cover a smaller range of RT when the mapping is compatible. Christensen, Ivkovich, and Drake (2001) obtained results consistent with this implication. In different trial blocks, subjects pressed a left key with the left index finger or a right key with the right index finger in response to the word LEFT or RIGHT using a compatible or incompatible mapping. They performed under three speed–accuracy instructions: speed emphasis, speed–accuracy emphasis, and accuracy emphasis. Although the tradeoff in accuracy was similar for the two mappings, the tradeoff in RT was 48 ms for the compatible mapping (526 ms with speed emphasis and 574 ms with accuracy emphasis) compared with 107 ms for the incompatible mapping (601 ms with speed emphasis and 708 ms with accuracy emphasis). Incompatibility and accuracy emphasis not only slowed RT but also increased the prominence of a late positive peak of the event-related potential, which Christensen et al. called P4, and slowed its onset, suggesting that P4 latency is an indicator of response selection.

Speed–accuracy relations are captured well by sequential sampling models of response selection, according to which, upon stimulus presentation, information starts accumulating in a later decision stage until a response–selection decision is reached (Ratcliff & Smith, 2004). One category of sequential sampling models, race models, assumes that there is a separate decision unit, or counter, for each response (e.g., Van Zandt, Colonius, & Proctor, 2000). The response that is selected is the one represented by the counter that "wins the race" and reaches its threshold first. Sequential sampling models account for the difference in performance with speed emphasis versus accuracy emphasis by setting the thresholds for all responses lower for speed and higher for accuracy. By varying the relative threshold levels across the different alternatives, biases toward one response or another can be captured as well.

One of the major trends in modeling S-R compatibility effects in recent years has been to characterize them in terms of the dynamic pattern of response activation that is produced during the period between stimulus presentation and response. Compatibility manipulations are typically presumed to affect primarily the rate and quality of this dynamic activation process. However, performance patterns across different conditions are a consequence of both the information activation and the response thresholds. For example, an emphasis on response speed induced by instructions or response deadlines may reduce the size of compatibility effects in RT while increasing their size in error rates (Dutta & Proctor, 1993).

2.3 UNCERTAINTY AND NUMBER OF ALTERNATIVES

Merkel (1885; described by Woodworth, 1938) provided the initial demonstration that RT increases as the number of alternative S-R pairs increases. He had subjects respond, in left to right order, to the Arabic numerals 1 to 5 with fingers on the left hand and the Roman numerals I to V with fingers on the right hand. Across different blocks of trials, the number of possible S-R alternatives varied between 2 and 10. RT increased from about 300 ms when there were only 2 alternatives to more than 600 ms when there were 10. This important finding of Merkel's was not systematically pursued until the now-classic studies of Hick (1952) and Hyman (1953).

2.3.1 THE HICK–HYMAN LAW

Hick (1952) and Hyman (1953) conducted similar studies in which the increase in RT with number of alternatives was explained in terms of information theory (Shannon, 1948). This theory, which was popular in the 1950s and continues to be used for certain purposes nowadays, provides a metric of information in bits, with the number of bits conveyed by an event being a function of uncertainty. The average number of bits for a set of N equally likely stimuli is $\log_2 N$. Because uncertainty also varies as a function of the probabilities with which individual stimuli occur, the average amount of information for a set of stimuli that occur with unequal probability will be less than $\log_2 N$. The more general metric for the average amount of information conveyed by a stimulus for a set of size N is $-\sum_{i=1}^{N} p_i \log_2 p_i$, where p_i is the probability of stimulus i. Across trials, all of the information in the stimulus set is transmitted to the responses if performance is perfectly accurate. However, when errors are made, the amount of transmitted information (H_T) will be less than the average information in the stimulus set.

In Hick's (1952) experiments, the stimuli were 10 lamps arrayed in an irregular circle, and subjects made responses by pressing with 1 of the 10 fingers, each of which was placed on a key. Hick examined his own performance under two

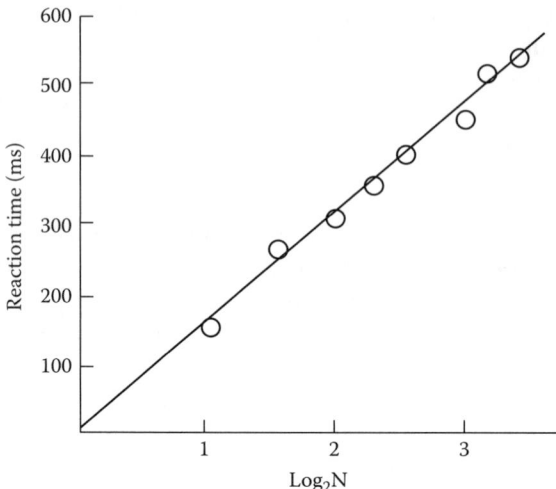

FIGURE 2.1   The Hick–Hyman law: Reaction time increases as a function of the amount of information transmitted ($\log_2 N$ for equally likely alternatives).

speed–accuracy conditions. In one, Hick performed blocks of trials with set sizes ranging from 2 to 10 in ascending and descending order, maintaining a high level of accuracy. In the other, he used only the set size of 10 but adopted various speed–accuracy criteria in different trial blocks. In both cases, RT increased as a linear function of the average amount of information transmitted (see Figure 2.1).

The stimuli used by Hyman (1953) were eight lights corresponding to the eight corners of inner and outer squares, and the response to each stimulus was a one-syllable spoken name, with the names arbitrarily assigned to the stimulus location. In agreement with Merkel's (1885) and Hick's (1952) studies that used keypress responses, RT was a linear function of $\log_2 N$, with the correlation being .983. In two other experiments, Hyman manipulated the probabilities of occurrence of the alternative stimuli and the sequential dependencies between them. In the former case, the correlation between mean RT and information transmitted was .975, and in the latter case it was .936. As predicted by information theory, RT increased as a linear function of the information transmitted in both studies. Thus, Hyman established that RT was not simply a function of the number of S-R alternatives, but of the average amount of information in a condition. This relation between RT and the stimulus information that is transmitted in the responses is known as Hick's law or the Hick–Hyman law:

$$RT = a + bH_T$$

where a is basic processing time and b is the amount that RT increases with increases in the amount of information transmitted (H_T, which is $\log_2 N$ for equally likely S-R pairs with no errors).

The Hick–Hyman function is obtained in a wide variety of situations. As Delaney, Reder, Staszewski, and Ritter (1998) noted, "Psychology's search for quantitative laws that describe human behavior is long-standing, dating back to the 1850s. A few notable successes have been achieved, including ... the Hick–Hyman law (Hick, 1952; Hyman, 1953)" (p. 1). Moreover, the Hick–Hyman law applies to species other than humans. For example, Vickrey and Neuringer (2000) tested pigeons with a circular display/response arrangement similar to that used by Fitts and Seeger (1953; see Figure 1.2, top row, in Chapter 1). When a centered black square was pecked, one of up to eight squares equally spaced in a virtual circle around the centered square turned black, and the pigeon was to peck that square. Response times increased as a linear function of the number of bits of information, as the Hick–Hyman law predicts.

2.3.2 INFLUENCE OF S-R COMPATIBILITY

Although the Hick–Hyman law is highly generalizable, the slope of the function is not a constant value but varies according to several factors (Teichner & Krebs, 1974), one of which is S-R compatibility. The fact that high S-R compatibility increases the efficiency of response selection implies that it can reduce the slope of the Hick–Hyman function. Fitts and Seeger's (1953) and Fitts and Deininger's (1954) findings for eight-choice spatial tasks, described in Chapter 1, are consistent with this implication. In their experiments, RT varied greatly as a function of the display and control arrangements and the element-level mappings, even though the number of S-R alternatives was the same in all cases. As described by Leonard (1955), "In both studies the same amount of information was presented to subjects, but the manner in which it was presented and the kind of responses which had to be made differed" (p. 307). Because the inefficiency of an incompatible S-R mapping relative to a compatible one is greater for large S-R sets of the type used in the studies of Fitts and colleagues than for smaller S-R sets, the slope of the Hick–Hyman function will be much lower for a highly compatible S-R mapping than for an incompatible one.

Fitts (1964) reported this to be the case for visual–spatial stimuli and responses. The slope of the function for a task in which the subject moved a finger from a centered starting position to one of a circular set of N lights was 17 ms/bit. This increased to 40 ms/bit when the display of lights was separate from the response array and the hand and response apparatus were hidden from view. Leonard (1959) used the even more compatible task of having subjects respond by pressing a key with a finger of each hand in response to a tactile vibration to it. Although choice RT was about 40 ms longer than simple RT, there was no increase in RT as the number of S-R alternatives increased. Leonard's results suggest that when the relation between stimuli and responses is maximally compatible, there is essentially no uncertainty about which response to make to an identified stimulus.

Similar results occur for digit naming. Fitts and Switzer (1962) showed that when subjects had to name digits in a choice RT experiment, RT was only slightly shorter when a familiar subset of two digits was used (e.g., the digits 1 and 2) than when there were four or eight alternative digits. Alluisi, Strain, and Thurmond (1964) manipulated both the compatibility of the S-R mapping and set size in a similar digit-naming study. The three set sizes were crossed with three levels of S-R compatibility, with a different group of subjects assigned to each of the nine combinations.

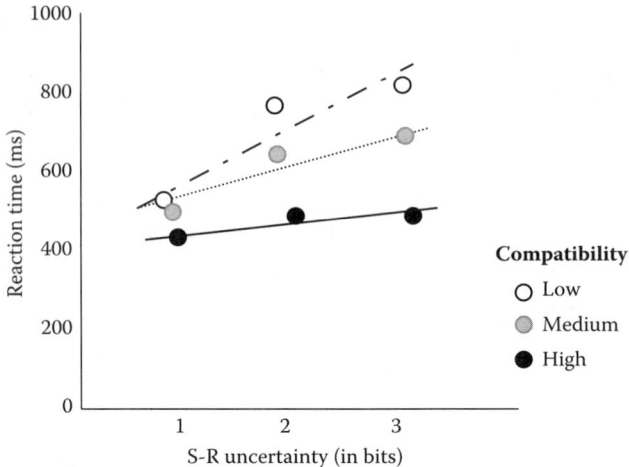

FIGURE 2.2 RT data from Alluisi, Strain, and Thurmond (1964), as a function of S-R compatibility and S-R uncertainty in a digit-naming study.

For set size 2 the digits were 1 and 2, for set size 4 the digits were 1 to 4, and for set size 8 they were 1 to 8. For the high-compatibility mapping, the response to each digit was its name, as in Fitts and Switzer's study; for the intermediate compatibility mapping, the response to each digit was the name of the number plus 2 (e.g., "four" to 2); for the low-compatibility mapping, the response to each digit was a randomly paired numeral name. As shown in Figure 2.2, the slope of the Hick–Hyman function was shallow for the high-compatibility S-R mapping and increased sharply as compatibility decreased.

Other studies have shown relatively small set-size effects for reading words and letters. Pierce and Karlin (1957) had subjects read lists of 60 words each, with the words for different lists selected randomly from vocabularies of different sizes. Although the reading rate was slightly faster for a two-word vocabulary than for the other vocabulary sizes, there was essentially no change in reading rate as the vocabulary size increased from 4 to 256 words. Moreover, Davis, Moray, and Treisman (1961) found that RT to name auditorily presented letters, for which there are 26 possibilities, was only about 20 ms slower than RT to name auditorily presented digits, for which there are 10 possibilities.

2.3.3 INFLUENCE OF PRACTICE

Another important factor influencing the slope of the Hick–Hyman function is practice. Because the benefit of practice is typically greater for more difficult conditions than for easy conditions, the slope of the Hick–Hyman function will decrease with practice. Davis et al. (1961) found that RT to repeat auditory nonsense syllables increased by about 40 ms as set size increased from two to eight alternatives in the first of three sessions, but by the third session, the function was relatively flat. Mowbray and Rhodes (1959) had a subject perform a total of 7,500 trials for two- and four-choice spatial tasks in which index-finger responses or index- and middle-finger responses were made to corresponding light positions. Whereas RT was approximately 20 ms slower for the four-choice task than for the two-choice task in the first three sessions, by the last three sessions there was no difference.

In one of the most dramatic demonstrations of the influence of practice, Seibel (1963) and two assistants performed reaction tasks with 1,023 alternatives (all combinations of 10 lights and 10 keypresses) for more than 75,000 trials each. As would be expected, RT decreased drastically with practice, being much longer than 1 s for the first 3,000 responses but reaching an asymptote of between 350 and 450 ms for each subject by the end of practice. At this time, the subjects were also tested using only the right hand and the 31 possible light combinations that could be made with it. RT for the 1,023-choice task after practice was only slightly slower (20 to 30 ms) than that for the 31-choice task.

2.3.4 An Indicator of Intelligence?

The slope of the Hick–Hyman function is negatively correlated with measures of intelligence. Several researchers have claimed that this correlation reflects an ability of people who score higher on the intelligence tests to process information more rapidly than those who score lower (e.g., Jensen, 1980). In other words, the negative correlation is proposed to arise from a link between the micro-level processes of information processing and macro-level processes of intelligence (e.g., Lindley, Bathhurst, Smith, & Wilson, 1993). Although some researchers have questioned whether the methods used to measure the slope of the Hick–Hyman function are appropriate for establishing a relation between it and intelligence (e.g., Smith, 1989), the correlation seems to be a real phenomenon. Researchers have found correlations as high as –.30 to –.60 for information ranging from 0 to 6 bits (e.g., Beh, Roberts, & Prichard-Levy, 1994), with the correlations being of similar magnitude when responding with the dominant or nondominant hand (Gignac & Vernon, 2004).

However, the fact that the slope of the Hick–Hyman function is highly dependent on amount of practice and S-R compatibility limits any conclusions that can be drawn from the negative correlation with intelligence tests. This point is illustrated by Vickrey and Neuringer's (2000) previously mentioned study that examined the Hick–Hyman function with pigeons. Human subjects performed the same task on the same apparatus as the pigeons, both with unimanual finger movements and "pecks" made with a plastic animal beak placed over the nose. The Hick–Hyman function had a steeper slope for the human subjects than for the pigeons, regardless of the method that the people used for responding. If one equates steeper slopes with lower intelligence, this outcome implies that humans are less intelligent than pigeons. Alternatively, it could just be that pecking is a more compatible response for pigeons than either pecking or keypressing is for humans.

2.3.5 Alternatives to the Hick–Hyman Law

Some researchers have suggested that the function relating RT to the number of alternatives is not logarithmic. Kvälseth (1980) introduced a variety of laws, includ-ing a power law for the case of equally likely alternatives and an exponential law

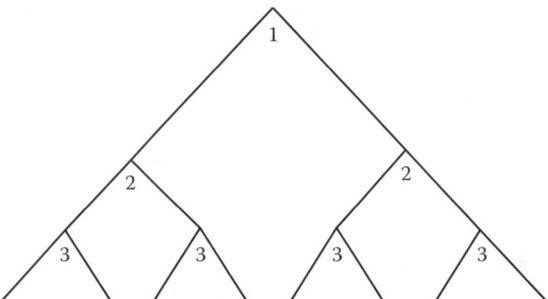

FIGURE 2.3 Hick's (1952) dichotomizing model, represented by a tree diagram. At each choice point at the three levels in the tree, a dichotomizing decision is made.

for cases in which the alternatives are not equally probable. Longstreth, El-Zahhar, and Alcorn (1985) claimed that the specific power function, RT $= a + b(1 - N^{-1})$, provides a better fit to data for equiprobable alternatives than does the logarithmic function. Longstreth et al.'s main argument for the power law is that as the number of alternatives increases beyond eight, the function is no longer linear with respect to the logarithm, but becomes curvilinear (see Longstreth, 1988). Although derived from a model of attention, Longstreth et al.'s power law is a special case of the more general power law proposed by Kvälseth. In addition, Kvälseth (1989) and Welford (1987) pointed out that Longstreth et al.'s power law has several problems. A quote by Kvälseth (1989) accurately describes the status of the Hick–Hyman law: "Although, on purely empirical grounds, Hick–Hyman's law may not be uniformly superior to other lawful relationships, it has been clearly established that it does provide a good summary description of a substantial amount of data" (p. 358).

2.3.6 A RACE MODEL

What mechanism is responsible for the Hick–Hyman law? Hick (1952) considered several conceptual models. The one that he directly related to information theory was a dichotomizing model, represented by a tree diagram (see Figure 2.3), in which the set of alternatives was partitioned into a series of equiprobable classes until the final classification was made. For eight choices, for example, the stimulus would first be classified as indicating one of a set of four alternatives, then as one of a set of two alternatives, and lastly as the identified alternative. This model predicts that the increase in RT from two choices (one decision) to four alternatives (two decisions) will be equal to that between four alternatives and eight alternatives (three decisions), as Hick found. However, as Hick noted, several of the details of his results did not fit this model well, and Leonard (1958) subsequently showed that the amount of time needed to benefit from a precue that halved the alternatives was much longer than the estimated time for a single binary decision.

    Usher, Olami, and McClelland (2002; see also Usher & McClelland, 2001) proposed that the Hick–Hyman law is a consequence of subjects attempting to maintain a constant level of accuracy at all set sizes. They provided evidence for this claim within the context of a race model that has N stochastic accumulators. In race models of this type, as mentioned earlier, a separate accumulator exists for each S-R alternative. When a stimulus is presented, activation relevant to each alternative builds up dynamically within the appropriate accumulators. When the activation in one accumulator attains a criterion, or threshold, value, that response is selected. Lower thresholds produce faster responses than higher thresholds because a threshold is reached sooner after stimulus presentation. But, because the activation process is noisy, the threshold for an incorrect alternative is more likely to be reached, resulting in a higher error rate.

Each additional S-R alternative adds another accumulator and thus provides an additional opportunity for an incorrect response to be selected. Consequently, if a subject is to keep error rate approximately constant as the size of the S-R set increases, then the response threshold must be adjusted upward. Thus, Usher et al. (2002) attribute the Hick–Hyman law to adjustments made to the response thresholds of the S-R accumulators in order to maintain a given level of accuracy.

Usher et al. (2002) showed that if the increase in the criterion as N increases is logarithmic, the probability of an incorrect response remains approximately constant. This logarithmic increase in criterion results in a logarithmic increase in RT. Based on their model fits, Usher et al. conclude regarding the regularity demonstrated by the Hick–Hyman law, "A major determinant of this regularity is the intrinsic increase in likelihood of spuriously reaching threshold as a function of an increase in the number of alternatives in conjunction with the attempt to maintain a fixed level of performance" (p. 712).

2.4 RESPONSE-PRECUING EFFECTS

In a response-precuing task, advance information is provided about a subset of possible responses (and stimuli), allowing the participant to prepare for just the subset of cued alternatives. By varying which subset of responses is cued, it is possible to determine the nature of the processing in which subjects engage to select and prepare for the cued responses. Furthermore, varying the interval between the precue and the imperative stimulus allows determination of whether different amounts of time are required under different conditions to achieve the maximal precuing benefit. Inclusion of a small percentage of trials on which the precue is invalid enables evaluation of the costs associated with being prepared for the wrong response alternatives.

Leonard (1958) introduced the partial advance information technique to test Hick's (1952) model, described earlier, according to which the response on a trial is selected by making successive binary classifications. Leonard had subjects perform a six-choice or three-choice task in which responses were made by using three fingers on one or two hands. He compared performance for a pure six-choice task, a pure three-choice task, and a precued six-choice task for which the precue signaled the responding hand, reducing the number of alternatives to three. The question of interest was whether the six-choice precued task would yield performance comparable with that of the pure three-choice task when the interval between the precue and imperative stimulus was equal to the difference in RT between the pure three- and six-choice tasks. The answer was negative: Whereas the difference in RT

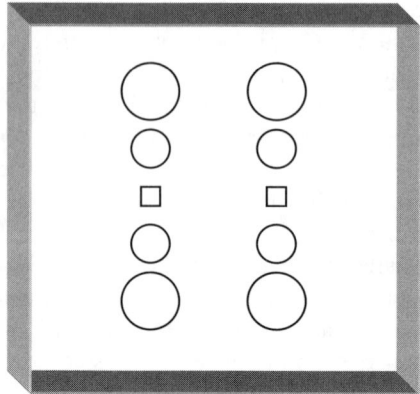

FIGURE 2.4 Illustration of the response panel for Rosenbaum's (1980) response-precuing task.

between the two pure tasks was less than 50 ms, a precuing interval of 300 ms was needed for the precued six-choice RT to be comparable with that of the pure three-choice task. So, subjects were able to use the precue information to prepare for the smaller set of alternatives, but the time required to do so was much longer than predicted by Hick's binary classification model.

2.4.1 Precuing Aimed Movements

One widely used response-precuing task, developed by Rosenbaum (1980), involves aimed movements of the left or right arm. In the prototypical aimed-movement task (see Figure 2.4), participants place their left and right hands on left and right home keys (depicted by squares); upon presentation of an imperative stimulus, one of the two fingers is moved to a response key (depicted by circles). For each hand, two response keys are located farther away from the participant's body in the horizontal plane than the home key and two are located closer. Precues can designate the hand, direction, or extent of the response to the imperative stimulus that will be required. In Rosenbaum's study, precues were letter stimuli signaling hand ("L" for left or "R" for right), direction ("F" for forward movement or "B" for backward movement), and extent ("N" for near the home keys or "D" for distant from the home keys), or any combination. The imperative stimulus was a colored dot presented at the center of the screen that matched the color of a response key. Precues effectively reduced RT: The more information that was given about the imperative stimulus in the precue, the more RT decreased. Moreover, the decrease in RT was largest when the precue designated the hand, intermediate when it designated the direction, and smallest when it designated the extent. Rosenbaum interpreted these differences in RT to the imperative stimulus in terms of specification of movement parameters in a fixed, serial order of hand, direction, and extent.

Subsequent research indicated, however, that the major factor determining the precuing patterns is the ease with which the cued subset of responses can be selected and prepared (e.g., Goodman & Kelso, 1980; Larish, 1986; Larish & Frekany, 1985).

Goodman and Kelso replicated Rosenbaum's (1980) method and found similar results using his stimuli. However, because the precues were letters and the imperative stimuli were colors, the mapping of precue and imperative stimuli to responses was not highly compatible. Consequently, subjects had to perform cognitive transformations in order to narrow down the subset of responses and select the final response itself. In their Experiment 2, Goodman and Kelso used a display consisting of eight light-emitting diodes arranged in the same manner as the response configuration. The corresponding diodes were activated to precue all parameters, and one diode was lit as the imperative stimulus to signal the corresponding response. For example, to precue the right hand, all four lights to the right were turned on for 3 s, and 500 ms later the imperative stimulus was lit. With this spatial precue display, results did not replicate Rosenbaum's findings of systematic effects of precuing particular parameters. Instead, precuing was equally beneficial for all parameters. Thus, when spatially compatible precues are used, there is no indication of fixed, serial specification of movement parameters. The differences in RT across precue conditions obtained in Rosenbaum's original version of this task reflect primarily S-R compatibility effects associated with response–selection processes (e.g., Larish, 1986), rather than motor-programming processes.

Stelmach and colleagues conducted a series of experiments using Goodman and Kelso's (1980) precuing method to compare performance of older and younger adults on the aimed-movement precuing task. Stelmach, Goggin, and Garcia-Colera (1987) and Stelmach, Goggin, and Amrhein (1988) used the two-limb, eight-choice aimed-movement task described above. Although mean RT was longer for the older adults than for the younger ones, the older adults benefited at least as much as the younger adults from precues for all parameters (arm, direction, and extent), indicating that they could use the precue information to prepare any subset of the responses. Similar results were obtained in studies that used a simpler two-choice task variation in which aimed movements of the right hand and arm were made in left or right directions (Larish & Stelmach, 1982), and a four-choice version in which hand and extent varied (Chua, Pollock, Elliot, Swanson, & Carnahan, 1995). Thus, these studies show that older adults have little or no deficiency in using precue information to select and prepare a subset of responses (see Proctor, Vu, & Pick, 2005, for a review).

2.4.2 Precuing Discrete Keypress Responses

Another widely investigated variation of the response-precuing task is one in which the responses are discrete keypresses (Miller, 1982). The prototypical procedure for this task has four stimulus and response alternatives, with the responses made by the index and middle fingers of the left and right hands (see Figure 2.5). A row of four plus signs is presented as a warning signal, followed shortly thereafter by two or four plus signs presented as precues in a row immediately below the warning row. After a variable interval, a single plus sign appears immediately below one of the precue plus signs, and the participant is to respond by pressing the corresponding response key as quickly as possible. Miller used precuing intervals of up to 1 s and found a benefit for precuing two responses only when they were on the same hand.

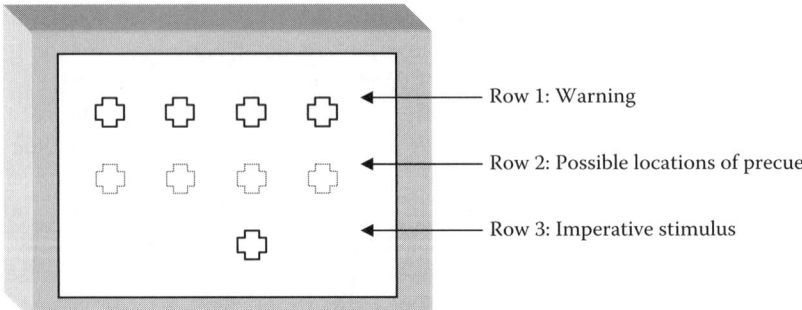

FIGURE 2.5 Procedure for precuing discrete keypress responses. A warning row appears, followed by stimuli in two or four locations of the precue row. After a variable interval, the imperative stimulus occurs below one of the cued locations.

He called this phenomenon the same-hand advantage and interpreted it as reflecting a characteristic of the motor system in which the hand for responding had to be specified prior to the finger (index or middle) that was to be used. However, Reeve and Proctor (1984) demonstrated that all pairs of responses benefit approximately equally from being precued if the interval between the precue and the imperative stimulus is made longer than 1s.

2.4.2.1 A Response–Selection Basis

In Reeve and Proctor's (1984) Experiment 1, the precuing interval was extended beyond the 1-s interval used by Miller (1982) to include values of 1,500 and 3,000 ms. As in Miller's study, when the interval was 750 ms or less, precuing was beneficial only for pairs of responses on the same hand. However, at 1,500 ms, all of the informative precue conditions showed some benefit, and by 3,000 ms, the precuing benefits were of similar magnitude for all of the pairings of fingers. This finding indicates that the hand does not have to be specified prior to the finger on that hand.

More importantly, Reeve and Proctor's (1984) Experiment 3 showed that the apparent "same-hand advantage" was really an advantage for precuing the two left or two right locations. Half of the subjects performed with a normal, adjacent-hands placement, as in their Experiment 1. However, the other half of the subjects performed with an overlapped-hands placement for which one hand was placed on top of the other, with the fingers from each hand alternated on the keys. With the overlapped placement, the left-to-right ordering of fingers on keys is right-index, left-middle, right-middle, and left-index. Thus, the two leftmost and two rightmost response keys are operated by different fingers (index, middle) from different hands. Nevertheless, the advantage remained with the precues for the two left and two right locations, with this precue condition showing a 49-ms benefit relative to the uninformative precue condition (for which all four responses remained possible) when the hands were adjacent and 50-ms when they were overlapped.

When averaged across short and long precuing intervals, the precuing benefit is greatest for the left and right pairs, intermediate for the inner or outer pairs, and

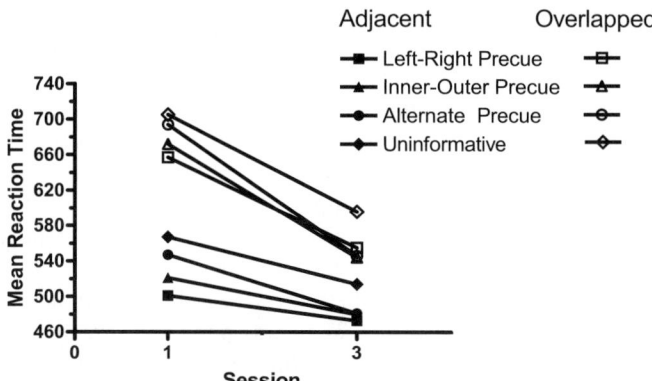

FIGURE 2.6 Mean reaction time as a function of precue condition, hand placement, and session in Proctor and Reeve's (1988) Experiment 1. For the informative precue conditions the precue designated two of the four locations (left or right pairs, inner–outer pairs, or alternate-location pairs), whereas for the uninformative condition the precue designated all four locations.

smallest for the pairs of alternate locations. This pattern of differential precuing benefits is primarily a spatial compatibility effect reflecting the time to translate the precue into a subset of responses to prepare. Proctor and Reeve (1988) found that the precuing functions changed with practice in a manner consistent with this interpretation. Specifically, in their Experiment 1, subjects practiced for three sessions of 310 trials each using either the adjacent- or overlapped-hands placement. The pattern of differential precuing benefits was evident for both hand placements in the first session, but by the third session, the inner–outer and alternate location conditions showed precuing benefits of similar magnitude to those of the left–right precue condition (see Figure 2.6). This equivalency was produced by a progressive shift in the functions relating RT to precuing interval, such that the asymptotic level of RT was reached at much shorter intervals.

It is important to note that, although the differences in precuing benefits for different pairs of responses reflect primarily differences in the time to select the cued responses, the benefit for precued responses relative to the case in which all responses remain possible reflects response preparation. This is implied by the fact that when precues designate responses on the same hand, a lateralized readiness potential is produced, indicating activation in the cerebral hemisphere that controls the cued hand, that is, the contralateral hemisphere (e.g., De Jong, Wierda, Mulder, & Mulder, 1988; Leuthold, Sommer, & Ulrich, 1996; Osman, Moore, & Ulrich, 1995). This response preparation may even extend to the latest stages of motor processing involving the recruitment of motor units (Possamaï, Burle, Osman, & Hasbroucq, 2002).

In contrast to the results of Stelmach et al. (1987, 1988) showing little or no precuing deficit for older adults with the aimed-movement method, Adam, Paas, et al. (1998) obtained results with the keypress method that they interpreted as showing "a substantial age-related deficit in preparing 2 fingers on 2 hands, but not on 1 hand"

(p. 870). This conclusion was based on two experiments in which subjects of different ages performed with only a normal, adjacent-hands placement. In Adam, Paas, et al.'s Experiment 1, a precue occurred 100 or 2,000 ms before the target stimulus, with the precue condition being either uninformative (plus signs in all four positions) or "hand cued" (either the two leftmost or two rightmost positions were cued). Six age groups with mean ages of approximately 25, 35, 45, 55, 65, and 75 years were tested. As is customary, mean RT was an increasing function of age; more importantly, older adults showed precuing benefits at least as large as those shown by younger adults at both precuing intervals.

Adam, Paas, et al.'s (1998) Experiment 2 tested only younger adults (mean age 24 years) and older adults (mean age 71 years) but added the remaining precue conditions (the "finger-cued" condition in which either the index or middle fingers were cued, and the "neither-cued" condition in which the index finger of one hand and the middle finger of the other were cued), with five precuing intervals (100, 500, 1,000, 1,500, and 2,000 ms). Averaged across all precuing intervals, the younger adults showed the typical pattern of differential precuing benefits: The hand-cued condition showed the most benefit, the finger-cued condition an intermediate benefit, and the neither-cued condition the least benefit. In contrast, the older adults showed a precuing benefit for only the hand-cued condition. For both age groups, the hand-cued condition had an advantage over the other precue conditions at the shortest precuing interval. But, whereas the younger adults showed the typical pattern of benefits appearing for the other precue conditions at longer intervals, the older adults did not.

Proctor, Vu, and Pick (in press) resolved this apparent inconsistency between Adam, Paas, et al.'s (1998) findings obtained with discrete keypress responses and those obtained with aimed movements. Proctor et al. extended the precuing interval beyond 2 s (intervals of 2, 3, 4, or 5 s) and used both the normal adjacent- and overlapped-hands placements to dissociate the hand distinction from spatial locations. With both hand placements, older adults benefited at least as much as younger adults from precuing any pair of responses. Although older adults may require longer than younger adults to select and prepare some subsets of cued responses, they show no deficit in the ultimate level of preparation that can be achieved, regardless of whether the responses are on the same or different hands, as long as sufficient time is allowed.

2.4.2.2   A Salient Features Coding Account

Proctor and Reeve (1986) proposed what they called the *salient features coding principle*, according to which "the stimulus and response sets are coded in terms of the salient features of each, with response determination occurring most rapidly when the salient features of the respective sets correspond" (p. 278). With regard to linear arrays of the type used in the spatial precuing studies, the salient feature for both the stimulus and response arrays is the left and right half. According to the salient features coding account, then, the precuing advantage for the left and right pairs of locations arises from fast determination of the cued responses because the left–right distinction is salient for both the stimulus and response sets.

Proctor and Reeve (1986) demonstrated that, consistent with this hypothesis, similar patterns of precuing benefits are obtained when the stimulus and response sets are oriented vertically. In their Experiment 1, the stimuli were three columns (warning, precue, target) of plus signs, and the response keys were aligned vertically on a tabletop at the body midline. To position their fingers on the keys, subjects turned their hands inward and placed the index and middle fingers from one hand on the top two positions and those from the other hand on the bottom two positions, or placed the fingers from each hand on alternating keys. Analogous to horizontal arrays, a top–bottom advantage was evident for which precuing the top two or bottom two locations produced larger benefits than precuing other pairs, for both the adjacent- and alternating-hands placements. In this case, there is no overall slowing of RT for the alternating-hands placement, as there is for the overlapped placement with horizontally arrayed responses. This finding indicates that the precuing benefit for the left and right pairs with the overlapped-hands placement in previous studies was not an artifact of the overall slower responses.

An implication of the salient features coding principle is that a similar advantage for precuing a pair of extreme locations should be evident in the four-choice task when the stimulus and response sets are oriented orthogonally. Proctor, Reeve, Weeks, Campbell, and Dornier (1997) tested this implication using all combinations of vertical and horizontal stimulus and response arrays. The pattern of differential precuing benefits was evident for orthogonal arrays as well as for parallel ones. Responses were slower overall when the arrays were orthogonal than when they were parallel, suggesting that a transformation such as mental rotation was needed to align the coordinate systems for the two arrays. In addition, subjects who practiced for three sessions with a vertical stimulus array and horizontal response array showed the same benefits of practice when the stimulus array was changed to horizontal in the fourth session as did subjects who practiced with the horizontal S-R arrays all along.

Reeve, Proctor, Weeks, and Dornier (1992) examined whether manipulations of salience of display and response features affect the relative precuing benefits. In their Experiment 1, the four locations in the display rows were either equally spaced, or the two left and two right positions were separated into distinct groups by a large central gap. Some subjects performed with their hands placed on four adjacent response keys and others with their hands on keys separated by a central gap. The equally spaced display showed the pattern of precuing effects typically obtained, which is that the left–right precue condition showed the largest benefit, followed by the inner–outer condition and the alternate-location condition. However, with the separated display, for which the stimulus locations for the inner–outer cue condition were in different display groupings, the precuing benefit for the inner–outer condition was eliminated. In Reeve et al.'s Experiment 2, a partitioned display (+ ++ +), for which the stimulus locations were grouped to make the distinction between the inner and outer locations more salient, was used in addition to the separated display, and responses were made with the hands placed in a normal manner on four adjacent keys. The precuing benefit for the left–right cues was reduced to about 20 ms with the partitioned display compared with more than 40 ms with the other two displays, whereas that for the inner–outer cues was increased to about 40 ms compared with about −5 ms with the separated display. Hence, by manipulating the relative salience

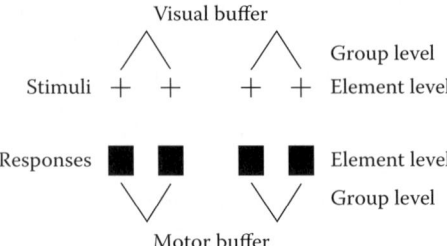

FIGURE 2.7 Adam, Hommel, and Umiltà's (2003) grouping model of response precuing. From "Preparing for perception and action (I): The role of grouping in the response–cuing paradigm," J. J. Adam, B. Hommel, and C. Umiltà, 2003, *Cognitive Psychology, 46*, p. 314. Copyright 2003 by Elsevier, Inc. Reprinted with permission.

of the features in the display, different patterns of precuing benefits were obtained. As Adam, Hommel, and Umiltà (2003) note, "These findings demonstrate that the spatial organization of the stimulus set is a powerful mediator of the pattern of precuing benefits" (p. 310).

The effect of response organization was weaker in Reeve et al.'s (1992) Experiment 1, showing no significant effects for RT. However, the error data showed a correspondence effect for the display and response groupings: For the uninformative condition, the percentage of errors was larger when the groupings did not correspond (2.5%) than when they did (1.6%), and this pattern held for all informative precue conditions except for the left–right condition. In this case, for which the precue indicated the left or right side of the arrays and responses on one hand, there was no difference. Beyak, Weeks, and Chua (1999) also found a small benefit for display–response correspondence using the same left–right display salience manipulation as Reeve et al. but manipulating response salience with textured and untextured equally spaced keys.

Adam et al. (2003) proposed a grouping model (see Figure 2.7) that builds on the salient features theme to account for response-precuing effects. Adam et al. describe the fundamental idea of the grouping model as follows:

> If the cue indicates a strong, good subgroup of stimuli that corresponds closely with a strong and similar subgrouping of responses, then a fast, automatic selection (activation) of the cued responses occurs. If, on the other hand, the cue indicates stimuli belonging to different subgroups, or if there is a mismatch between the grouping of the stimuli and the grouping of the responses, then a slower, effortful process is needed to create a good, finely tuned subgroup. (p. 308)

The model distinguishes a visual buffer and motor buffer, each of which is created on every trial. The stimulus and response sets are represented in the respective buffers with a default organization into subgroups based on Gestalt principles. When the stimulus and response organizations are in conflict, the response buffer is reorganized to conform to the organization of the visual buffer. With linear arrays of four positions, the natural subgroups are the two pairs of positions to either side of center.

Precues consistent with this organization allow the cued responses to be selected with little effort, whereas the precues that include one location from each of the natural subgroups require more effortful processing.

2.5 SEQUENTIAL EFFECTS

2.5.1 REPETITION BENEFIT

In many situations, choice reactions are faster when the stimulus and response from the previous trial are repeated than when they are not. Bertelson (1961) showed that when left and right stimuli were mapped compatibly to left and right response keys, RT was shorter overall when the proportion of repetitions was .75 than when it was .25. This repetition benefit was evident when the response–stimulus interval (RSI) was 50 ms but not when it was 500 ms. Bertelson (1963) also had subjects perform with an incompatible mapping of the stimulus and response locations, using a 50-ms RSI, and found that the repetition benefit was larger for the incompatible mapping (110 ms) than for the compatible mapping (70 ms).

Since Bertelson's (1961, 1963) studies, numerous investigations of sequential effects in choice–reaction tasks have been conducted. First-order sequential effects are those that involve the relation of the current trial to the immediately preceding trial. The most common first-order effect is that the response to a stimulus is faster when the S-R pair for a trial is a repetition of the preceding S-R pair than when it is not. In two-choice tasks of the type studied by Bertelson, this repetition benefit is obtained only when the RSI was short. At RSIs of 500 ms or longer, a benefit for alternations over repetitions is often found, instead. In tasks with more than two choices, the repetition benefit is larger, and a repetition benefit is found even at long RSIs (Schvaneveldt & Chase, 1969; Soetens, 1998).

The first-order sequential effects have been attributed to two processes. At short RSIs, residual activation from the preceding trial produces automatic facilitation when the current trial is identical to it; at long RSIs, expectancy regarding the nature of the next trial produces faster responses for expected than for unexpected stimuli (Soetens, 1998). This expectancy, at long RSIs, is for the alternative S-R pair in two-choice tasks because participants tend to anticipate a switch in the response. However, for tasks with more than two alternatives, the expectation at long RSIs is for repetition of the same pair. This repetition expectancy when there are several alternatives may be attributed to the fact that participants cannot prepare for the "alternative" response because there is more than one possible alternative.

Pashler and Baylis (1991) evaluated the locus of the repetition benefit for tasks in which two stimuli were assigned to each of three response keys (left, middle, and right keys operated by index, middle, and ring fingers of the right hand). Two of the stimuli were digits, two were letters, and two were nonalphanumeric symbols (e.g., & and #). Stimuli were mapped to responses in a categorizable (e.g., digits to left response, letters to middle response, and symbols to right response) or uncategorizable (e.g., a digit and a letter to the left response) manner. For both mappings, the repetition benefit occurred primarily when the same stimulus was repeated, but not when only the response was repeated. This repetition benefit for the same stimulus was not found when responses on alternate trials were vocal (the stimulus name) and manual. Consequently, Pashler and Baylis concluded that the repetition effects were at the stage of response selection, with the normal response–selection process being bypassed when the stimulus and response were repeated.

In Pashler and Baylis's (1991) experiments, a benefit for response repetition alone tended to occur with categorizable but not uncategorizable S-R mappings. Campbell and Proctor (1993) verified this effect, showing a benefit of approximately 40 ms for response repetition alone with categorizable mappings but not uncategorizable mappings. Their remaining experiments showed that this response repetition benefit, as well as the additional benefit for repeating the same stimulus, could be obtained when the responses on successive trials were made with different hands. In the critical conditions, the stimuli were presented to the left or right of fixation on alternate trials, with responses to the left stimulus made with the three fingers on the left hand and responses to the right stimulus made with the three fingers on the right hand. A cross-hand repetition benefit was obtained when either spatial information (relative position) or finger information (which finger, e.g., index) was consistent across hands, but not when both consistencies were eliminated. These results imply that the response sets can be coded in terms of locations or effectors, and that response selection benefits from repetition of the stimulus category when it maps onto a salient feature of the response set.

Soetens (1998) examined sequential effects for tasks in which subjects responded to four stimuli located at the corners of an imaginary square by pressing a left key if the stimulus was to one side and a right key if it was to the other. When left–right stimulus locations were mapped compatibly to the left–right responses, the repetition benefit at the short RSI (50 ms) was primarily associated with the response (i.e., the benefit was evident when the stimulus side was the same as on the previous trial, even when the location at the top or bottom of the square was different). At the long RSI (1,000 ms), a small alternation benefit was evident. With an incompatible S-R mapping (i.e., left side to right response), the results were similar, but with an increased benefit for repeating the same stimulus, particularly at the short RSI. When up–down responses were made to the left–right stimulus locations, response and stimulus repetition benefits of similar magnitudes were found at the short RSI. At the long RSI, the only effect was a repetition benefit for the same stimulus. Soetens concluded that automatic facilitation shifted toward stimulus-related processes as the mapping became less compatible. Together, the studies of Pashler and Baylis (1991), Campbell and Proctor (1993), and Soetens indicate that response repetition, without stimulus repetition, is beneficial when there is a structural relation between the stimulus and response sets and that repetition of the stimulus is more important when the mapping is arbitrary.

Although first-order sequential effects have been most widely studied, higher order repetition effects involving the sequence of the preceding N stimuli are larger and more consistent (Soetens, 1998). For two-choice tasks, at short RSIs, RT is benefited by preceding multiple repetitions, regardless of whether the present trial is a repetition or an alternation (called a benefit-only, or cost-only, pattern). For example, responses on the current trial tend to be faster if the three preceding trials were repetitions than if they were alternations. At long RSIs, however, a prior string of repetition trials is beneficial if the current trial is also a repetition, but a prior string of alternation trials is beneficial if the current trial is an alternation (called a cost-benefit pattern). These patterns of higher order effects, which were also apparent in Soetens's four-choice tasks, have also been attributed to automatic facilitation at the short RSI and subjective expectancy at the long RSI.

Jentsch and Sommer (2002) obtained the above patterns of higher order sequential effects for two-choice spatial tasks in which two vertically aligned stimulus locations were mapped compatibly to two vertically aligned response keys. The cost-benefit pattern was evident when the RSI was 700 ms, whereas the benefit-only pattern was evident at a short RSI of 50 ms. With a 700-ms RSI, a noncompatible mapping of a red- or green-centered stimulus to vertically arrayed response keys yielded an intermediate result pattern, suggesting that for less compatible mappings the cost-benefit component still contributes in addition to the benefit-only component at long RSIs. Using analyses of intertrial interval, based on the length of the immediately preceding RT, and analyses of the lateralized readiness potential, Jentsch and Sommer concluded that the cost-benefit pattern has a premotoric basis, whereas the benefit-only pattern has its basis in motoric processes.

Soetens and Notebaert (2005) recently provided evidence that at short RSIs, the first-order repetition benefit can be dissociated from the higher-order benefit-only pattern, implying that they are not due to the same mechanism. Soetens and Notebaert suggest that whereas the first-order benefit is due to automatic activation, or a priming mechanism, the higher-order benefit-only pattern is due to monitoring of response processes. In accordance with previous accounts, the third process, subjective expectancy, dominates at long RSIs.

2.5.2   Is the Hick–Hyman Law an Artifact of Repetition Effects?

Kornblum (1967, 1968) noted that, unless explicitly controlled, the proportion of repetition trials decreases as set size increases. Therefore, he proposed that the Hick–Hyman law is an artifact of repetition effects. Kornblum (1968) used a four-choice task in which four lights were mapped to four response keys, and information was varied by manipulating stimulus probabilities. For three levels of information, conditions were constructed in which the probability of repetition was high or low. RT was shorter for the high-repetition conditions than for the corresponding low-repetition conditions, and these latter conditions showed only a nonsignificant effect of information on RT. Kornblum (1967) conducted a similar experiment in which the number of alternatives was two, four, or eight. For the four- and eight-choice tasks, RT was shorter on repetition than on nonrepetition trials, with the slope being less for repetition trials. Within these tasks, RT for repetition trials increased as the amount of stimulus information increased, but RT for nonrepetitions did not.

Hyman and Umiltà (1969) noted that the RSI in Kornblum's (1967, 1968) experiments was approximately 140 ms, a relatively short interval that would maximize automatic repetition benefits and minimize preparation for the subsequent trial. They replicated three of Kornblum's (1968) conditions, but used an average RSI of 7.5 s. Though RT was faster for repetition than nonrepetition trials, the slopes of the two functions were approximately equal. Hyman and Umiltà concluded, "There seems little doubt that the information hypothesis is much more compatible with our results than those of Kornblum's" (p. 47). In other words, the Hick–Hyman function is not an artifact of the proportion of repetition trials when there is adequate time to prepare for the next trial.

2.6 INFLUENCE OF PRACTICE ON SET-SIZE EFFECTS AND SEQUENCE LEARNING

A general finding in choice–reaction studies is that RT becomes shorter and the proportion of errors decreases as subjects become practiced at the task. Whereas the task is initially effortful and responses tend to be slow and frequently incorrect, with practice performance becomes much more efficient and less effortful. The speedup in responding in choice–reaction tasks and many others follows what has come to be called the power law of practice (Newell & Rosenbloom, 1981):

$$RT = A + BN^{-\beta},$$

where N is the number of practice trials, B is performance time on the first trial, β is the learning rate, and A is the asymptotic RT. Although several authors have recently challenged whether the power law provides the best fit to individual subject practice functions in some tasks (e.g., Delaney et al., 1998; Heathcote, Brown, & Mewhort, 2000), it provides a reasonable approximation to the changes in RT that occur in many choice–reaction tasks.

Welford (1968, 1976) and Teichner and Krebs (1974) proposed that the largest effects of practice are on response selection, or what Welford calls S-R translation processes. Welford (1976) proposes that this increase in efficiency of S-R translation occurs at least in part because the "translations become 'built in' with practice" (p. 72). As described in Chapter 1, as might be expected if practice exerts a strong effect on response selection, the benefit of practice is greater for situations that are S-R incompatible than for ones that are more compatible. Recollect that Fitts and Seeger (1953) had subjects perform an eight-choice task with three stimulus arrays (panels A, B, and C in Figure 1.2) and a circular response array (panel A in Figure 1.2) for 32 sessions spread over 2½ months. RT for all display–control configurations decreased with practice, with the greatest benefit being for the display–control configuration that was initially of lowest compatibility (stimulus panel C and response panel A). However, the additional practice benefit for this condition was not sufficient for performance with it to attain the level of performance for the most compatible configuration (Display A and Response Array A), a finding that is shown in the next chapter to have broad generality.

Hick (1952) noted that practice at a choice–reaction task may be more beneficial when a fixed sequence is repeated than when the order of stimuli is random. In comparing Merkel's (1885) earlier study to his own, Hick observed that Merkel did not indicate how the sequence of stimuli in his experiment was determined, but "the very large practice improvement, even with ten alternatives, suggests that the sequence was easily learnt, since the present writer, using an irregular sequence, found very little improvement with practice" (pp. 11–12). Nissen and Bullemer (1987) confirmed Hick's speculation in an influential study showing that practice effects in a four-choice spatial choice task were indeed much larger when a fixed sequence of stimuli and responses was repeated periodically than when the order was randomly determined.

In Nissen and Bullemer's (1987) study, subjects responded to an asterisk presented in one of four horizontally aligned locations by pressing a corresponding response key, which triggered the next stimulus. Each subject performed 800 trials with a random order or a repeating sequence (DBCACBDCBA, with the letters A, B, C, and D designating the locations from left to right). RT decreased from the first block of 100 trials to the last by only about 20 ms for the random order but by about 160 ms for the repeating sequence. Nissen and Bullemer presented evidence that this sequence learning did not seem to require explicit awareness of the sequence, but this is a point that has been, and continues to be, hotly debated (see, e.g., Destrebecqz & Cleeremans, 2001; Wilkinson & Shanks, 2004).

Hoffmann and Koch (1997) examined the influence of S-R compatibility on sequence learning. Specifically, they compared performance using Nissen and Bullemer's (1987) task and sequence with a task variation in which the stimuli were the digits 1, 2, 3, and 4 (presented at the center of the screen), which designated the four response locations (in left-to-right order). Thus, the repeating stimulus sequence in the digit condition was 4231324321. Responses averaged more than 50 ms slower overall with the digit stimuli than with the location stimuli, but this factor did not interact with the decrease in RT across practice blocks. Also, when a random order was introduced in a ninth block, RT increased approximately 140 ms for both conditions. Because the response sequences were the same across conditions, the authors concluded that the observed learning was primarily response-based, another issue that continues to be debated.

Although healthy adults show similar sequence learning both when the stimuli are spatially compatible with the responses and when they are not, patients with Parkinson's disease show a dissociation between the two situations. Werheid, Ziessler, Nattkemper, and von Cramon (2003) examined sequence learning for Parkinson's patients and healthy subjects in two experiments. In one experiment, two letters were assigned to each of four response keys, and the eight letters occurred in a repeating sequence. Both subject groups showed improvement with practice, but the overall improvement was less for the Parkinson's patients than for the control group. However, the increase in RT on interspersed trial blocks in which the stimuli were presented in random order was similar for the two subject groups, indicating no deficit in sequence learning for the Parkinson's patients. The second experiment used a similar procedure, except that the stimulus set was an upper and lower row of four locations each, with the position in each row mapped compatibly to the corresponding response position. Although the overall decrease in RT with practice was less than in Experiment 1, it was of the same magnitude for both subject groups. However, the Parkinson's patients showed a significantly smaller increase in RT for the interspersed random trial blocks, indicating less serial learning than that evidenced by the healthy control subjects. Werheid et al. suggest that patients with Parkinson's disease are able to respond relatively fast with a spatially compatible mapping because attention is captured by the stimulus, and the response is executed automatically. However, this relatively fast responding on a particular trial occurs at the expense of generating an expectancy for the forthcoming trial.

2.7 CHAPTER SUMMARY

S-R compatibility is only one of several factors that influence the speed and accuracy of response selection. In this chapter, we reviewed several robust phenomena that also are thought to have their basis primarily in response–selection efficiency. As indicated by the Hick–Hyman law, response time increases logarithmically as a function of uncertainty. The degree of increase is moderated by many factors, with S-R compatibility and practice being two of the most important ones. Evidence suggests that this lawful relation is a consequence of trying to retain a given accuracy level when each additional alternative provides more opportunity for selecting an incorrect response.

Performance typically benefits from precues that reduce uncertainty, but the ease with which various precues can be used is itself subject to compatibility effects such that, unless the interval to prepare is long, some S-R combinations may show more benefit than others. Older adults' choice reactions are typically much slower than those of younger adults. However, the older adults benefit at least as much from advance information that reduces uncertainty, as long as they are allowed sufficient time to process that information. Similar to precuing effects, performance on a particular trial is influenced by preceding events, with the advantage of an immediate repetition of the preceding trial being more beneficial for an incompatible S-R mapping than for a compatible one. S-R repetitions benefit from automatic facilitation at short response–stimulus intervals, and both repetitions and alternations can benefit at long intervals from conscious expectancies induced by a prior sequence of repetitions or alternations, respectively. Higher order sequential effects occur that also vary as a function of whether the response–stimulus interval is short or long.

Some authors have attributed practice effects primarily to response–selection processes. Practice not only reduces the slope of the Hick–Hyman function, but it also results in a gradual reduction in RT that follows a power law. Practice is more beneficial for incompatible than compatible situations, but the compatibility effect does not entirely disappear. Benefits of practice are much more evident when there is a repeating sequence of trials than when there is not. The exact nature of the changes in response selection that occur with practice is covered in more detail later in Chapters 3 and 6 when we examine practice and transfer effects for S-R compatibility proper and the Simon effect.

3 Basic Stimulus–Response Compatibility Effects

3.1 INTRODUCTION

The initial studies of Fitts and Seeger (1953) and Fitts and Deininger (1954), described in Chapter 1, sparked considerable interest in S-R compatibility effects. Numerous variations of compatibility effects have been the subject of a substantial amount of research since that time. There are many reasons for the continued interest in S-R compatibility effects, including that they (a) are pervasive phenomena evident in performance of both simple laboratory tasks and complex human–machine interactions, (b) are perhaps the purest measures of the basic cognitive processes that intervene between perception and action, and (c) yield considerable theoretical insight into the nature of those processes. The research since the early 1950s has built on the foundation provided by Fitts and colleagues to develop many new compatibility phenomena of interest as well as models of the processes that underlie S-R compatibility effects.

The primary goal of the present chapter is to describe research on the fundamental nature of S-R compatibility through examining effects of S-R compatibility proper, that is, those effects that involve the mapping of relevant stimulus information to responses. The next chapter continues exploring the fundamental nature of S-R compatibility, but focuses on variations of the Simon effect, or those effects involving correspondence of irrelevant stimulus information with responses.

3.2 SPATIAL STIMULI AND RESPONSES

Following in the tradition of Fitts and Seeger (1953) and Fitts and Deininger (1954), much work on S-R compatibility effects has focused on spatial compatibility. The detailed investigations of spatial compatibility provide insight into the nature of compatibility effects more generally.

3.2.1 BASIC PHENOMENA

3.2.1.1 Set-Level Compatibility with Complex Displays and Controls

Contemporaneous with the research of Fitts and colleagues, Knowles, Garvey, and Newlin (1953) extended Fitts and Seeger's (1953) demonstration of set-level spatial compatibility effects, that is, effects of pairing different display and control arrangements to "somewhat more complex systems than those used by Fitts" (p. 65). In Knowles et al.'s Experiment 1, subjects performed self-paced tasks with combinations of three different displays and three different control arrangements (see Figure

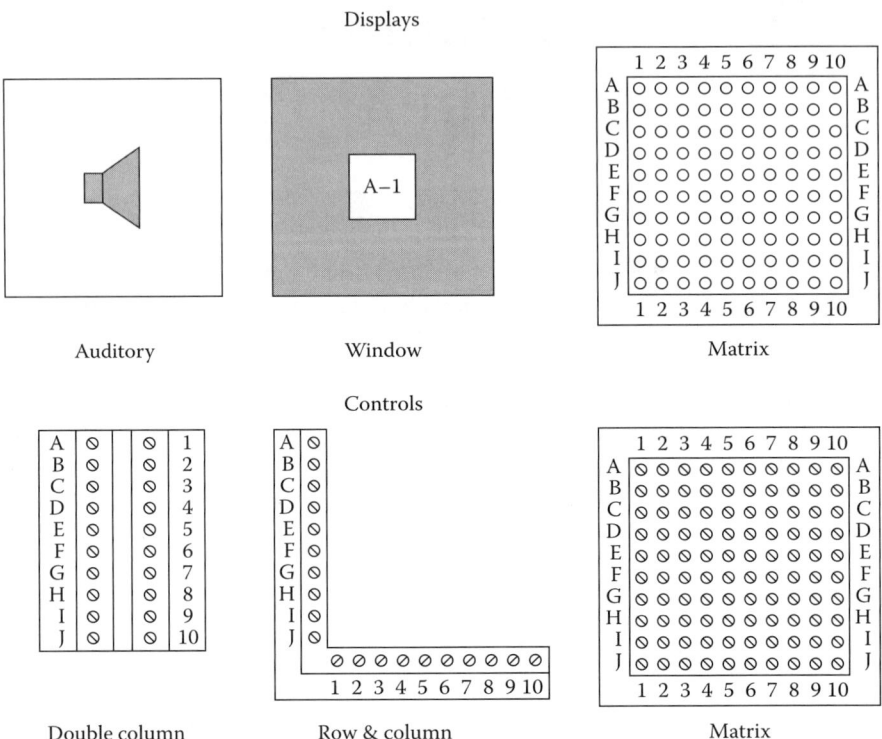

FIGURE 3.1 Display and control arrangements used by Knowles, Garvey, and Newlin (1953). From "The effect of speed and load on display–control relationships," by W. B. Knowles, W. D. Garvey, and E. P. Newlin, 1953, *Journal of Experimental Psychology, 46*, p. 67. Copyright 1953 by the American Psychological Association. Reprinted with permission.

3.1). The tasks involved entering values for letter-number pairs (composed from the letters A through J and the numbers 1 through 10; e.g., F-6) by pushing buttons. The letter and digit were presented auditorily (e.g., "F, 6"), displayed visually in a window (e.g., F-6), or designated by which light in a 10 (column of letters) × 10 (row of digits) matrix was lit. Responses were made by pressing a single button on (a) a matrix control panel corresponding to the matrix display, or two buttons simultaneously, one for the letter and one for the digit, (b) a double column array, or (c) a row and column array. In agreement with Fitts and Seeger's findings, response times to the matrix display were considerably shorter with the matrix control array than with the other two arrays, whereas for the auditory and window displays, performance tended to be worst with the matrix control array.

In a second experiment, Knowles et al. (1953) varied the presentation rate of the stimuli between the values of 0.27 and 0.45 stimuli per second, with the intent of inducing errors through speed stress. Only the matrix and window displays were used, coupled with the matrix and double column response arrays. As expected, the number of errors increased with increases in presentation rate, but the number of errors was least for the highly compatible matrix–matrix combination and most for the matrix–double column combination. Garvey and Knowles (1954) examined the same display–control configurations but added both matrix and column arrangements in which each stimulus light was placed immediately above its corresponding button (see Figure 3.2). These arrangements led to considerably faster self-paced performance than that for the compatible pairings of matrix–matrix and column–column display–control arrangements. Across 24 sessions of 100 trials of practice each day with each of the arrangements, the difference between the most and least compatible combinations was reduced. However, the relative ordering of the combinations remained the same, and substantial differences in performance were still evident in the last sessions.

Garvey and Knowles (1954) also examined the response speeds for the individual elements of the arrays. For the superimposed display–response arrangements, there was little difference between elements, but for the other arrangements, response time showed an increase from periphery to center of the matrix. This pattern was most pronounced for the matrix–matrix display–control combination. Garvey and Mitnick (1955) showed that providing horizontal and vertical lines as frames of reference that divided the matrix into four quadrants could reduce the response times for those elements in the center. Thus, they concluded not only that "the degree of isomorphic relationship between S-R sets is an important determinant of display–control efficiency," but also "the efficiency of a display–control system can be further enhanced by providing the operator with additional spatial references on the system where the internal interference among stimuli is greatest" (p. 282). The importance of frames of reference for spatial coding demonstrated by Garvey and Mitnick is a theme that is repeated often in the literature on spatial compatibility.

3.2.1.2 Element-Level Mapping Effects

Although set-level compatibility has continued to be investigated, the majority of research on S-R compatibility has focused on element-level mapping effects. Fitts and Deininger's (1954) first demonstration of mapping effects for spatial tasks, described in detail in Chapter 1, used an eight-choice task for which the locations of stimulus lights arranged in a circle were mapped to target locations for responses arranged in a corresponding manner. The effects were large, with the highly compatible, corresponding mapping yielding response latencies that were 154 ms shorter than a mirror-opposite mapping and 724 ms shorter than a random mapping. Using a similar arrangement, but with responses made to the actual locations of the stimuli rather than on a separate response panel, Griew (1964) showed that the difference between corresponding and mirror-opposite mappings was larger for the eight-choice task (180 ms) than for a two-choice task in which only two of the alternatives were possible (65 ms).

Based on Griew's (1964) findings and others', Fitts and Biederman (1965) remarked, "Results from recent studies … indicate that compatibility effects are negligible for two-alternative choices, but increase rapidly as the number of alternatives is increased" (p. 409). Although S-R compatibility effects do in fact increase with increasing numbers of alternatives, the typical effect obtained in two-choice tasks is sufficiently large (e.g., RT for the incompatible mapping in Griew's study

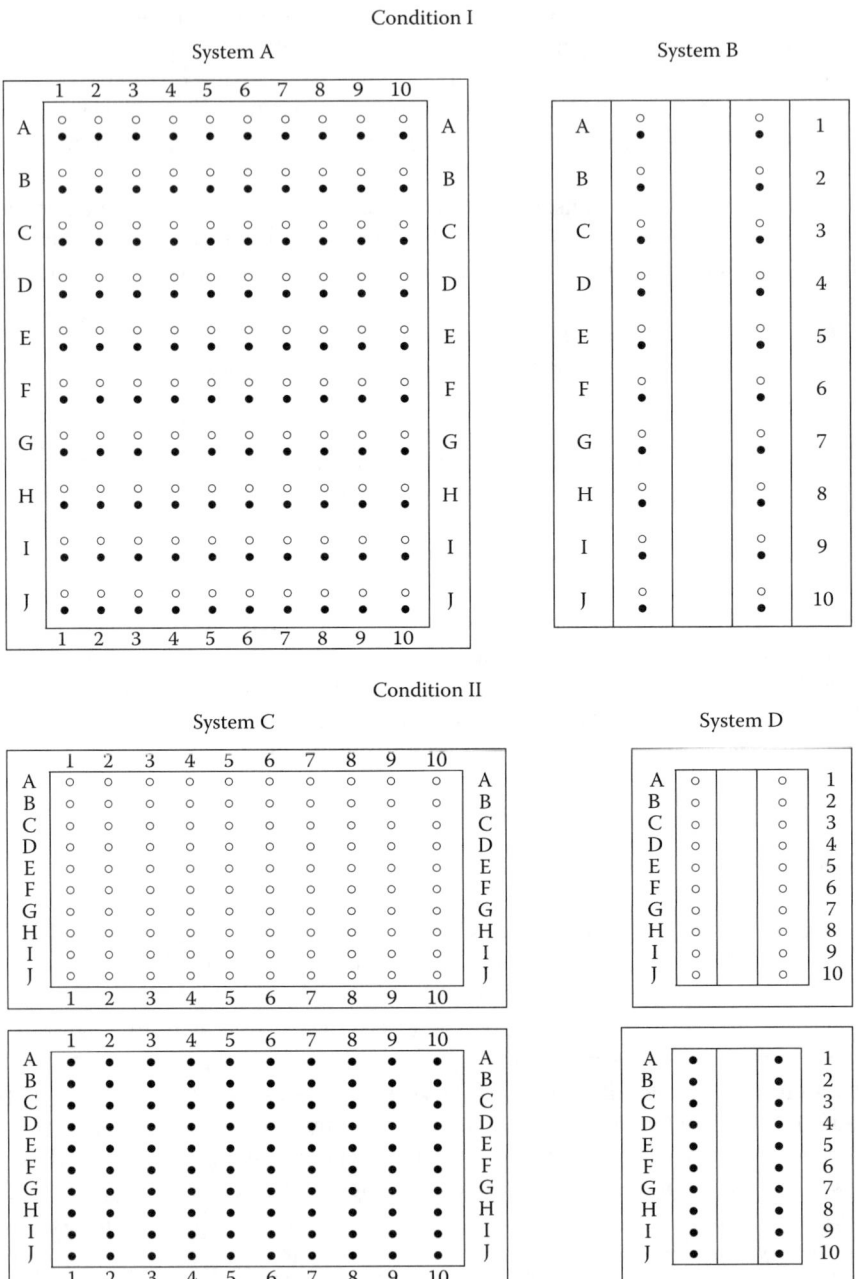

FIGURE 3.2 Additional display–control arrangements used by Garvey and Knowles (1954). Unfilled dots are display elements; filled dots are control elements. From "Response time patterns associated with various display–control relationships," by W. D. Garvey and W. B. Knowles, 1954, *Journal of Experimental Psychology, 47*, p. 316. Copyright 1954 by the American Psychological Association. Reprinted with permission.

Condition III

System E

	1	2	3	4	5	6	7	8	9	10	
A	o	o	o	o	o	o	o	o	o	o	A
B	o	o	o	o	o	o	o	o	o	o	B
C	o	o	o	o	o	o	o	o	o	o	C
D	o	o	o	o	o	o	o	o	o	o	D
E	o	o	o	o	o	o	o	o	o	o	E
F	o	o	o	o	o	o	o	o	o	o	F
G	o	o	o	o	o	o	o	o	o	o	G
H	o	o	o	o	o	o	o	o	o	o	H
I	o	o	o	o	o	o	o	o	o	o	I
J	o	o	o	o	o	o	o	o	o	o	J
	1	2	3	4	5	6	7	8	9	10	

A	•	•	1
B	•	•	2
C	•	•	3
D	•	•	4
E	•	•	5
F	•	•	6
G	•	•	7
H	•	•	8
I	•	•	9
J	•	•	10

System F

A	o	o	1
B	o	o	2
C	o	o	3
D	o	o	4
E	o	o	5
F	o	o	6
G	o	o	7
H	o	o	8
I	o	o	9
J	o	o	10

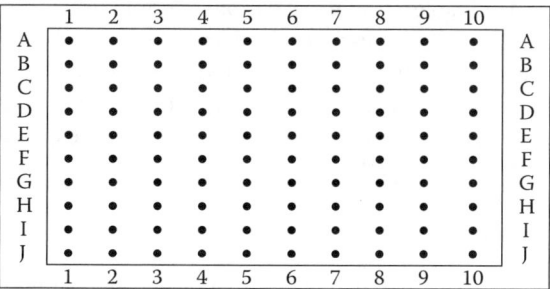

FIGURE 3.2 (continued)

increased by 18% compared with the compatible mapping) that much of the subsequent research on S-R compatibility has used tasks with only two choices. Two-choice tasks typically yield reliable effects for spatial S-R compatibility in the range of 50 to 125 ms and can yield larger effects for other stimulus and response sets. For example, in one of the first studies to examine S-R compatibility solely with two alternatives, Broadbent and Gregory (1962) obtained a 107-ms S-R compatibility effect for keypress responses to tactile stimulation of the left or right index finger and an even larger 164-ms verbal compatibility effect for saying "bid" or "did" in response to the speech stimuli BID or DID.

Subsequent studies have shown similar S-R compatibility effects for two-choice tasks with visual and auditory stimuli. Shaffer (1965) had subjects make keypress responses with the left and right index fingers to left and right stimulus locations. RT was 54 ms faster for the compatible mapping than for the incompatible one. Similarly, Simon (1967) showed a 95-ms S-R compatibility effect for tones presented in the left or right ear mapped to left and right keypresses. Two-choice compatibility effects in this general range have also been reported for aimed movement responses (Griew, 1964), left–right joystick or handle-movement responses (Simon, 1969), and left (counterclockwise) versus right (clockwise) wheel rotations (Proctor, Wang, & Pick, 2004; Stins & Michaels, 1997). As should be evident from the studies described in this chapter and others, versions of two-choice tasks can be used to investigate a range of issues.

As with S-R compatibility effects in general, those obtained in two-choice spatial tasks can be attributed primarily to response–selection processes. Shulman and McConkie (1973) varied both spatial mapping and whether the responding fingers were on the same or different hands (the index and middle fingers of the right hand or the index fingers of each hand). Responses were almost 50 ms slower when executed by different fingers on the same hand than by the same finger on different hands (a response–response compatibility effect first demonstrated by Kornblum, 1965), but the 47-ms compatibility effect did not differ significantly for the two response sets. Similarly, Adam (2000) manipulated stimulus discriminability by having small or large distances between the two stimulus locations and response locations (left index and right middle fingers; index and middle fingers of the right hand), along with spatial mapping. All three variables produced main effects: shorter RT with more discriminable stimulus locations, shorter RT for the between-hand response set than for the within-hand one, and shorter RT for compatible than incompatible mappings. More important, neither the stimulus discriminability nor response-set factor interacted with compatibility, or with each other. This finding is consistent with the view that S-R compatibility affects a response–selection stage that is neither perceptual nor motor.

3.2.2 SPATIAL CODING

Simon and Wolf (1963) conducted an experiment using two stimulus lights mounted on a circular panel that, in different blocks of trials, were rotated to different orientations relative to the horizontal. For the 0° condition, the subject was to respond to the right light with the right key and left light with the left key. For the remaining

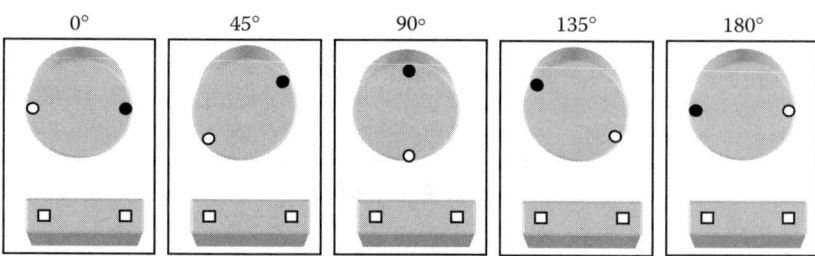

FIGURE 3.3 Depiction of Simon and Wolf's (1963) display and control arrangement with the stimuli rotated 0, 45, 90, 135, and 180 degrees. The small black circle indicates the stimulus location to which a right response is to be made, and the small white circle indicates the stimulus location to which a left response is to be made.

conditions, the panel was rotated to 45°, 90°, 135°, and 180°, but each light remained assigned to the same response (e.g., at 90°, the right response was made to the upper stimulus and left response to the lower stimulus, and at 180°, the left response was made to the right stimulus and right response to the left stimulus; see Figure 3.3). RT was approximately 120 ms longer for the incompatible 180° condition than for the compatible 0° condition, with RT for the 90° condition being intermediate. Additionally, RT was only slightly longer for the 45° condition than for the 0° condition, and slightly shorter for the 135° condition than for the 180° condition. This outcome implies that the crucial factor is whether the stimuli appeared in distinct left–right positions, allowing them to be coded as left or right.

Umiltà and Liotti (1987) showed that two-choice S-R compatibility effects are a function of the relative location of the stimuli. In Experiment 1, two boxes were presented to the left or right of a fixation cross and, after a delay of 0 or 500 ms, the imperative stimulus occurred in one of the boxes (see Figure 3.4a). In the compatible condition, subjects were to respond to a stimulus in the right box of the pair with a right button-press and to a stimulus in the left box of the pair with a left button-press; in the incompatible condition, this mapping was reversed. An S-R compatibility effect of 65 ms was obtained that did not vary as a function of whether the boxes were in the left or right hemispace. In their Experiment 2, all four boxes were presented, two in dashed lines to mark relative positions and two in solid lines, and the target stimulus occurred in one of the latter two boxes (see Figure 3.4b). Again, an S-R compatibility effect was obtained, which Umiltà and Liotti interpreted as evidence for coding by left or right hemispace as well.

For two-choice tasks in which keypress responses are made with the left and right hands, a confound exists between the spatial relationships for (a) the stimulus locations and the response mechanisms or locations, (b) the stimulus locations and the responding effectors (left or right hand), and (c) the responding effectors and the response locations. Brebner, Shephard, and Cairney (1972) dissociated these relationships by having subjects perform a visual two-choice task using compatible and incompatible spatial mappings, with the hands held in a normal, uncrossed placement or a crossed placement for which the right hand operated the left response key and the left hand the right response key (see Figure 3.5). Responses were

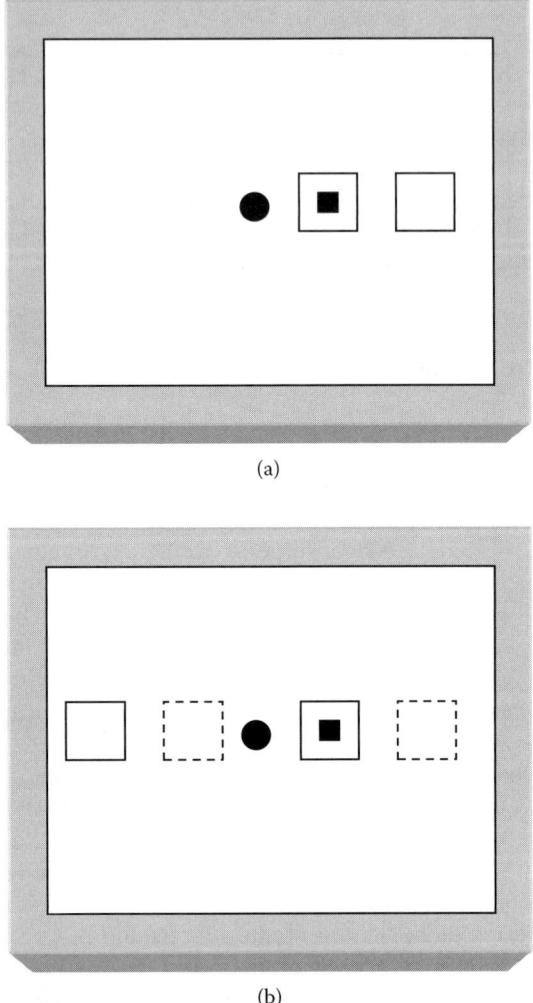

FIGURE 3.4 Illustration of the display arrangements used by Umiltà and Liotti (1987) in their Experiment 1 (a) and Experiment 2 (b).

approximately 30 ms slower for the crossed-hands placement than for the uncrossed placement, indicating a compatibility effect for the hand–key relation. However, this crossed-hands effect did not interact significantly with the spatial mapping effect, which was 33 ms for the crossed-hands placement and 43 ms for the uncrossed-hands placement. This outcome indicates that the primary factor contributing to the difference between the compatible and incompatible mappings is the relation between the stimulus location and response key (or location), with the contribution of the stimulus-hand relation being negligible. Another way of characterizing the results is that the element-level S-R mapping effect is a function of the spatial relationships, and there is an additional slowing with the hands crossed due to having to press the left key with the right hand and vice versa.

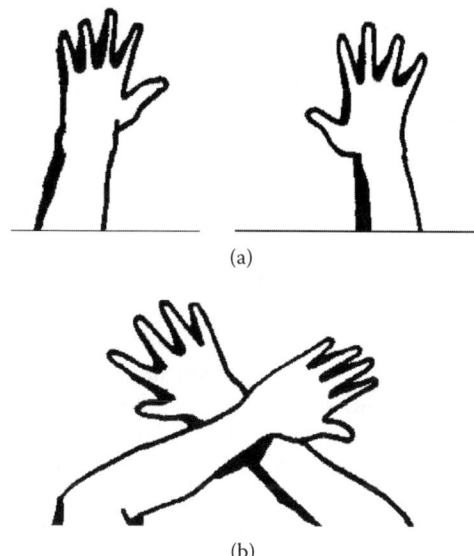

FIGURE 3.5 Uncrossed- (a) and crossed- (b) hands placements.

Several authors, including Anzola, Bertolini, Buchtel, and Rizzolatti (1977) and Aglioti, Tassinari, and Berlucchi (1996), have obtained similar additive effects of S-R mapping and crossed- versus uncrossed-hands placement for normal adults performing two-choice visual tasks. Aglioti et al. found that persons with complete defects of the corpus callosum, which normally connects the two cerebral hemispheres, similarly showed additive effects of mapping and hand placement, but also evidenced faster responding with the anatomically ipsilateral hand. They concluded, "These findings suggest that rules for coding the stimulus and response sets in the performance of a typical spatial compatibility task are similar in the normal and acollosal brain; yet, unlike normal subjects, in acollosal subjects there is a 'neural pathway' effect which appears to be additive with those of spatial stimulus–response compatibility and hand placement" (p. 627). Aglioti et al. proposed that, for normal subjects, both hemispheres are involved in the response–selection decision, whereas for acollosal subjects, the decision is primarily restricted to the hemisphere receiving the visual stimulus.

Heister, Schroeder-Heister, and Ehrenstein (1990) showed that spatial relations are also most important when responding with the index and middle fingers from the same hand. Subjects responded in different sets of trials with the left hand and the right hand, in a prone or supine hand position, using compatible or incompatible spatial mappings. An S-R compatibility effect was obtained of approximately 55 ms (see also Katz, 1981) that did not interact significantly with hand (left or right) or hand posture (prone or supine). In other words, the S-R compatibility effect was primarily a function of the relation between the stimulus and response locations and not of the fingers used for responding.

Nicoletti, Anzola, Luppino, Rizzolatti, and Umiltà (1982) had subjects perform using both compatible and incompatible spatial mappings with their hands crossed and uncrossed, but positioned both visual stimulus locations in the right visual field or in the left visual field in different trial blocks. Subjects were to maintain their gaze on a fixation point, and the stimulus for a trial could be presented 5° or 10° away from fixation in the appropriate visual field. As typically found, responses were 37 ms slower with the hands crossed than with them uncrossed. Most important, spatial compatibility effects were obtained as a function of the relative location of the stimulus, with the effects being of similar magnitude for the uncrossed- (45 ms) and crossed- (52 ms) hands placements. Thus, relative position of the stimuli was the critical factor determining the spatial S-R compatibility effect.

Nicoletti et al.'s (1982) Experiment 2 was similar to their Experiment 1, except that subjects performed the task with the hands uncrossed but with both hands placed to the same side of the body in the right or left hemispace. Their results showed a 40-ms benefit for mapping the right stimulus to the right response and the left stimulus to the left response that did not differ significantly as a function of whether the hands were located to the left or right of the subject. Thus, similar to the effects of relative location for stimuli, the relative position of the response locations is sufficient to produce a spatial compatibility effect. Nicoletti, Umiltà, and Ladavas (1984) showed that the slowing of RT due to crossing the hands could also be attributed to relative position. They had subjects respond with a spatially compatible mapping to two stimulus locations placed to the left and right of the center of the display or both in one hemispace (the left for one condition and the right for another). The response keys were aligned with the stimulus locations, and subjects executed the responses with their hands crossed or uncrossed. A crossed-hands effect of about 50 ms was obtained, and this effect did not differ reliably as a function of the location of the two responses relative to body midline. Nicoletti et al. (1984) attributed the effect to conflict between a spatial response code and a code for the relative position of the response hand.

In contrast to the results obtained with visual stimuli, Simon, Hinrichs, and Craft (1970) reported an effect of ear–hand correspondence for an auditory S-R compatibility task. Subjects in their experiment performed only two of the four possible conditions resulting from crossing spatial mapping and hand placement, with the assignment of tone locations to response hands being the same for both conditions. One group of subjects performed the compatible mapping with their hands uncrossed and the incompatible mapping with their hands crossed, and another group performed the compatible mapping with their hands crossed and the incompatible mapping with their hands uncrossed. An S-R compatibility effect of 129 ms was obtained with the uncrossed-hands placement, but the effect was only 6 ms with the crossed-hands placement, suggesting that both the relation between tone and response locations and that between tone location and response hand contributed to performance.

Roswarski and Proctor (2000) examined the discrepancy between Simon et al.'s (1970) results and those of the studies using visual stimuli. Their Experiment 1 was similar to Simon et al.'s, with subjects performing with uncrossed- and crossed-hands placements using either a compatible or incompatible mapping of tone locations to responses. The compatibility effect was 112 ms with the hands uncrossed and 96 ms with them crossed, which was a nonsignificant difference. The slight tendency toward a smaller compatibility effect for the crossed-hands placement was due to the error rate for the spatially incompatible mapping being particularly high for that hand placement. Roswarski and Proctor's Experiment 2 replicated this pattern of results and showed that the pattern did not differ from that obtained for visual stimuli. In sum, their experiments provided no evidence that the relation between tone and responding hand influences performance.

3.2.3 ACTION GOALS AND REFERENCE FRAMES

Riggio, Gawryszewski, and Umiltà (1986) noted that, in the crossed-hands experiments, the position of the effector is confounded with the position of the response goal. They conducted two experiments that dissociated the location at which an action was effected from the location of the goal, or response key, that was operated. In Experiment 1, instead of crossing the arms so that the left hand was located to the right of the right hand, subjects crossed only the index fingers. The spatially compatible mapping showed an advantage of 53 ms over the spatially incompatible mapping with this crossed-fingers placement, compared with a 58-ms advantage with a normal finger placement for which the left finger operated the left key and the right finger the right key. Thus, a "crossed-hands" compatibility effect was obtained even though only the fingers were in opposite positions from normal. As with performing with crossed hands, though, crossing the fingers slowed RT by about 68 ms compared with when the fingers were not crossed.

More fascinating was Riggio et al.'s (1986) Experiment 2, in which subjects responded holding two sticks whose tips lay on the response keys. In the uncrossed condition, the stick on the left operated the left key and the one on the right the right key, whereas in the crossed condition, this relation was reversed. Thus, in the crossed condition, the hand that moved the stick in order to press the key was on the opposite side of the key. Nevertheless, a spatial compatibility effect with respect to the key was obtained, with the compatibility effect being 51 ms in the crossed condition and 52 ms in the uncrossed condition. Consequently, Riggio et al. concluded, "The results were clear in showing that spatial compatibility depended on the location of the response goal whereas the location of the effector had no effect whatsoever" (p. 99).

Wheel-rotation responses are not explicitly left–right, but consist of several features that could provide a basis for left–right coding. Clockwise rotations are typically made for right turns and counterclockwise rotations for left turns, and this association is sufficiently strong that it is more natural to refer to turning the wheel left or right than to turning it counterclockwise or clockwise. Additionally, the top of the wheel, which is the part nearest to the line of sight, moves rightward for clockwise rotation and leftward for counterclockwise rotation, and when the wheel is held at the top or bottom, the hands also move leftward or rightward. With one or two hands placed at the top of the wheel, all three frames of reference are in agreement in favoring clockwise rotation for a right response and counterclockwise rotation for a left response. Not too surprisingly, a substantial element-level compatibility effect is obtained in this case. For example, Stins and Michaels (1997) found a 46-ms advantage for the mapping of right stimulus to clockwise response and left stimulus to counterclockwise response than for the opposite mapping when the wheel was operated by a single hand placed at the top.

More interesting is the situation in which the wheel is held at the bottom, for which the direction of hand movement (or the hand-referenced frame) is opposite that of the wheel-referenced frames. In this case, the optimal strategy would be to code the responses with respect to the wheel-referenced frame for the right–clockwise/left–counterclockwise mapping but with respect to the hand-referenced frame for the right–counterclockwise/left–clockwise mapping, which would yield no S-R compatibility effect. Stins and Michaels (1997), in fact, found a nonsignificant S-R compatibility effect of –13 ms (i.e., a tendency favoring the right–counterclockwise/left–clockwise mapping) when a single hand at the bottom operated the wheel. However, across subjects there were large individual differences: Nine subjects showed negative S-R compatibility effects averaging –50 ms, four exhibited positive effects averaging +64 ms, and three displayed no effect at all. Because of these individual differences, Stins and Michaels concluded that, for both S-R mappings, some subjects used hand-referenced response coding (which would yield a negative S-R compatibility effect with respect to the wheel rotation) and others used wheel-referenced response coding (which would yield a positive effect).

Proctor, Wang, and Pick (2004, Experiment 1) replicated the presence of a large overall positive S-R compatibility effect when the wheel was held at the top with both hands and the absence of an overall effect when the wheel was held at the bottom. However, their data suggested that subjects who performed with their hands at the bottom position coded their responses with respect to the direction of hand movement when instructed to "turn away" from the tone. Relative to this frame of reference, the S-R mapping would be compatible rather than incompatible. In other words, subjects seemed to adopt the frame of reference that would make the S-R mapping compatible. Merz, Kalveram, and Huber (1981) obtained similar results, finding that subjects could code an incompatible mapping of left–right knob pressure to cursor movement in a tracking task as compatible with the direction in which a visible steering wheel would turn when the knob was located at the bottom of the wheel.

Whereas the instructions used in Proctor et al.'s (2004) Experiment 1 were neutral with respect to using the hands or wheel as a reference frame, the instructions for Experiment 2 specified the responses in terms of moving the hands left or right in response to the stimuli. In that experiment, a negative S-R compatibility effect was obtained that was of similar magnitude to the positive effect obtained with the hands at the top of the wheel in Experiment 1. This negative S-R compatibility effect indicates that most if not all subjects coded the responses in terms of the instructed hand movements even when this coding resulted in an incompatible S-R mapping. In Proctor et al.'s Experiment 3, clockwise rotation of the wheel caused a visual cursor to move to the right and counterclockwise rotation caused the cursor to move to the left. A large positive S-R compatibility effect occurred even though the wheel was held at the bottom, indicating that the responses were coded relative to the same frame, in this case, direction of cursor movement, regardless of whether the resulting mapping was compatible or incompatible.

Heister et al. (1990) compared anatomical and spatial reference frames for keypresses made with the same hand using a procedure that allowed them to distinguish the spatial distance between the two responding fingers from the anatomical distance. Subjects performed a task similar to that in the previously mentioned study with the left hand on one day and the right hand on another. In each session, they responded with (a) the index and ring fingers operating narrowly separated buttons, (b) the thumb and little finger operating more widely spaced buttons, and (c) the thumb and little finger operating narrowly spaced buttons. All three conditions showed spatial S-R compatibility effects. Most importantly, the two conditions for which the narrow buttons were operated showed similar effect sizes, even though the fingers were separated by only one finger in the case of the index and ring fingers, which yielded a 51-ms S-R compatibility effect, and by three fingers in the case of the thumb and little fingers, which yielded a 46-ms effect. The effect was smaller, 26 ms, when the thumb and little fingers were widely spaced, though, suggesting that spatial distance may matter though anatomical distance does not.

Schroeder-Heister, Heister, and Ehrenstein (1988) had subjects respond with different head positions: head upright, tilted 90° to the right, or tilted 90° to the left. Subjects performed the spatial S-R compatibility task for which keypresses were made with two fingers on the same hand while using the three different head positions in separate blocks of trials. They performed on two days, one with the hands uncrossed and the other with them crossed, using both spatially compatible and incompatible mappings. The results for the upright head position were consistent with the crossed-hands studies described above: Responses were slower with the hands crossed than with them uncrossed, but a spatial S-R compatibility effect of 42 ms was obtained that was independent of the hand placement. With the head tilted to the left or right, an effect of 41 ms was obtained with the uncrossed-hands placement, but with the crossed-hands placement, the effect was reduced to 19 ms. Schroeder-Heister et al. suggested that when the head is tilted, a normal hand placement helps to maintain the task as one of left–right discrimination. However, when the hands are crossed in addition to the head being tilted, "neither the head midline nor the anatomically right and left hands can guide spatial right/left orientation" (p. 43). They proposed the resulting greater uncertainty in coding positions as left or right leads to more reliance on what they call *spatio-anatomical* mapping (whether the response effector is right or left).

Heister et al. (1990) provided a hierarchical model of spatial compatibility to account for effects such as those described in the previous paragraph. According to this model, three main factors contribute to S-R compatibility: spatial coding of the response keys' positions, spatial coding of the responding effectors' positions, and spatio-anatomical mapping. In some descriptions of the model, coding is based on a lower level only when a higher level provides no basis for coding the responses, whereas in other descriptions it is assumed that the three factors jointly determine compatibility effects, with the degree to which each factor influences the effects being a decreasing function of its ranked order. For many situations, though not all, the patterns of results seem more consistent with the first description of the model.

Klapp, Greim, Mendicino, and Koenig (1979) obtained strong evidence for spatio-anatomical mapping for situations in which left–right coding of response locations was difficult. Their first experiment replicated the crossed-hands effect with subjects using the left or right thumb to press a single switch button located between them. With a crossed-hands placement, the spatial compatibility effect was

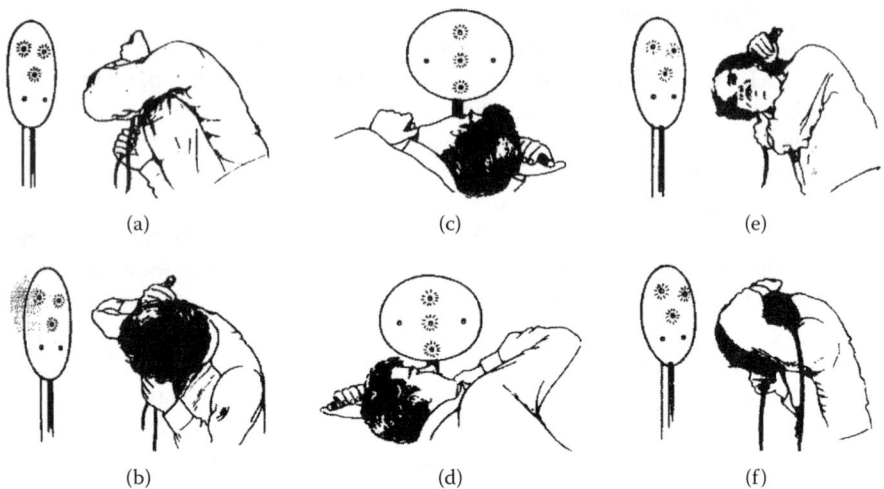

FIGURE 3.6 Illustration of conditions in Ladavas and Moscovitch's (1984) experiments. From "Must egocentric and environmental frames of reference be aligned to produce spatial S compatibility effects?" by E. Ladavas and M. Moscovitch, 1984, *Journal of Experimental Psychology: Human Perception and Performance, 10*, p. 207. Copyright 1984 by the American Psychological Association. Reprinted with permission.

61 ms, compared with 79 ms when the hands were not crossed. Of most interest is a third condition in which the hands were placed one above another, with the thumbs pointing vertically toward each other and resting just above the single response button. The right hand was above the left hand for half of the trials for this condition, and the left hand was above the right for the other half. For this up–down condition, a 66-ms advantage for mapping the right stimulus to the right thumb and left stimulus to the left thumb was obtained, suggesting that anatomical coding was being used.

In Klapp et al.'s (1979) Experiment 3, subjects lay on their sides, with the head resting at an angle approximately 45° from horizontal. This allowed all aspects of the anatomy associated with the top arm (i.e., the shoulder, upper arm, etc.) to be located entirely above that for the lower arm. By varying the side on which the subject lay in different trial blocks, the right thumb/arm was on top for half of the trials and the left thumb/arm for the other half. The stimuli were still two horizontal locations, although their orientation relative to the observer would vary on both horizontal and vertical dimensions. Even though this experiment eliminated other environmental cues to left and right that would have been present in Experiment 1, a compatibility effect with respect to the mapping to left or right thumb of 56 ms was still evident, implicating spatio-anatomical coding.

Ladavas and Moscovitch (1984) investigated in more detail the effects of mis-aligning egocentric and environmental reference frames (see Figure 3.6). As in Schroeder-Heister et al.'s (1988) experiment, subjects performed with the head upright or tilted 90° to the left or right. Each hand held a plastic cylinder with a pushbutton that was operated by the thumb. The cylinders were held to the left or right side of the head, aligned with the vertical axis of the display screen when the head was in a tilted position. With the head upright, the S-R compatibility effect was 43 ms; it was 49 ms and 32 ms with the head tilted to the right and left, respectively, in relation to the anatomical effector. Experiment 2 obtained similar results for situations in which the stimulus locations were up and down on the screen and the hands were placed above the top of the head and below the chin. Experiments 3 and 4 used crossed-hands conditions but in other respects were similar to Experiments 1 and 2, respectively. The results again indicated that responses were coded with respect to the hand being left or right. Ladavas and Moscovitch concluded that their results indicate "that at least under conditions of head rotation, spatial compatibility effects are based on a right–left classification of the responding hand" (p. 213).

3.2.4 RULE-BASED RELATIONS

Fitts and Deininger's (1954) finding that performance was much better with a mirror-reversed mapping for their eight-choice task than with a random mapping provided an initial demonstration that response selection can benefit from systematic rules. Morin and Grant (1955) provided an even more compelling demonstration. For their experiment, the stimulus display was a row of eight red lights, each of which had a green response light located immediately below it. The response panel was a row of eight response keys, operated with the four fingers of each hand. The stimulus for a given trial was a pattern of red lights, to which the subject was to respond by pressing the keys that would turn on the corresponding green lights. Across nine groups of subjects, Morin and Grant varied the degree of systematicity in the mapping of response keys to stimulus (and response light) locations, as measured by Kendall's τ, an indicator of rank correlation. For three groups, the correlations were +1.00 (perfect spatial correspondence), 0.00 (completely random assignment), and –1.00 (perfect reversed assignment). For the remaining six groups, the response and stimulus positions had a positive or negative correlation that was less than perfect: +0.86, +0.57, +0.29, –0.29, –0.57, and –0.86.

As shown in Figure 3.7, response time was fastest with a perfect positive correlation and next fastest with a perfect negative correlation, consistent with Fitts and Seeger's (1953) findings. In addition, performance benefited from partial correlations, with response time for the positive correlation conditions decreasing monotonically as τ decreased from +1.00 to 0.00. For the negative correlation conditions, only the groups for which τ was –1.00 or –0.86 showed a substantial benefit of the correlation. However, it is also important to note that even a slight change from perfect correspondence greatly increased response time. When τ was +0.86, either only two pairs of lights were displaced from direct correspondence or one light was displaced two positions away from correspondence. Yet, mean response time was more than twice that when τ was +1.00.

Duncan (1977b) obtained evidence for rule-based response selection in four-choice tasks for which the responses were made with the index and middle fingers of each hand. Subjects were tested with one of four S-R mapping conditions: pure corresponding, for which each stimulus was assigned to its corresponding response; pure opposite, for which each stimulus was assigned to its mirror opposite response;

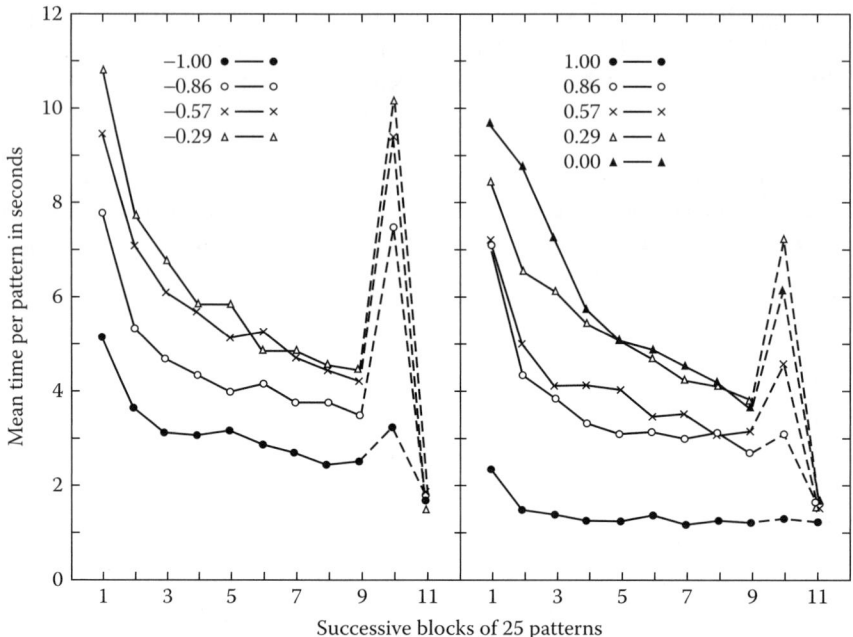

FIGURE 3.7 Results from Morin and Grant's (1955) experiment. From "Learning and performance on a keypressing task as a function of the degree of spatial stimulus–response correspondence," by R. E. Morin and D. A. Grant, 1955, *Journal of Experimental Psychology, 49*, p. 42. Copyright 1955 by the American Psychological Association. Reprinted with permission.

and two mixed mappings, one for which the two inner stimulus positions were mapped to the corresponding responses and the two outer stimulus positions to the opposite responses and the other for which the corresponding and opposite relations were reversed. The finding of most importance in Duncan's experiment was that RT for both corresponding and opposite responses was faster in the pure mapping conditions than in the mixed ones. An analysis of incorrect responses showed that for the mixed mapping conditions, the most common error was the response that would be correct for the alternative mapping rule. Duncan interpreted his results as indicating that response selection can be based on rules and not only on individual S-R associations. He attributed the longer RTs in mixed mapping conditions to the need to select the appropriate transformation rule prior to applying it.

3.2.5 Role of Practice

It seems logical to think that S-R compatibility effects would be eliminated with practice. They are typically thought to involve intentional response–selection processes, possibly coupled with an interfering tendency to activate the corresponding response when the mapping is incompatible, both of which should be possible to overcome with practice. However, this has repeatedly been shown not to be the case. As indicated, Fitts and Seeger (1953) and Knowles et al. (1953) found that the differences in set-level compatibility in their 8- and 100-choice tasks decreased with practice but showed no sign of disappearing entirely. Morin and Grant (1955) found similar results for their task using combinations of eight stimulus and response locations, with large differences evident after nine 25-trial blocks of practice (see Figure 3.7). Consequently, they concluded, "Practice will compensate very little for loss of correspondence or compatibility" (p. 45). Morin and Grant conducted two additional trial blocks after the original nine practice blocks, one in which the green response lights were turned off and one in which the mapping was spatially corresponding for all subjects. Performance degraded in these blocks for all groups except the one for which the stimulus and key locations had a perfect positive correlation, and the degradation was severe for all of the remaining groups except for the group with the perfect negative correlation and the one with a high, but not perfect, positive correlation of +0.86. Thus, subjects in all except these easiest conditions apparently were relying extensively on the feedback provided by the action effects of turning on the green lights corresponding to the locations of the red lights.

Even the smaller S-R compatibility effect obtained in two-choice tasks is robust across practice. Brebner (1973) had subjects practice for 10 sessions of 120 trials each with the four conditions created by combining compatible and incompatible spatial mappings with crossed- and uncrossed-hands placements. RT was reduced with practice for all conditions, with the greatest reduction occurring for the incompatible–crossed condition, which yielded the worst initial performance. However, a benefit for the spatially compatible mapping remained evident for both hand placements.

Dutta and Proctor (1992) reasoned that the most likely situation in which performance with the incompatible mapping might become as efficient as that with the compatible mapping would be when each subject used only a single mapping throughout practice. Consequently, they had subjects perform a two-choice task, with uncrossed hands, for eight sessions of more than 300 trials each with only a compatible or incompatible mapping. The S-R compatibility effect decreased from 73 ms in session 1 to 48 ms in session 5, but was still 46 ms in session 8. In a further attempt to eliminate the S-R compatibility effect with practice, Dutta and Proctor (1993) provided summary feedback about mean accuracy and mean reaction time after each block of 40 trials in one experiment and, in another experiment, imposed a response deadline that was reduced as the experiment progressed. In neither case did 2,400 trials of practice eliminate the compatibility effect. Using a PET scan to measure cerebral blood flow while subjects practiced a two-choice task, Iacoboni, Woods, and Mazziotta (1996) found a bilateral increase in blood flow in the posterior parietal lobule for the incompatible mapping compared with the compatible mapping. However, this difference did not interact with practice. Practice influenced blood flow in the dorsolateral prefrontal cortex, the premotor cortex, and the motor cortex, but similarly for the compatible and incompatible mappings, which Iacoboni et al. attributed to procedural learning on a process shared by the two mapping conditions.

Proctor and Dutta (1993) had subjects practice for three sessions of 300 trials each with one of four conditions involving compatible or incompatible spatial mapping and crossed- or uncrossed-hands placement, and transferred a fourth of the subjects in each group to one of the four mapping conditions in a fourth session.

The practice sessions showed that practice yielded a larger benefit for incompatible than compatible mapping and for crossed- than uncrossed-hands placement. In the transfer session, performance was best in all cases for subjects who continued performing the same condition with which they had practiced. For all practice conditions except the incompatible–crossed one, subjects showed a benefit from maintaining the same spatial mapping that did not depend on whether the hand placement remained the same or changed from the practice session. This outcome indicates that the improvement in performance of maintaining a constant spatial mapping was due to processes that did not depend on the specific effector used for responding. For the difficult, incompatible–crossed mapping, it was as if the subjects had to start entirely anew even when the hands placement was now changed to uncrossed and the incompatible spatial mapping was maintained.

In Proctor and Dutta's (1993) Experiments 2 and 3, subjects practiced for four sessions, each of which contained 10 blocks of 42 trials. Some subjects in each experiment practiced with one of the four possible mapping/hand-placement conditions for all blocks, others with a second condition, and still other subjects switched between those two conditions on alternate blocks of trials. In Experiment 2, the conditions were compatible–uncrossed and incompatible–crossed, for which the relation between stimulus locations and effectors remained constant when switching between conditions. In Experiment 3, the conditions were incompatible–uncrossed and incompatible–crossed, for which the spatial S-R mapping remained constant when switching between conditions. In Experiment 2, the compatible–uncrossed condition showed no performance cost for switching conditions compared with practicing only that condition, but the incompatible–crossed condition showed a large switching cost. In Experiment 3, in contrast, neither that condition nor the incompatible–uncrossed condition with which it was alternated showed any cost compared with performing only the respective condition. These data imply that subjects benefit little if any from a constant relation between stimulus locations and effectors, but do benefit from a constant relation between stimulus and response locations, as if the specific effectors used for responding were immaterial.

Proctor and Dutta's (1993) findings are in general agreement with results obtained by Wickens (1938) for conditioned responses of the middle finger of the right hand. Wickens had subjects place their right hands palm down in an apparatus through which an electric shock could be administered. It was possible to avoid or escape the shock by making an extension movement of the finger. Conditioning trials, on which a buzzer sounded about 500 ms prior to onset of the shock, were administered until the response became conditioned to the buzzer. After conditioning, the hand position was changed to palm up, requiring a flexion movement if the shock was to be avoided. The conditioning transferred to this situation in which the avoidance response was executed with the antagonistic muscle group: Ten of eighteen subjects made the flexion response to the buzzer on the first trial, without receiving a shock, and all but one subject showed the conditioned response after receiving only a small number of shocks. Wickens (1943) demonstrated that this result did not depend on whether flexion or extension was conditioned initially and also showed that responding for subjects transferred to the new response was just as strong as that for subjects who continued making the original response.

Although Proctor and Dutta's (1993) results indicate that considerable positive transfer from practice occurs when the spatial mapping remains constant across practice and transfer conditions, this is not the case when the mapping remains the same for only one of two spatial dimensions. Healy, Wohldmann, and Bourne (2005) conducted a study in which subjects practiced a task that required using a mouse to move a cursor from a start location to one of several possible target locations of a clock-face display. The mapping of mouse movement to cursor movement was varied between subjects in a practice session of 400 trials. In the up–down reversal condition, upward movement of the mouse produced downward cursor movement and downward movement of the mouse produced upward cursor movement, but the left–right movement was normal. In the left–right reversal condition, left–right movement of the mouse produced noncorresponding right–left movement of the cursor, but up–down movement was normal. In the combined reversal condition, both left–right and up–down movements of the mouse produced noncorresponding cursor movements along the respective dimensions. Subjects returned a week later for another session, in which they performed with either the same mapping condition or a different one.

Healy et al. (2005) found that practice produced substantially shorter RT for all three mapping conditions, and these benefits of practice were retained well across the week interval. However, practice showed little positive transfer from one mapping condition to another. There was only a small benefit of practicing with one of the conditions in which there was incompatibility along only one dimension when transferred to the combined reversal condition, and practice with the combined reversal condition yielded no benefit to performance with the individual reversal conditions. Thus, in this task, what is learned appears to be the relation between specific mouse and cursor movements and not relations along the individual dimensions.

3.3 VERBAL AND NONVERBAL S-R MODES

3.3.1 ROLE OF PERCEPTUAL SIMILARITY

Spatial S-R compatibility effects can be obtained with stimuli other than physical locations that convey spatial information, although the magnitude of effects depends on perceptual similarity to the response set. Beller (1975) demonstrated differences in set-level compatibility using arrows that pointed up, down, left, and right or the letters U, D, L, and R as stimuli to convey directions. Sixty-four arrows or letters were presented in an 8×8 matrix, and the task was either to proceed through the matrix from left to right and top to bottom naming the stimuli or to start at one location and trace a path through the matrix moving in the direction indicated by each stimulus in each successive box along the path. The mean time to name the elements in a matrix was 12.8 s for letters versus 17.8 s for arrows, whereas the mean time to trace the path was 26.7 s for letters and 22 s for arrows. That is, letters afforded better performance when the responses were verbal–vocal, and arrows afforded better performance when the responses were visuospatial.

Wang and Proctor (1996; see also Proctor & Wang, 1997a) performed experiments using two-choice tasks in which stimulus modes were crossed with response modalities. Subjects performed eight blocks of trials composed from a combination of stimulus mode (left–right physical locations or centered words LEFT and RIGHT), response modality (left–right keypresses or vocal "left"–"right" responses), and mapping (compatible or incompatible). All sets of stimuli and responses yielded mapping effects, but the effects were larger for the sets with high set-level compatibility (spatial–manual, 67 ms; verbal–vocal, 152 ms) than with those with low set-level compatibility (spatial–vocal, 41 ms; verbal–manual, 74 ms). Another way of characterizing these results is that the element-level mapping effect is larger when the stimulus and response sets share both perceptual and conceptual similarity than when they share only conceptual similarity. This effect was due entirely to the compatible mapping, with the incompatible mapping showing no reliable interaction of stimulus mode and response modality, implying that the additional perceptual similarity acts primarily to facilitate responding for compatible mappings.

Wang and Proctor (1996) obtained comparable results when the physical location stimuli were replaced with left- and right-pointing arrows, the mapping effects being larger for the sets with high set-level compatibility (arrow–manual, 81 ms; verbal–vocal, 143 ms) than with those of low set-level compatibility (arrow–vocal, 51 ms; verbal–manual, 81 ms). Again, this interaction was due primarily to effects on RT with the compatible mappings and not the incompatible ones. A similar, though smaller, interaction was obtained with the physical location and arrow stimuli when aimed movements made on a touchscreen from a home location to a left or right target location were contrasted with vocal responses. Wang and Proctor confirmed this finding in a final experiment, showing that the spatial–keypress and verbal–movement combinations yielded larger mapping effects (61 and 95 ms, respectively) than the spatial–movement and verbal–keypress combinations (38 and 68 ms, respectively).

Proctor and Wang (1997b) conducted experiments that analyzed in more detail the reason for the differences between the on-screen aimed movement responses and the keypresses in Wang and Proctor's (1996) study. The results suggest that the primary determinant of the differences in set-level compatibility is whether the responses are made with one or two effectors. In Experiment 1A, both types of responses were performed on the computer keyboard, whereas in Experiment 1B, both were performed on the display screen (for the press responses, each index finger was held just above the appropriate response box and a "press" response was made to touch the screen). Regardless of whether responding on the keyboard or screen, responses were faster overall to spatial location stimuli than to the location words. This advantage was larger for the press responses than for aimed movement processes, which can be interpreted as the spatial–press and verbal–movement combinations having higher set-level compatibility than the alternative combinations. Although this evidence for a set-level compatibility difference was obtained, there was no reliable difference in size of the element-level mapping effect between the high and low set-level conditions when the confound between screen and keyboard was removed. Other experiments conducted by Proctor and Wang showed (a) no such difference in set-level compatibility for these stimulus sets with press and aimed movement responses when the left and right aimed movements were made with the left and right hands, respectively; (b) higher set-level compatibility of bimanual movements to spatial stimuli and unimanual movements to verbal stimuli than for the other combinations; and (c) higher set-level compatibility of keypresses made by two fingers on the same hand with spatial stimuli and of unimanual aimed movements with verbal stimuli rather than vice versa.

3.3.2 IDEOMOTOR COMPATIBILITY

The concept of ideomotor action has a long history in psychology (Stock & Stock, 2004). The basic idea is that actions are represented in terms of images of the sensory feedback they produce and that actions can be initiated by "nothing other than the idea of the *sensory consequences* that typically result from them" (Stock & Stock, p. 176). Greenwald (1970a, 1970b) applied this concept to S-R compatibility, reasoning, "If images of feedback do mediate voluntary responding, then one should be able to demonstrate this in an RT procedure" (1970a, p. 20). Greenwald (1970a) made two assumptions: "The feedback modality most essential for spoken responses is audition, while for written responses vision would be most important," and "sensory images in a given modality are most rapidly activated by their corresponding stimuli in that same modality" (p. 20).

Greenwald (1970a) tested this hypothesis in several experiments in which he used auditory and visual stimulus modalities paired with spoken or written responses. For one of the experiments, the stimuli were the letters A, E, I, K, O, P, and the four combinations of stimulus and response modalities were performed in distinct trial blocks. In four other experiments, which differed in only minor respects, the stimuli were the digits 1, 2, 4, 6, 8, 9, and visual and auditory stimuli were randomly intermixed in trial blocks using each of the response modalities. In all cases, an interaction between stimulus and response modality was evident, with the interaction contrast showing shorter RT for the ideomotor compatible combinations (visual–written and auditory–spoken) than for the nonideomotor compatible conditions (visual–spoken and auditory–written). It is important to note that all of these conditions would be regarded as being S-R compatible, indicating that the modality combinations classified as ideomotor compatible are more compatible than the others. Greenwald favored an ideomotor interpretation, "which asserts that encoding is easier for the visual-written and auditory-spoken combinations because these allow selection of the proper response (via its controlling image) without necessity for translating the stimulus code into another modality" (p. 24).

Ten Hoopen, Akerboom, and Raymakers (1982) similarly favor an ideomotor compatibility account of results they obtained for a vibrotactile choice–reaction task. In different conditions, subjects received vibratory stimuli to the fingertips that varied in frequency (40 and 150 Hz) and amplitude (15 and 135 microns), with the number of S-R alternatives being two, four, or eight. The results with the 150 Hz, 135 micron stimuli replicated those reported by Leonard (1959), described in Chapter 2, for which increasing the number of S-R alternatives from two to eight produced no increase in RT. However, the 40 Hz, 15 micron stimuli showed an increase in RT of about 50 ms per bit of information, and the other two combinations showed increases of about 20 ms per bit. Ten Hoopen et al. noted that the slope size of the Hick–Hyman function was inversely related to subjective intensity and that the lowest intensity vibration stimulated only cutaneous receptors (Meissner corpuscles), whereas the highest intensity vibration stimulated both cutaneous and subcutaneous receptors (Pacinian corpuscles). Because the force required to depress the vibrating button was sufficient to stimulate both Meissner and Pacinian corpuscles, Ten Hoopen et al. concluded that ideomotor compatibility was only high for the highest intensity vibration.

Brass, Bekkering, and Prinz (2001) recently argued that there is a close relationship between ideomotor compatibility and action imitation. They provided evidence for this argument with an experiment in which subjects made a single finger movement (a lift or tap of the right index finger) within a block of trials in response to onset of movement of a finger of an animated hand. Even though the response was known in advance, a compatibility effect of 55 ms was obtained: Responses were faster when the observed movement (tap or lift) corresponded with the executed movement. Brass et al. also showed that the compatibility effect was approximately half that size when the stimulus was a square that moved up and down as the tip of the finger would, providing only a movement direction feature. The effect was also smaller when the observed hand was upside down so that the viewed direction of movement was incompatible with that of the direction of response movement. They interpreted these results as implicating contributions of both dynamical spatial S-R compatibility and ideomotor compatibility to performance.

Brass et al. (2001) concluded, "From the ideomotor point of view, it can be argued that imitation of action is a special case of S-R compatibility such that the stimulus prespecifies certain aspects of the perceivable consequences of the action" (p. 20). Heyes and Ray (2004) agreed with Brass et al. that imitation is mediated by the same processes as S-R compatibility, but saw no need to invoke ideomotor compatibility. Heyes and Ray instructed their subjects to respond as fast as possible to one of six possible movements of a computer graphic representation of a female by imitating the movement with the same side of the body as the model. The model could be facing the subject or facing away, with her back turned. Responses took longer to initiate and included more errors of using the wrong side of the body when the egocentric locations of the stimulus and response did not correspond than when they did (i.e., when the model was facing the subject than when her back was turned). In a second experiment, subjects were instructed to imitate with the opposite side of the body. In this case, subjects were slower to respond and made more errors when the egocentric locations of the stimulus and response were compatible than when they were incompatible (i.e., again, when the model was facing the subject rather than having her back turned). Based on the importance of the spatial relations in their study, Heyes and Ray concluded, "Our results are entirely consistent with the hypothesis that common processes are responsible for imitation and for responding to inanimate stimuli on the basis of arbitrary S-R mappings" (p. 708).

As these latter two studies suggest, the concept of ideomotor action has become increasingly prominent in recent years, and ideomotor compatibility has been the subject of debate. Because ideomotor compatibility has been investigated extensively in the context of dual-task performance, we discuss it in more detail in Chapter 10. We also delve more into the topic of ideomotor action when discussing the theory of event coding in Chapter 11.

3.3.3 STIMULUS–CENTRAL PROCESSING–RESPONSE COMPATIBILITY

Wickens, Sandry, and Vidulich (1983) proposed that the concept of S-R compatibility should be elaborated to incorporate a mediating central processing (C) component, introducing the concept of stimulus–central processing–response (S-C-R) compatibility. They provided two reasons for this expansion. The first is that with increasingly complex systems, operators often have to incorporate the stimulus information into a mental model of the system and make an action in response after some delay. The second reason, and more relevant for present concerns, is that research has suggested that spatial codes are distinct from verbal codes. Wickens et al. argued

> There appears to be a unique compatibility relation between modalities of input (auditory, visual, or A, V) and output (manual, speech, or M, S) and codes of central processing (spatial versus verbal). The benefits of A/S modalities will be most realized when they are associated with the verbal task, whereas those of V/M modalities are best realized when associated with the spatial task. (p. 228)

Wickens et al. (1983) and Wickens, Vidulich, and Sandry-Garza (1984) tested the predictions derived from the concept of S-C-R compatibility in several experiments. Wickens et al.'s (1983) Experiment 2 contained single-task conditions involving combinations of auditory and visual stimulus modalities and speech and manual response modalities for spatial and verbal tasks. The verbal task consisted of a series of communication, navigation, and identification tasks, whereas the spatial task required target localization. Each part of the verbal task included a command, presented visually on a screen or auditorily through headphones, to which a series of actions was to be carried out. These actions could involve entering commands manually on a control panel or giving spoken commands. For the spatial task, a stimulus was presented, auditorily or visually, to designate the identity of one of three target stimuli that was to be localized. The subject moved a circular cursor to the target in order to "designate" it. Manual control was accomplished by using a finger control to locate the cursor on the target and pressing a button. Speech control entailed saying a command with a clock-like direction (e.g., 3 o'clock) to indicate the direction of cursor movement, and the command "designate" when the cursor reached the target. Response times showed no main effect of task, indicating that the verbal and spatial tasks were of similar difficulty. However, task interacted with condition. As predicted by the S-C-R compatibility concept, response time for the verbal task was shortest with the combination of auditory input and speech output and longest with the combination of visual input and manual output, and response time for the spatial task showed the reverse relation.

S-C-R compatibility, like ideomotor compatibility, emphasizes that relations between stimulus and response modalities are important contributors to compatibility effects. Because the concept of S-C-R compatibility has had considerable influence on applied research, particularly in the area of dual-task performance, we return to it in Chapter 12.

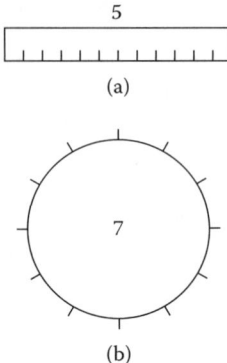

(a)

(b)

FIGURE 3.8   The schematic ruler and clockface used by Bächtold, Baumüller, and Brugger. (1998). From "Stimulus–response compatibility in representational space," by D. Bächtold, M. Baumüller, and P. Brugger, 1998, *Neuropsychologia, 36*, p. 732. Copyright 1998 by Elsevier Science Ltd. Reprinted with permission.

3.4   OTHER ELEMENT-LEVEL S-R COMPATIBILITY EFFECTS

Although we have focused in this chapter on spatial compatibility effects for which physical stimulus locations, location words, or arrow directions are mapped to physical or verbal spatial response alternatives, S-R compatibility effects occur in virtually any situation for which there is overlap between the mental representations of stimuli and responses on some dimension. We end this chapter with several examples of S-R compatibility effects that differ from the typical spatial compatibility effects.

3.4.1   MENTAL IMAGES

Spatial S-R compatibility effects are not restricted to the mental representations of stimulus and response sets with spatial properties. They can also occur for mental images of stimuli that are not physically present. This point is illustrated by two studies.

Bächtold, Baumüller, and Brugger (1998) conducted two experiments in which subjects classified the numbers 1 to 11, presented in the center of the visual field, with a left or right keypress. In one experiment, subjects first received practice with a schematic ruler (see Figure 3.8a) in which they were trained to conceive of the indicators as distances in centimeters. After practice, the schematic ruler was removed and subjects were to classify the centrally presented digit by aligning it with the corresponding location on the imagined ruler. They then performed two blocks of 100 trials in which the numbers were classified as "less than 6" or "greater than 6," with less than assigned to the left response and greater than to the right response for one block, and this mapping being reversed for the other block. Responses were about 150 ms faster for the mapping of < 6 to left response and > 6 to right response than for the opposite mapping. In their second experiment, the procedure used was similar to the first, except that a schematic clockface (for which the small numbers are to the right side and the large numbers to the left; see Figure 3.8b) was used instead of the ruler. In this case, a mapping effect at least as large as that of Experiment 1 was obtained, but with response time being shorter for the mapping of > 6 to left and < 6 to right than for the other mapping. Bächtold et al. concluded, "Taken together, our two experiments thus point to an intriguing property of a mental image, that is, its division into a left and right 'representational hemi-field'" (p. 734).

Tlauka and McKenna (1998) conducted two experiments in which subjects studied a simple visual map or read a verbal description of the same map. In both cases, subjects responded with the hands crossed or uncrossed, as described for several studies earlier in the chapter. In the first experiment, prior to performing, subjects were shown the map with two letters, one located in a location to the left side and one in a location to the right side. Once subjects had memorized the map, they performed several blocks of trials without the map being visible. In those blocks, subjects were to respond to the letter stimuli, and "before each response subjects were to imagine the map which they had previously studied. It was emphasized that subjects should press the appropriate key only once they had a 'clear' image of the map" (p. 71). Responses were about 100 ms faster on average when the letter location in the map and the response location corresponded than when it did not, and this effect did not interact with whether the hands were crossed or uncrossed. Tlauka and McKenna's second experiment was similar to the first, except that subjects were instructed verbally to imagine a map with one letter positioned in the middle left section and the other in the middle right section. Again, a compatibility effect of at least 100 ms was obtained that did not depend on hands placement.

The studies by Bächtold et al. (1998) and Tlauka and McKenna (1998) emphasize the point that S-R compatibility effects depend on mental representations. They also are consistent with the view that mental imagery and perception are functionally equivalent in many respects.

3.4.2 INTENSITY–FORCE COMPATIBILITY

Romaiguère, Hasbroucq, Possamaï, and Seal (1993) had subjects respond to the intensity of a stimulus (low or high) by making a response with a certain amount of force (weak or strong). They found that responses were faster when the low-intensity stimulus was mapped to the weak response and high-intensity stimulus to the strong response than with the opposite mapping. Mattes, Leuthold, and Ulrich (2002) replicated the intensity–force compatibility effect with visual stimuli and with auditory stimuli as well. For the auditory condition, subjects responded to low- or high-intensity tones by making a weak or strong response on a force-sensitive key. Similar to the visual condition, faster responses were obtained when the weak response was made to the low-intensity tone and strong response to the high-intensity tone than with the alternative mapping.

Moreover, Mattes et al. (2002) found that the intensity–force compatibility effect was larger for auditory (120 and 95 ms for their Experiments 2 and 4, respectively) than visual stimuli (35 and 59 ms for their Experiments 1 and 3, respectively), and that the larger effect obtained with auditory stimuli was not a result of different intensity differences between the two modalities. Because the magnitude of the intensity–force compatibility effect varies as a function of stimulus modality, Mattes et al. suggested that the codes involved in intensity-to-force translation are analog representations of the physical and sensory properties of the stimulus rather than conceptual or linguistic categories such as large versus small.

3.4.3 NUMEROSITY COMPATIBILTY

Miller, Atkins, and Van Nes (2005, Experiment 1) had subjects respond to one or two short tones with one or two taps of a response key. They found a numerosity compatibility effect in which response initiation times were 152 ms shorter when the number of responses generated by the participant matched the number of stimuli presented than when it mismatched. Because the numerosity compatibility effect obtained with auditory stimuli could be associated to making rhythmic responses to them, Miller et al. used visual stimuli in their second experiment. For the visual condition, one or two simultaneously presented rectangles appeared above or below a fixation point and subjects were instructed to respond to the number of rectangles with a compatible or incompatible mapping. They obtained a smaller, but highly significant, 95-ms numerosity compatibility effect. Miller et al. also showed that the numerosity compatibility effect is obtained with bimodal stimuli (auditory, visual, or one of each) and with digit (i.e., "1" and "2") stimuli.

In addition to measuring RT in terms of response initiation (time from stimulus onset to pressing the response key), Miller et al. (2005) examined response execution times by measuring the interval between the successive taps for the two-tap responses. A compatibility effect or tendency toward one was also obtained for the inter-response interval between the first and second taps, in which the mean interval was shorter for the compatible mapping of stimulus number to response taps. This finding suggests that numerosity compatibility influences response production as well as response initiation.

3.5 CHAPTER SUMMARY

Spatial S-R compatibility effects occur in many situations. Compatibility effects for different spatial stimulus sets mapped to spatial response sets extend to larger arrays than those studied initially by Fitts and Seeger (1953). Although S-R compatibility effects are largest when many stimuli are mapped to many responses, an effect of at least 50 ms is typically obtained in two-choice tasks. Because two-choice tasks allow isolation of the factors involved in compatibility effects, they have been used extensively to investigate the nature of such effects. Spatial compatibility effects depend on the frames of reference with respect to which stimuli and responses are coded. Adding reference frames to a complex array can improve performance, and humans seem to be quite flexible in their ability to code stimuli and responses not only with regard to environmental frames of reference but also with regard to egocentric frames of reference. S-R compatibility effects occur for verbal stimuli and responses, mentally imaged spatial relations, and so on. However, the magnitude of the effects varies as a function of set-level relations, being largest when the stimulus and response sets are themselves highly compatible. Although S-R compatibility effects are reduced with practice, the effects typically remain quite large even after extensive practice.

4 Correspondence of Irrelevant Stimulus Information and Responses: The Simon Effect

4.1 INTRODUCTION

One of the more important facts about compatibility effects is that they occur for stimulus dimensions that are irrelevant to the task at hand. That is, even though a dimension along which stimuli vary may be defined as irrelevant, responses are often faster and more accurate when the value of that stimulus dimension corresponds with the response signaled by the relevant stimulus dimension than when it does not. As with S-R compatibility proper, correspondence effects for an irrelevant stimulus dimension with a response dimension have been studied most extensively using spatial location, in which case responses are faster when the irrelevant stimulus location corresponds with the location of the response that is to be made. The *Simon effect*, introduced in Chapter 1, is the name given to this advantage in response selection for corresponding S-R locations for tasks in which the stimulus location is irrelevant and the relevant dimension (e.g., color) is not spatial in nature. The Simon effect has received increasing emphasis in recent years in studies of perception–action relations, as indicated by the fact that entering "Simon Effect" as a keyword into the PsycINFO index in the first week of January 2006, yielded 151 entries, 87 of which were published in the year 2000 or later. Correspondence effects of this type for other dimensions such as stimulus and response duration are often called Simon-type effects.

As noted in Chapter 1, Simon and Small (1969) had subjects respond to high- and low-pitch tones, presented to the left or right ear, by pressing a left or right key. Responses were faster and more accurate when the pitch assigned to the right response was presented in the right ear and pitch assigned to the left response was presented in the left ear (see Figure 4.1). In fact, with auditory stimuli, the magnitude of the Simon effect (approximately 40 to 60 ms) was comparable to the magnitude of the S-R compatibility effect obtained when location was relevant. Simon and Small attributed the Simon effect to a natural tendency to react toward the source of stimulation.

Another version of the Simon effect, sometimes referred to as the spatial Stroop effect, occurs when the stimulus location is irrelevant to the task but the relevant dimension also conveys location information (e.g., the word LEFT or RIGHT). In

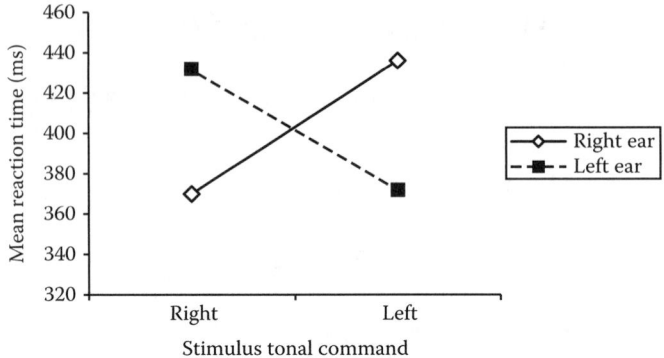

FIGURE 4.1 Results of Simon and Small's (1969) experiment showing an auditory Simon effect.

this case, responding is faster if a left keypress is to be made to the word LEFT when it appears in the left position than when it appears in the right position, and vice versa for the word RIGHT. Simon and Rudell (1967), in a study also mentioned in Chapter 1, in fact demonstrated an effect of irrelevant location with this task prior to Simon and Small's (1969) study for which the relevant stimulus dimension did not involve location. In addition to finding a 42-ms correspondence benefit for stimulus location with response location (and the relevant word name) when the ear in which each word would occur was uncertain, Simon and Rudell found a 26-ms correspondence benefit even when all stimuli within a trial block were presented to the same ear. Although a distinction has been made between the Simon and spatial Stroop effects (Kornblum, 1992), and separate lines of research have been devoted to each, both types of effects seem to have a similar processing basis (Lu & Proctor, 1994, 1995). Consequently, for the purpose of the present chapter, this distinction will not be emphasized, and both types of effects will be referred to as Simon effects.

4.2 SPATIAL STIMULI AND RESPONSES

4.2.1 Basic Phenomena

Like S-R compatibility proper, the Simon effect is a highly reliable finding that is also evident with tactile stimuli (Hasbroucq & Guiard, 1992) and, more importantly, with visual stimuli. For example, Umiltà and Nicoletti (1985) had subjects respond to red or green stimuli with left or right keypresses and obtained a 21-ms Simon effect. The Simon effect also occurs when the relevant visual dimension is geometric form (Nicoletti & Umiltà, 1989), letter identity (Proctor & Lu, 1994), bright versus dim stimulus intensity (Proctor, Lu, & Van Zandt, 1992), and so on. As noted in Chapter 1, the Simon effect is typically attributed to automatic activation of the response code that corresponds with the stimulus location. Consistent with this view, analysis of the lateralized readiness potential on trials for which stimulus and response location do not correspond shows an early component (peaking around 200 to 250 ms after stimulus onset) favoring activation of the corresponding response, which then shifts to a later component favoring the correct noncorresponding response (e.g., Sommer, Leuthold, & Hermanutz, 1993; Valle-Inclán, de Labra, & Redondo, 2000). Also, as expected if the Simon effect were due to automatic activation of the corresponding response code, the effect is obtained with nonhuman animals, such as pigeons, that are sensitive to spatial locations and can discriminate between symbolic cues (Urcuioli, Vu, & Proctor, 2005).

For humans, the Simon effect obtained with visual stimuli is usually smaller (typically in the range of 15 to 30 ms) than that obtained with auditory stimuli (typically in the range of 40 to 60 ms). One reason auditory stimuli yield larger Simon effects than do visual stimuli is that, as originally suggested by Simon (1968), they tend to elicit an automatic orienting toward the stimulus location. This is evident in a study by Simon, Craft, and Small (1970) in which auditory noise stimuli were presented in addition to the relevant tones of two different pitches. Subjects made a left or right keypress to a 96-dB high- or low-pitch tone presented to the left or right ear. In their Experiment 2, a noise tone of lower (74 dB) or higher (106 dB) intensity was present in the ear opposite the relevant tone on two-thirds of the trials. The Simon effect was 65 ms when there was no accompanying noise tone, 25 ms when the lower intensity noise occurred in the opposite ear, and –15 ms when the higher intensity noise occurred in the opposite ear.

Noise not only can counteract the Simon effect for the location of the relevant tone stimulus, but it can also increase the effect size. In Simon et al.'s (1970) Experiment 3, an 89-dB noise burst was presented in the ear opposite the relevant tone on one third of the trials and in the same ear as the relevant tone on another third of the trials. The 50-ms Simon effect when there was no noise reversed to –10 ms when the noise was in the opposite ear, but increased to 75 ms when the noise was in the same ear as the relevant tone. As described later, a visual noise stimulus in the location opposite the relevant visual stimulus does not reduce the magnitude of the visual Simon effect (e.g., Proctor & Lu, 1994), emphasizing the point that the auditory Simon effect has an orienting component to it that the visual Simon effect does not.

As might be expected from the results suggesting that attention is automatically directed to an auditory stimulus, an irrelevant auditory stimulus can also produce a Simon effect when the relevant stimulus dimension is visual. Simon and Craft (1970) had subjects make a unimanual aimed movement from a start position to the position of a left or right key when it was lit. On some trials, an accompanying tone was presented to the left or right ear. A 36-ms Simon effect was obtained, with responses faster when the tone occurred in the ear corresponding to the response than when it occurred in the opposite ear. Simon and Pouraghabagher (1978) obtained a similar effect for the location of an irrelevant tone when the visual discrimination involved a centered letter. This form of the Simon affect is sometimes called the accessory–stimulus version, with reference to the auditory tone being an accessory to the relevant visual stimulus.

It is of interest to note that a similar effect appears to occur when the auditory stimulus is relevant and the visual stimulus irrelevant. Ragot, Cave, and Fano (1988) presented a brief visual stimulus in a left or right location simultaneous with a brief tone presented from a left or right loudspeaker. In agreement with the findings of

Simon and colleagues, when visual location was relevant and tone location irrelevant, a 55-ms Simon effect for correspondence of the irrelevant tone with the visual stimulus (and response) location was obtained. When auditory location was relevant and visual location irrelevant, a 58-ms Simon effect as a function of visual location was obtained. Due to the fact that stimulus location was both the relevant and irrelevant dimension in this study, the apparent Simon effect for visual stimuli could be due to correspondence with tone location rather than with the response signaled by the tone.

4.2.2 Spatial Coding

As with S-R compatibility proper, most accounts of the Simon effect have emphasized spatial coding. The Simon effect is obtained primarily when stimuli and responses share spatial codes (e.g., convey left–right location information). For example, Simon and Craft (1970) found that providing visual stimulation to the left or right eye without having the perceived location of the stimulus varying in left–right locations did not produce a Simon effect. Thus, their results suggested that an anatomical difference in the eye stimulated is not sufficient to yield a Simon effect, but rather that the stimuli must be perceived and coded as being at distinct locations for an effect to occur (however, see Valle-Inclán, Hackley, & de Labra's, 2003, results described later in the chapter).

Umiltà and Liotti (1987, Experiment 3) conducted an experiment much like that described for S-R compatibility proper in Chapter 3, in which two boxes occurred to the left or right of fixation and a target stimulus was presented in one of the boxes. In this case, the task was a Simon task in which the subject was to press one response key for a square and the other for a rectangle of greater width than height. A Simon effect of 21 ms with respect to relative stimulus position was obtained when the target appeared 500 ms after the squares, but no effect was evident when their onsets were simultaneous. A comparable Simon effect was obtained for stimulus side in their Experiment 4, in which all four boxes were presented, two with dashed borders to provide referent objects and two with solid borders, one in which the stimulus occurred.

As for S-R compatibility proper, Simon effects are obtained regardless of whether the limbs remain in their natural left–right positions or are crossed to the contralateral sides of the body. Simon, Hinrichs, and Craft (1970) were the first to demonstrate this point. In their Experiment 1, a high- or low-pitch tone was presented to the left or right ear, and subjects were to make a left keypress to a high-pitch tone and a right keypress to a low-pitch tone, or vice versa. In some trial blocks, the hands were uncrossed, and in others they were crossed. Responses were 28 ms slower with the hands crossed than with the hands uncrossed, but the Simon effect was a function of spatial location and was of similar magnitude for the two hand placements: 65 ms with the hands uncrossed and 56 ms with the hands crossed. Roswarski and Proctor (2000, Experiment 2) confirmed this result, finding a 46-ms auditory Simon effect that did not interact with hand placement.

In a widely cited study, Wallace (1971) showed similar results for the visual Simon task, using shape as the relevant stimulus dimension. Wallace had subjects make a left or right keypress to the stimulus form (square or circle), with the hands uncrossed in one trial block and crossed in another. He found a Simon effect of 54 ms when the hands were uncrossed and 51 ms when they were crossed, indicating that the effect was due entirely to spatial correspondence. Wallace (1972) obtained similar results when subjects were brought into the room blindfolded and a halter was placed over the response keys, so that the coding of response location could not be based on visual cues.

With color as the relevant stimulus dimension, Umiltà and Nicoletti (1985, Experiment 1) obtained a 23-ms Simon effect when the hands were crossed, compared with a 21-ms effect when they were uncrossed. Effects of similar size were obtained in their Experiment 2, which used only the uncrossed placement, when both stimuli and responses were in the left hemispace or right hemispace. This result emphasizes the point that relative location of the stimuli is crucial, regardless of whether the hands are crossed or uncrossed.

Wascher, Schatz, Kuder, and Verleger (2001) compared the crossed–uncrossed hands manipulation for the visual and auditory Simon tasks, varying whether instructions specified responses in terms of the fingers on the left or right hand (hand mapping) or in terms of the left or right response location (key mapping). Their Experiments 1 and 2, which were conducted as a single within-subjects design, examined the visual and auditory Simon effects, respectively. With visual stimuli, the key- and hand-mapping instructions yielded similar magnitude Simon effects of 24 and 25 ms with the hands uncrossed and 32 and 23 ms with the hands crossed. The interaction of instructions with hand placement did not approach statistical significance. With auditory stimuli, although the key- and hand-mapping instructions yielded similar magnitude Simon effects of 46 and 41 ms with the hands uncrossed, they did not with the hands crossed: In that case, the key-mapping instructions yielded a Simon effect of 35 ms and the hand-mapping instructions only a nonsignificant 3-ms effect. However, the latter difference, suggesting an effect of tone–hand correspondence in the auditory Simon task with overlapped hands, was based on only nine subjects who experienced all conditions of both Experiments 1 and 2. Moreover, in contrast to all other studies, neither instruction condition showed overall slowing when subjects performed with crossed hands versus uncrossed hands.

Because of the unusual nature of Wascher et al.'s (2001) results, Roswarski and Proctor (2003) attempted to replicate them in distinct experiments in which each subject performed with only a single mapping, using a crossed-hands placement in one trial block and an uncrossed-hands placement in another. Their Experiment 1 did not replicate the nonsignificant tendency toward an interaction reported by Wascher et al. for visual stimuli, with the Simon effect being 19 and 18 ms for key- and hand-mapping instructions with the hands uncrossed and 24 and 28 ms with the hands crossed. Likewise, their Experiment 2 showed no indication of an interaction of the type Wascher et al. reported for auditory stimuli: The Simon effect was 49 and 38 ms for key- and hand-mapping instructions with the hands uncrossed and 46 and 33 ms with the hands crossed. In Roswarski and Proctor's Experiment 3, in which subjects received considerably more trials (two sessions of 400 trials each), as would have been the case across conditions in Wascher et al.'s study, a shift of the auditory Simon effect in the direction of Wascher et al.'s results was obtained. For the first session, the Simon effect tended to be larger with the hands crossed

(49 ms) than with them uncrossed (31 ms), but in the second session, it tended to be smaller with the hands crossed (27 ms) than with them uncrossed (38 ms). Thus, the data suggest that although the most critical factor in both the visual and auditory Simon effects is the correspondence between stimulus and response locations, the correspondence between stimulus location and response hand may come to play a role in the auditory Simon effect after extensive practice.

Consistent with the point that response location, and not the hand distinction, is the critical factor for producing the Simon effect, Heister, Ehrenstein, and Schroeder-Heister (1987) demonstrated the Simon effect when responses were made with two fingers on the same hand. Subjects performed using the left or right hand in a prone or supine posture, making keypresses with the index or middle finger to the red or green color of visual stimuli. A Simon effect as a function of correspondence of spatial positions was obtained with the hands in the prone posture (49 ms) and the supine posture (73 ms).

Relative location coding is also important for producing the Simon effect with tactual stimuli. Hasbroucq and Guiard (1986) had subjects perform tactual Simon tasks in which the index and middle fingers of each hand were placed on response keys that could deliver vibratory stimuli. One of two fingers could be stimulated with high- or low-intensity vibration, and one of the other two fingers was used to execute the response. In Experiment 1, the vibrated fingers were either the two middle or two index fingers, and the responding fingers were the alternative pair. In Experiment 2, the stimulated fingers were on one hand and the responding fingers on the other, and in Experiment 3, the leftmost or rightmost finger on each hand served for stimuli and the others for responses. Simon effects averaging 55 ms were obtained in all three conditions, indicating that each member within the pair of stimulus fingers and the pair of response fingers was coded in terms of relative location, regardless of the specific fingers involved in the pairs.

4.2.3 ACTION GOALS AND REFERENCE FRAMES

4.2.3.1 Action Effects

Hommel (1993a) demonstrated that spatial coding in the Simon task is a function of task goals, as Riggio, Gawryszewski, and Umiltà's (1986) results suggested for S-R compatibility proper. In his study, subjects were presented with high- and low-pitch tones to the left or right ear and responded by pressing a key that turned on a light located in a left or right position (see Figure 4.2a). In one condition, the left key operated the left light and the right key operated the right light. In two other conditions, the left key operated the right light and the right key operated the left light: For one of those conditions, subjects were told to press one key in response to the high-pitch tone and the other key in response to the low-pitch tone, whereas for the other condition, subjects were told to turn on one light in response to the low-pitch tone and the other light to the high-pitch tone. When the instructions were in terms of the keys, the Simon effect was 73 ms for the condition in which the key operated the ipsilateral light and 52 ms for the condition in which it operated the contralateral light. The smaller Simon effect in the latter condition suggests that the light was coded as part of the response action. More interesting, the Simon effect

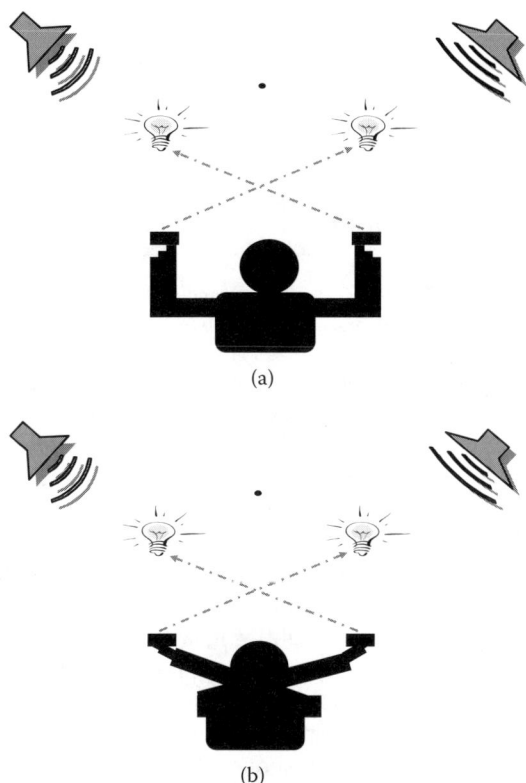

(a)

(b)

FIGURE 4.2 Depiction of Hommel's (1993a) study, with (a) normal and (b) crossed-hands placements. Pressing a left or right key in response to the pitch of a tone presented from a left or right speaker turned on a light on the opposite side from the keypress.

reversed to a –30 ms for the contralateral condition when the instructions were in terms of turning on the lights, indicating that the direction of the Simon effect followed the location of the light, even though the keypress was made on the other side. The fact that this reversed effect was smaller than the positive effect obtained with key instructions indicates that the key location could not be ignored entirely.

In Hommel's (1993a) Experiment 2, instructions were always in terms of turning on the left or right light, and a crossed-hands manipulation was added to the light-key manipulation of the previous experiment to allow dissociation of stimulus correspondence with the key and hand from that with the light (see Figure 4.2b). As for the contralateral condition in Experiment 1 when instructed in terms of turning on the lights, the instructed stimulus-light relation determined the direction of the Simon effect, but the other response features influenced performance as well, with their effects combining additively.

Hommel (1996b) provided a stronger test of whether an action effect could automatically affect performance. Subjects performed a visual Simon task in which

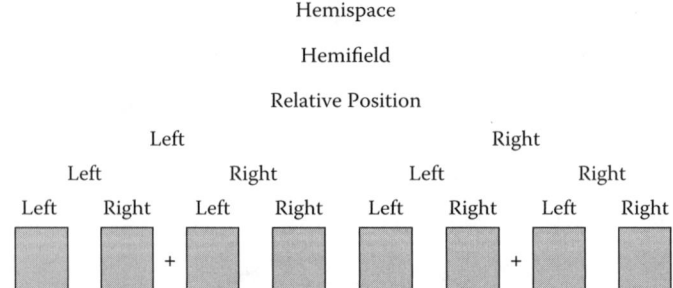

FIGURE 4.3 Depiction of Lamberts, Tavernier, and d'Ydewalle's (1992) stimulus array, with left and right hemifields, hemispaces, and relative positions indicated.

they made a left or right keypress to a red or green stimulus presented to the left or right. For the first or second half of the experiment, when a response was made, a completely irrelevant tone was presented through a loudspeaker on the opposite side from the key. The Simon effect was 28 ms without the tone and 18 ms with it, indicating that the auditory–action effect was coded in opposition to the key location.

Grosjean and Mordkoff (2002) confirmed Hommel's (1996b) results and provided evidence that postresponse stimuli have both facilitatory and inhibitory effects. Their subjects performed a normal Simon task, responding to the letter H or S, presented at fixation or to the left or right of it, with a left or right keypress. Visual and auditory postresponse stimuli (an expanding plus sign and a rising tone) occurred to the same side as the response in one set of trials, to the opposite side in another, and a third block contained no postresponse stimulus. When the postresponse stimuli occurred to the opposite side of the response, the Simon effect was reduced from 30 ms to 23 ms, whereas when they occurred on the same side, the effect was increased from 30 ms to 47 ms.

4.2.3.2 Multiple Reference Frames

Several studies have provided evidence that stimuli are coded relative to multiple frames of reference. Lamberts, Tavernier, and d'Ydewalle (1992, Experiment 2) used a display for which the stimulus could occur in any of eight possible locations (see Figure 4.3). A trial began with presentation of a fixation sign to the left or right side of the display for 500 ms. This event acted as a precue designating the hemispace in which the stimulus would appear. At the end of that time, two outline boxes appeared to one side or the other of the fixation sign. One of the boxes contained the target stimulus, a square or circle, to which a left or right keypress was to be made. A Simon effect occurred with respect to all three reference frames. That is, responses were approximately 10 ms faster when hemispace corresponded with response location than when it did not, 15 ms faster when the hemifield for the two boxes corresponded with the response than when it did not, and 20 ms faster when the relative position of the stimulus within the two boxes corresponded with response location than when it did not.

Roswarski and Proctor (1996) replicated Lamberts et al.'s (1992) experiment, but in addition to using the circle–square stimulus set used by Lamberts et al., they used a red–green stimulus set for which the relevant discrimination was easier and a rectangle–square stimulus set for which it was more difficult. The overall Simon effects, collapsed across the three reference frames of hemispace, hemifield, and relative position, were decreasing functions of the difficulty of the relevant discrimination and, consequently, overall RT, being 16 ms for red/green color, 10 ms for circle/square, and 6 ms for rectangle/square. This outcome is consistent with the hypothesis that location codes are activated only transiently, with their strength decreasing across time, as described in Section 4.2.4.

Recently, Lleras, Moore, and Mordkoff (2004) reported results similar to those of Lamberts et al. (1992) and Roswarski and Proctor (1996) for displays containing only a fixation cross. An exogenous cue appeared for 67 ms at the middle of three locations to the left or right of fixation, and then offset 133 ms before stimulus onset at one of the locations on the side. Simon effects of 11 ms relative to the fixation cross and 12 ms relative to the cue location were obtained.

Logan (2003) provided evidence that multiple response codes are activated when subjects are engaged in a typing task. In his experiments, experienced typists rested their hands on the home row of a computer keyboard, as they would when typing, and then typed the displayed stimulus as quickly as possible using the fingers that they normally use in typing. In Experiments 1 and 2, the stimuli were single letters presented to the left and right of fixation or above and below fixation, respectively. Simon-type spatial correspondence effects were observed in both cases, with RT shorter when the stimulus location corresponded with the keyboard location of the letter to be typed than when it did not.

For Logan's (2003) Experiment 3, the stimuli were three-, four-, and five-letter words, presented to the left or right of fixation, to which the entire sequence of letters was to be typed in response. Some of the words were composed of letters typed entirely with one hand, whereas others included letters typed by both hands. Previously, Inhoff, Rosenbaum, Gordon, and Campbell (1984) demonstrated S-R compatibility effects for movement sequences. RT to initiate a sequence of keypresses made with the fingers of one hand was shorter with a compatible mapping of stimulus location to hand than with an incompatible mapping, and this compatibility effect was smaller when the last two presses of a sequence of four were made with a different hand than the first two presses. Logan obtained similar results for the Simon task, showing a Simon effect for RT to the first letter for words typed with one hand but not for words for which the first letter was typed by one hand and the remaining letters by the other hand. Logan's findings, along with those of Inhoff et al., suggest that the occurrence of a stimulus in a left or right location activates the first two or more responses in the response sequence. Logan's results also indicate that correspondence of irrelevant stimulus location with the response that is to be made can affect performance of operators performing tasks at which they are highly skilled.

4.2.3.3 Wheel Rotation Responses

Wheel rotation responses also provide a valuable means for examining the influence of multiple reference frames. As noted in Chapter 3, clockwise and counterclockwise wheel rotations can be coded as left or right based on several relations. When the hands are positioned at the top of the wheel, the direction of hand movement is the same as the direction of the wheel movement, but when both hands are positioned at the bottom, the direction of hand movement is opposite that of the wheel movement.

Guiard (1983) used a Simon task to examine factors that influence whether wheel rotation responses are coded as left and right with respect to a wheel-referenced or hand-referenced frame. In his experiments, one response was to be made to a high-pitch tone and the other to a low-pitch tone, with the ear in which the tone occurred being irrelevant. Guiard showed that a positive Simon effect (faster responses when tone location corresponds with the typical direction of rotation) occurred when the wheel was held at its sides or at the top. However, no overall Simon effect occurred when the hands were at the bottom of the wheel and, consequently, the hand- and wheel-referenced frames were in opposition.

Guiard (1983) found, though, that this absence of Simon effect in the mean RT data when the hands were placed at the bottom was not due to individual subjects showing little or no Simon effect. Instead, approximately half of the subjects showed a positive Simon effect of substantial magnitude and half a negative Simon effect (slower responses when the tone location corresponds with the typical direction of rotation than when it does not) of similar magnitude. The reverse Simon effect implies that the hands were used as the frame of reference because the trials for which direction of movement corresponded with the stimulus location yielded faster responding than those for which it did not. These individual differences suggest that, when wheel- and hand-referenced frames are in opposition, some people rely primarily on wheel-referenced coding and some on hand-referenced coding. Wang, Proctor, and Pick (2003) replicated this finding and showed that when the sign of the Simon effect was disregarded by taking the absolute value for each subject, the mean Simon effect size with the bottom hand placement was similar to that with the hands at the top or sides of the wheel. This outcome implies that, although some people used a hand-referenced frame and others a wheel-referenced frame when responding with their hands at the bottom of the wheel, the group as a whole showed Simon effects that were just as strong as those for people who responded with their hands at the sides or top of the wheel.

For subjects using the bottom hand placement, Wang et al. (2003, Experiment 2) also varied whether the instructions described responses in terms of hand movement or wheel movement. With hand-movement instructions, a negative Simon effect was obtained for all subjects. This outcome indicates that these instructions were effective in getting the subjects to code the responses relative to the direction of hand movement. However, with wheel-movement instructions, only about half of the subjects showed a positive Simon effect indicative of wheel-referenced coding, with half still showing a negative Simon effect indicative of hand-referenced coding. Thus, subjects who tend to use wheel-referenced coding when instructions are neutral apparently use hand-referenced coding when the hands are emphasized, but those who tend to use hand-referenced coding with neutral instructions apparently continue to use that reference frame even when the instructions emphasize wheel movement.

Although wheel-movement instructions are not sufficient to cause all subjects to code responses with respect to a spatial reference frame when responding with their hands at the bottom, both Guiard (1983) and Wang et al. (2003) found that all subjects do so when the instructions are neutral but the wheel controls a cursor that moves in a direction consistent with the wheel movement. In this case, the cursor movement is coded as the intended action effect, and a positive Simon effect is obtained. Wang et al. also found that the cursor movement had to be controlled directly by the wheel. When movement of the cursor was triggered at completion of an 8° wheel rotation, the cursor movement did not influence the Simon effect. As found without a cursor, approximately half of the participants showed a positive Simon effect and half a negative effect. Wang, Proctor, and Pick (2006) reduced the amount of rotation at which the cursor movement was triggered to 3°, to place the cursor motion in closer temporal proximity to the response, but still found no influence of the triggered movement on the Simon effect.

Although a triggered cursor does not in itself serve as an action effect, Wang et al. (2005) found evidence that it could do so if the subject had some experience controlling the cursor. In their Experiment 1, subjects performed three trial blocks (no cursor, triggered cursor, controlled cursor) with their hands at the wheel bottom and were instructed in terms of direction of hand movement. The triggered cursor condition again yielded a large negative Simon effect indicative of the instructed hand coding, with an exception occurring when that condition immediately followed the controlled cursor condition. In that case, the Simon effect shifted to a slightly positive 6 ms, indicating that the triggered cursor was serving to some extent as an action effect. In Experiment 4, the wheel controlled the cursor during three different periods of each trial for different groups of subjects: (a) between onset of a warning signal and the auditory imperative stimulus, (b) from onset of the imperative stimulus until completion of the response, and (c) during the interval between response completion and onset of the warning for the next trial. Not too surprisingly, a positive Simon effect of 15 ms, indicative of the cursor movement being coded as an action effect, was obtained when the cursor was controlled during the response period itself. However, the Simon effect was even larger when the cursor was controlled only during the warning interval (24 ms) or intertrial interval (39 ms). Control of the cursor when the imperative stimulus was not present was sufficient to allow the triggered cursor movement to be coded as a consequence of the subject's actions.

4.2.4 TIME COURSE OF ACTIVATION

Because the spatial coding that produces the Simon effect is presumed to be automatic and irrelevant to the task, the possibility exists that its effect will be greatest shortly after stimulus onset and will dissipate with the passage of time. Simon, Acosta, Mewaldt, and Speidel (1976) suggested that the activation produced by a visual stimulus dissipates over a period of about 250 ms. In their study, subjects made a left or right keypress to the color of a red or green light that occurred in a left or right location. They were asked to withhold their response until a go signal, a binaural tone, was presented after a delay of 0, 150, 250, or 350 ms. The Simon effect was 35 ms at the 0-ms delay, 20 ms at the 150-ms delay, 12 ms at the 250-ms delay, and a nonsignificant 2 ms at the 350-ms delay.

In Simon et al.'s (1976) Experiment 2, the onset of the key labels (red or green) was manipulated. The mapping of tone pitch (high or low) to response key color was constant for each subject. In one condition, the key labels were fixed, but in the remaining five conditions, the key labels varied randomly from trial to trial, with the labels appearing 1 s prior to, simultaneously with, or 150, 250, and 350 ms after onset of the imperative auditory stimulus. When the key labels were fixed, the Simon effect was 62 ms. Large Simon effects were also obtained when the response-key labels were presented 1 s before the imperative stimulus (36 ms) or simultaneously (37 ms) with it. However, when there was a delay of 150 to 350 ms, the Simon effect was virtually eliminated, being 6 ms on average. In agreement with Simon et al.'s Experiment 1, this result indicates that there is an initial advantage for the corresponding S-R locations that decreases over time.

Hommel (1993b, 1994) developed a specific account to explain why the magnitude of the Simon effect decreases over time. According to Hommel's temporal overlap account, automatic activation of the corresponding response occurs quickly and then decays. As a result, the Simon effect can be reduced and eliminated when the temporal overlap between the coding of the relevant and irrelevant stimulus properties is reduced. Hommel (1993b) manipulated temporal overlap by varying the eccentricity at which stimuli were presented on the retina. Stimulus locations from 0.2° to over 6.0° from a central vertical line that divided the screen into left and right halves were used. When stimulus eccentricity (low, medium, or high) varied randomly from trial to trial, a Simon effect of 13 ms was obtained when eccentricity was low, and this effect was nonsignificantly reduced to 6 ms when eccentricity was high. However, when the stimulus eccentricity conditions were presented in pure blocks, the Simon effect was 23 ms when eccentricity was low, and this effect was −5 ms when eccentricity was high.

Similarly, temporal overlap can be manipulated by varying stimulus quality by partly masking the letter stimuli, as Hommel (1993b) did. When stimulus quality was high, a Simon effect of 16 ms was obtained, and the effect reversed to −24 ms when stimulus quality was low. Hommel also varied stimulus contrast and delay of stimulus formation to slow identification of the relevant stimulus dimension, and these manipulations produced results consistent with the notion that delaying the processing of the relevant stimulus attribute from the irrelevant spatial location reduces, eliminates, or reverses the Simon effect. Subsequent studies have also shown that the Simon effect is larger when the discriminability of the relevant dimension is high than when it is low (see, e.g., Lu & Proctor, 1994).

Consistent with the notion that automatic activation of the corresponding response is transient, De Jong, Liang, and Lauber (1994) showed that, within a condition, the Simon effect is smaller for those trials on which the RT is longer than for those on which it is shorter. De Jong et al. performed distribution analyses of the RT data in which the Simon effect was computed for separate quantiles of the RT distribution (see Chapter 1). They found that the Simon effect was positive at the shortest RT bins and then decreased in magnitude across RT bins, being negative at the longest bin.

Two accounts have been proposed for why the Simon effect is reduced, eliminated, or reversed when processing of the relevant stimulus dimension is delayed:

(a) The activation of the irrelevant dimension is actively suppressed, and (b) the activation of the irrelevant dimension decays over time. The active suppression account postulates that when the irrelevant code becomes available before the relevant code, inhibition can be initiated, and this inhibition will be stronger with time. The transient activation account postulates that the activation of the irrelevant spatial code decays over time, and a delay in processing the relevant code will result in weaker interference from the irrelevant code. It has not been determined which account is more accurate, but the transient activation account seems to be invoked more often as an explanation of the temporal properties for the irrelevant spatial code over the active inhibition account (see Hommel, 1993b, 1994; Proctor & Vu, 2002a; Simon et al., 1976).

Although numerous results with two-choice visual S-R sets arrayed along the horizontal dimension are consistent with the temporal overlap account, results from RT distribution analyses using different S-R sets are not. For visual stimuli, when the stimuli and responses are arrayed along the vertical dimension, distribution analyses show an increase in Simon effect size from the shortest RT bins to the longest ones (Proctor, Vu, & Nicoletti, 2003; Wiegand & Wascher, 2005). An increasing function is also found when the stimuli are arrayed along the horizontal dimension but the hands are crossed for responding (Wascher et al., 2001), and the effect size is relatively constant across the RT distributions for a Simon effect produced by left–right target location relative to the majority of flanking distractor stimuli (Ansorge, 2003). An increasing function is obtained for auditory stimuli that vary along the horizontal dimension, both in the standard auditory Simon task and in the accessory stimulus version (Proctor, Pick, Vu, & Anderson, 2005), when the hands are in a normal placement as well as when they are crossed (Wascher et al., 2001).

Although it is not clear at present exactly why the distribution analyses yield decreasing Simon effect functions for some conditions and flat or increasing ones for others, Wascher et al. (2001) proposed that the decreasing function is a consequence of automatic activation through privileged visual pathways that occurs only when responses are coded anatomically. Wiegand and Wascher (in press) obtained evidence consistent with this interpretation for the Simon effect obtained with vertical stimuli and responses. They noted that two previous studies using vertical arrangements had found decreasing Simon effects across the RT distribution (De Jong et al., 1994; Valle-Inclán and Redondo, 1998), and in both cases the relevant S-R mapping was designated randomly at the start of each trial. Wiegand and Wascher examined a task in which subjects responded to an A or B by pressing an upper or lower response key, comparing blocks in which a cue designated the mapping (in one experiment, a lowercase *a* and *b* located above and below two vertically arranged squares in which the target stimulus could appear; in another experiment, a centered circle or square) to blocks in which the mapping was held constant. Their results showed a decreasing function for the cued mapping condition but a slightly increasing function for the blocked mapping condition. Wiegand and Wascher attributed the decreasing function for randomized mappings to subjects coding the responses anatomically and thus receiving automatic activation by way of the visual pathways. Although Wiegand and Wascher's results are consistent with their two-process interpretation, another possibility is that the different functions

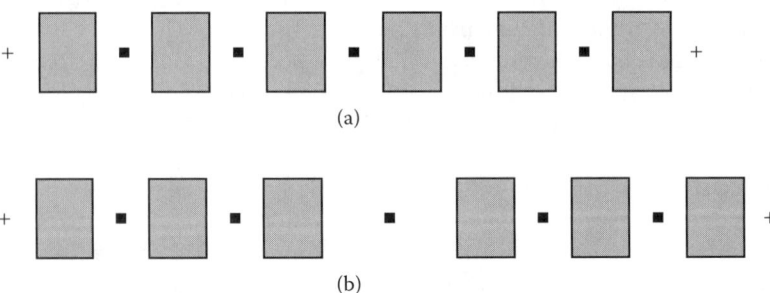

FIGURE 4.4 The display variants used by Nicoletti and Umiltà (1989) in their Experiment 3 (a) and Experiment 4 (b).

represent two parts of a single activation and decay function. That is, because responses were slower overall when the mappings were randomized than when they were blocked, as is customary (see Chapter 6), responses in the blocked mapping condition could have been made in the first half, and responses in the random mapping condition in the second half, of a single function of activation to a peak and then decay.

4.2.5 GENERATION OF SPATIAL CODES

The two types of spatial coding accounts for the Simon effect that have received the most emphasis are attention shifting and referential coding. The attention shifting account attributes spatial coding to the frame of reference provided by the location to which attention is shifted, whereas the referential coding account assumes that the spatial code is generated because the stimulus location is represented relative to the position of a reference object (e.g., a fixation cross) or objects.

4.2.5.1 Attention Shifting Account

Nicoletti and Umiltà (1989) were the first to provide evidence that "attention orienting plays a crucial role by yielding the spatial code of the stimulus" (Umiltà & Nicoletti, 1992, p. 339). They proposed that a shift of attention to a spatial location causes the spatial code to be formed that produces the Simon effect. Two implications of the attention shift explanation are that the spatial code will be relative to where attention is positioned at the onset of the stimulus and that if an attention shift does not occur, then no Simon effect should be found.

Nicoletti and Umiltà (1989, Experiment 3) tested the first implication using the display shown in Figure 4.4a, in which six empty boxes designated possible stimulus positions. Half of the subjects were to maintain fixation on a plus sign placed to the left of the leftmost box and half on a plus sign placed to the right of the rightmost box. At the start of a trial, a small solid square appeared at one of five locations between the boxes for 500 ms, and subjects were instructed to orient attention to this box while maintaining fixation on the plus sign. After another 500-ms interval, the imperative stimulus, a square or oblong rectangle, occurred in one of the boxes,

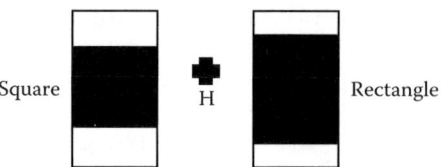

FIGURE 4.5   Stimulus displays used by Rubichi et al. (1997).

and the form was to be identified with a left or right keypress. A 47-ms Simon effect relative to the attentional focus (i.e., the location of the solid square) was obtained. Stimuli to the right of the square, regardless of where it was located, were responded to faster with the right response than the left response, and stimuli to the left of the square were responded to faster with the left response than the right response.

Nicoletti and Umiltà's (1989) Experiment 4 and Nicoletti and Umiltà's (1994) Experiment 1 yielded similar results for a display with a large central gap that grouped the boxes into two sets of three (see Figure 4.4b), indicating that locus of attention was more important to the Simon effect than the left–right grouping distinction created by spatial proximity. More important, Nicoletti and Umiltà's (1994) Experiments 3 and 4 provided evidence that the Simon effect does not occur in the absence of an attention shift. They used a display similar to that shown in Figure 4.4(b), with the large central gap, but without the solid squares used to direct attention. Instead, the fixation sign was presented at the center of the display, and the letter X, N, H, or W appeared immediately below it, simultaneous with the imperative stimulus. A response was to be made to the imperative stimulus if the letter was an X, N, or H but not if it was a W. The logic was that attention would have to remain at the centered location, rather than shifting to the location of the imperative stimulus, in order to detect the target. Consistent with the attention shifting hypothesis, no Simon effect was evident. Note that this contrasts with the large Simon effect obtained relative to the center under similar conditions, except for the letter stimuli, in Nicoletti and Umiltà's (1989) Experiments 1 and 2.

Rubichi, Nicoletti, Iani, and Umiltà (1997) used a similar procedure of presenting one of four letters at fixation, with the W signaling not to respond. Their display had only two rectangular boxes, one to the left and one to the right of fixation, and the imperative stimulus was a square that filled the box horizontally but not vertically or a rectangle that also filled the box horizontally but was slightly longer in the vertical dimension (see Figure 4.5). The idea was that this discrimination was difficult, which would allow more time for attention to shift. In Rubichi et al.'s Experiment 1, the letter was shown for 100 ms, and 500 ms later the square or rectangle was shown in one of the boxes for 100 ms. The rationale was that the interval between letter onset and onset of the imperative stimulus allowed time to process the letter before stimulus onset. Consequently, attention could be shifted to the location of the imperative stimulus. However, because the discrimination was difficult, attention would be shifting back to the fixation point prior to a response being made, resulting in a spatial code that was opposite that of the stimulus location. Consistent with this scenario, the results showed a 22-ms reverse Simon effect, for which the response that did not correspond to the stimulus location was faster than the one that did, and this effect was evident at all five equal RT bins except the one with shortest mean RT.

Rubichi et al.'s (1997) Experiment 2 was similar to their Experiment 1, except with the discrimination made easier by increasing the length of the vertical axis of the rectangle stimulus. In this case, a positive 18-ms Simon effect was obtained, with the effect decreasing across RT bins and even reversing by 16 ms at the longest one. Experiment 3 was also similar to Experiment 1, except that the stimulus duration was increased to 400 ms, with the reasoning that attention would not return to the center while the stimulus was still visible. This experiment also showed a positive Simon effect of 14 ms, which decreased across RT bins, reversing slightly to –7 ms at the longest bin. In a final experiment, Rubichi et al. used the easy discrimination of their Experiment 2 but had the letter indicating whether a response should be made appear 100 ms after offset of the 100 ms imperative stimulus. The letter could occur at fixation, as in the previous experiments, or to the left of the left box or the right of the right box. The Simon effect was a function of the position of the letter, which Rubichi et al. interpreted as occurring because an additional shift of attention had to be made after the shift to the imperative stimulus location because the letter occurred after it.

Stoffer (1991) distinguished a lateral shift of attention from attentional zooming, in which attention shifts from a more global level to a local level, and proposed that a spatial code is generated only when the last step in attentional focusing is a lateral attention shift. Stoffer derived his hypothesis from the results of Experiment 3 by Umiltà and Liotti (1987), mentioned earlier, in which two squares, one of which contained the imperative stimulus, appeared to the left or right of fixation to designate the side on which the stimulus would occur. When the boxes onset 500 ms before the imperative stimulus, a 21-ms Simon effect was obtained. However, when the boxes and stimulus appeared simultaneously, no Simon effect was evident. According to Stoffer's hypothesis, the 500-ms interval allowed attention to be shifted to the position between the boxes, resulting in a left or right shift of attention from that position to the imperative stimulus when it appeared. With simultaneous presentation, attention was first shifted to the side at which the boxes appeared and then zoomed down to the level of the imperative stimulus to identify it. Because the last step involved attentional zooming, no left–right spatial code was generated, and no Simon effect was obtained.

Stoffer (1991) replicated and extended Umiltà and Liotti's (1987) Experiment 3, adding a condition in which the two possible stimulus locations were cued by a large rectangle that encompassed both locations. The results for the two-box condition were similar to those reported previously, showing no Simon effect when the boxes were presented simultaneously with the imperative stimulus and a 51-ms Simon effect when they were presented 500 ms in advance. In contrast, the single-box condition showed no Simon effect even when it preceded the imperative stimulus by 500 ms. According to Stoffer's shift/zooming account, the absence of effect in this case is due to attention being at the level of the large rectangle when it is presented in advance, rather than at a location between the two possible stimulus locations, thus requiring that attention zoom in on the imperative stimulus.

Although the absence of a Simon effect in Stoffer's (1991) single-box condition is consistent with the hypothesis that the last operation must be a lateral attention shift if a Simon effect is to occur, Weeks, Chua, and Hamblin (1995) were not able to replicate this crucial aspect of his results. Their Experiment 1 was methodologically similar to Stoffer's experiment, and the results replicated the finding that both the two-box and single-box conditions showed negligible Simon effects when box onset was simultaneous with target onset. However, when box onset preceded target onset by 500 ms, the Simon effects for both conditions increased similarly in magnitude. Weeks et al. reported two other experiments that also showed comparable effect sizes for the single- and two-box conditions when box onset preceded onset of the target stimulus.

Hommel (1993c) also found Simon effects when a large box was presented 500 ms before the target stimulus. For each trial in his Experiments 2 and 3, the target stimulus (a red or blue rectangle) in one of two positions was accompanied by a reference stimulus (a dark green rectangle) in the other position. Simon effects of 29 and 24 ms, respectively, were obtained when the large box always occurred at the center of the display (Experiment 2) or appeared randomly in the left or right visual field (Experiment 3), as in Stoffer's (1991) study. Thus, the key finding in support of the zooming hypothesis, the absence of a Simon effect when the large box is presented in advance of the target stimulus, has not been supported in the studies of Hommel or Weeks et al. (1995).

Predictions of the attention shifting account have been tested by precuing stimulus location, with the general idea being that if a valid precue directs attention to the target location, then a shift in attention to the target stimulus at its onset should not occur and little or no Simon effect should be evident. Yet, the most widely obtained outcome is that the Simon effect does not vary in magnitude when stimulus location is validly precued compared with a neutral condition. Verfaellie, Bowers, and Heilman (1988) found that for a task in which the target stimulus had to be classified as bright or dim, only a nonsignificant Simon effect was evident both when stimulus location was precued and when it was not. Proctor, Lu, and Van Zandt (1992) obtained a small, but significant, Simon effect using a similar procedure, but replicated the lack of difference between the valid cue trials and the uncued trials. Zimba and Brito (1995) conducted a thorough study with a different methodology that used both endogenous and exogenous precues, and likewise found no influence of validly cuing stimulus location. The attention shifting hypothesis also predicts that a large Simon effect should be obtained on invalidly cued trials because attention must subsequently be redirected to the target stimulus location at its onset. Verfaellie et al. did not report results for their invalidly cued trials, but Proctor et al. and Zimba and Brito did report those results, and neither study found this predicted outcome. In Proctor et al.'s study, the Simon effect on invalidly cued trials was not statistically significant, and in Zimba and Brito's study, the effect size on those trials did not differ from that on validly cued trials.

Other evidence has suggested that a shift in attention is not necessary to produce a Simon effect. Valle-Inclán et al. (2003) presented subjects with monocular stimuli that were fused within binocular space to be perceived as located in the center of the visual field. The attention shift hypothesis predicts that no Simon effect should be obtained in this situation because attention should remain at the center of the visual field on all trials. As described earlier, Craft and Simon (1970) reported that a Simon effect was not obtained under such circumstances, but Valle-Inclán et al. note that the stimuli used by Craft and Simon were not optimal for generating a spatial code for each eye and no measurement was made as to whether subjects in that experiment could discriminate stimulation of the left eye from that of the right eye. In addition to having subjects perform a Simon task, Valle-Inclán et al. had them perform a discrimination task in which they had to indicate to which eye a stimulus was presented. The Simon task showed a Simon effect of 17 ms as a function of whether the left or right eye was stimulated. That a Simon effect can be obtained solely on the basis of which eye was stimulated under conditions in which the stimuli are perceived to occur at the same, central location suggests that a Simon effect can be obtained in the absence of an attention shift. Of some interest, the magnitude of this Simon effect did not correlate with how well subjects were able to discriminate stimulation of the left or right eye, suggesting that the eye-stimulated code producing the Simon effect may be outside of conscious awareness.

4.2.5.2   Referential Coding Account

Hommel (1993c) presented evidence against the attention shifting account and proposed instead the referential coding hypothesis, according to which "the stimulus code is not spatially coded in reference to, or depending on, the focus of spatial attention, but in reference to an intentionally defined object or frame of reference (or several of them)" (p. 209). An implication of the referential coding hypothesis is that the Simon effect should occur when a reference object is present in the display. In Hommel's aforementioned experiments that showed a Simon effect when the two stimulus locations were precued by a large surrounding box, contrary to the attention-zooming hypothesis, all trials included a reference object in the position opposite that in which the target stimulus occurred. Other studies have shown that, even without the surrounding box, Simon effects are obtained when a noise, or referent, item occurs in the position opposite the target. For example, Grice, Canham, and Boroughs (1984) presented a target S or H in a left and/or right position. For trials on which only a single instance of the target was present, an irrelevant noise letter, Y, occupied the opposite position. Simon effects of 45 and 23 ms were obtained when the spacing between letters was large and small, respectively. Proctor and Lu (1994) similarly obtained a 22-ms Simon effect with a small spacing, and this effect was larger than the nonsignificant 6-ms Simon effect obtained for trials on which no noise letter was present. Although Proctor and Lu reported other experiments that showed a larger Simon effect with the noise stimulus than without, this difference has not been consistently reported in other studies (e.g., Grice, Canham, & Gwynne, 1984; Wascher et al., 2001). Regardless of this detail, it is apparent that, unlike for auditory stimuli, an irrelevant visual stimulus to the side opposite the relevant stimulus does not reduce or eliminate the Simon effect.

Some of the strongest support for the referential coding hypothesis comes from the studies cited earlier in the chapter showing that the Simon effect can occur

FIGURE 4.6   Context of Marilyn Monroe's face rotated 90° to the right or left.

relative to multiple frames of reference. For example, as noted, Lamberts et al. (1992) and Roswarski and Proctor (1996) found that when eight stimulus positions could be divided into left–right hemispace, left–right hemifield, and left–right relative position within the hemifield, performance showed Simon effects relative to all three reference frames. Additionally, Hommel and Lippa (1995) provided evidence suggesting that stimulus position can be coded relative to an object-based frame of reference. They presented stimuli in top or bottom positions, at the locations of the eyes of a picture of Marilyn Monroe's face rotated either 90° clockwise or 90° counterclockwise (see Figure 4.6). In their Experiment 2, a left or right keypress was to be made to a black or white circle, and this situation yielded an 8-ms Simon effect relative to the eye that would be seen as left or right when viewed in the standard upright position. Finally, as described earlier, when a wheel is held at the bottom and the direction of hand movement is in opposition to the direction of wheel rotation, some people tend to code responses in terms of the hand movements and others in terms of the wheel's movements (Guiard, 1983; Wang et al., 2003). In sum, although attention is involved in the Simon effect, its contribution seems to come primarily from weighting some reference frames more strongly than others and not directly from shifting attention.

4.3 Verbal and Nonverbal S-R Modes

In Chapter 3, we presented evidence that S-R compatibility effects occur when spatial information is conveyed verbally and that the effects are larger when the stimulus and response modes are similar (i.e., the stimulus and response sets have perceptual similarity). Simon effects can also be obtained with location words as stimuli or responses, as well as with arrows that point to the left or right. For example, Lu and Proctor (2001) obtained a Simon effect of 17 ms from an irrelevant word LEFT or RIGHT on left–right keypresses made to the color of a red or green outline rectangle that enclosed the word.

When both the relevant and irrelevant stimulus dimensions concern location, it is possible to vary the match between their respective modes and those of the response dimension. Virzi and Egeth (1985) conducted an experiment in which the word RIGHT or LEFT was presented to the left or right of a fixation point. Some subjects responded to the meaning of the word as relevant and others to its position, making left–right keypresses in one trial block and vocal "left"–"right" responses in another. When responding to meaning, a Simon effect for irrelevant position of 42 ms was obtained with manual responses but only a nonsignificant 8-ms effect with naming responses. In contrast, when responding to position, a nonsignificant 8-ms Simon effect for irrelevant words was obtained with manual responses but the effect was 38 ms for naming responses. Virzi and Egeth interpreted this interaction in terms of a translational model that distinguishes parallel spatial and verbal systems. When the irrelevant stimulus dimension is coded in a system that differs from that for the relevant stimuli and responses, it is not translated into that system and therefore produces no significant Simon effect.

O'Leary and Barber (1993) obtained results similar to those of Virzi and Egeth (1985) in a study in which the four combinations of stimulus and response modes were varied between subjects. The main exception was that they found a significant 11-ms correspondence effect for irrelevant location when the task was to respond by saying "left" or "right" to the location word. The correspondence effect of the irrelevant word for the task of responding manually to stimulus position was only a marginally significant 8 ms, but this effect increased to 32 ms when the display also included a neutral color word on the side opposite the location word. O'Leary and Barber attributed this larger effect to increased salience of the word meaning due to having to discriminate the location word from the color word to know the side on which to respond.

Similar patterns of results are obtained when spatial information is conveyed by left- or right-pointing arrows and location words. Lu and Proctor (2001) had subjects make keypresses to stimuli composed of the word LEFT or RIGHT inside an outline arrow that pointed to the left or right, with the interval between onsets of the irrelevant and relevant stimulus dimensions varied between 0 and 500 ms. A consistent Simon effect for irrelevant arrow direction was evident when location word was relevant. When arrow direction was relevant, the irrelevant word produced a negligible Simon effect at short onset intervals, but it attained a size of 17 ms when the irrelevant word preceded the relevant arrow by 300 ms. This outcome suggests that activation in the spatial response system from an irrelevant word may take a while to build up and dissipate.

Baldo, Shimamura, and Prinzmetal (1998) presented an arrow and word simultaneously, one above the other, and required keypress or vocal responses to one type of stimulus or the other in different trial blocks. They found that arrow direction yielded an interaction with response modality similar to that shown by stimulus position. However, they did find small but significant Simon effects for irrelevant arrow direction when responding vocally to words and irrelevant words when responding manually to arrows. Of importance, in a second experiment, Baldo et al. established that the critical factor is the relation between the relevant and irrelevant modes. When the relevant and irrelevant stimuli were both arrows or both words, with location determining which stimulus was relevant, similar size Simon effects were obtained that did not depend on response modality.

4.4   SIMON-TYPE EFFECTS FOR OTHER STIMULUS AND RESPONSE DIMENSIONS

Simon effects are not restricted to the stimulus and response dimensions of location. They occur for a variety of other dimensions on which stimulus and response sets can overlap. We describe several such effects in this section, including ones based on motion, intensity, numerosity, duration, semantic relations, and affective relations.

4.4.1   MOTION AND MOVEMENT

Bosbach, Prinz, and Kerzel (2004) discovered a Simon effect for stationary moving stimuli. In their Experiments 1 and 2, the stimuli were drifting sine-wave gratings that conveyed left or right motion but did not actually shift position. Subjects were instructed to respond to the spatial frequency of the grating (i.e., whether the bars were narrow or wide) in Experiment 1 and to its color in Experiment 2. Small, though statistically significant, Simon effects of 7 and 4 ms, respectively, were obtained as a function of correspondence between direction of motion and response position. Experiment 4 used a point-light walker display that walked in the left or right direction, again without a change of relative positive over time. This situation yielded an 11-ms Simon effect for the direction in which the walker was walking. The authors concluded that direction of motion is sufficient to produce a Simon effect without requiring a change in position.

Kerzel and Bekkering (2000) demonstrated a Simon-type effect for visible speech and spoken responses. In their Experiment 2, subjects responded to the stimuli && or ## by saying "ba" or "da." The symbols were printed on the response-irrelevant mouth movements of a speaker pronouncing "ba" or "da." Responses were 42 ms faster when the visible mouth movements corresponded with the spoken response than when they did not.

4.4.2   INTENSITY–FORCE, NUMEROSITY, AND DURATION CORRESPONDENCES

Mattes, Leuthold, and Ulrich (2002) found that the intensity–force compatibility effect described in Chapter 3 also occurs in a Simon task for which force is irrelevant. Mattes et al. (Experiment 5) had subjects respond to the shape (circle or diamond) of a visual stimulus, while ignoring its intensity (dim or bright), by making a weak or strong response on a force-sensitive key. Responses were 8 ms faster for trials on which the intensities of the stimulus and response corresponded than on trials for which they did not. The intensity–force Simon effect was also evident when the stimuli tones for which pitch was relevant and high or low intensity irrelevant. In this case, a 21-ms intensity–force Simon effect was obtained.

Kunde and Stöcker (2002) demonstrated a similar Simon effect for S-R duration. They required subjects to make a short (< 120 ms) or long (between 121 and 300 ms) keypress to the color of a visual stimulus presented for a short (42 ms) or long (200 ms) duration. A Simon effect in the range of those typically obtained for visual location was found, with mean RT being 25 ms shorter when the duration of the stimulus corresponded with that of the response than when it did not. In a second experiment, Kunde and Stöcker had subjects make left or right keypresses of long or short duration, with the form of the stimulus (O or X) indicating the response location and the color indicating the response duration. Simon effects for both stimulus position and stimulus duration were found, and these effects were approximately additive.

Miller (2006) showed that numerosity correspondence for stimuli and responses also produces a Simon effect, as well as the element-level compatibility effect mentioned in Chapter 3. In his Experiment 1, subjects made one or two taps of a response key depending on the pitch of a tone. The tone could be presented as a single burst of 40 ms or two bursts of 40 ms each, separated by 20 ms. A 65-ms Simon-type effect was obtained for correspondence of stimulus numerosity and response numerosity. In his Experiment 2, visual stimuli showed a similar benefit for numerosity correspondence of 22 ms.

4.4.3 SEMANTIC AND AFFECTIVE SIMON EFFECTS

De Houwer (1998) found a semantic variant of the Simon effect based on word meaning. He presented bilingual subjects with Dutch or English words from the categories of animals or occupations. Subjects were to ignore the word meaning and indicate whether the word was in Dutch or English by saying "animal" or "occupation." Responses were 45 ms faster, with lower error rate, when the meaning of the word corresponded with the response (e.g., saying "animal" to *dog*) than when it did not (e.g., saying "occupation" to *dog*). Similar results were obtained in other experiments in which all words were Dutch (the native language of the subjects) and the response category was based on whether the word was in upper- or lowercase, or whether the word was a noun or an adjective.

De Houwer and Eelen (1998) extended this procedure to an affective variant of the Simon task. The stimuli in this study were nouns and adjectives with a positive, negative, or neutral affective meaning. Subjects were to respond to the grammatical category of the word by saying "positive" or "negative," with the affective meaning of the words being irrelevant. Yet, responses were 66 ms faster when the affective meaning of the word corresponded with the response than when it did not. This Simon-type effect was replicated even when the instructions stated explicitly to ignore the affective meaning of the words and when the words "flower" and "cancer" were used as the positive and negative responses, respectively. De Houwer, Crombez, Baeyens, and Hermans (2001) demonstrated that the affective Simon effect also generalizes to tasks in which semantic category or letter-case of words is relevant, as well as to ones in which the semantic category of photographed objects is relevant.

4.5 CHAPTER SUMMARY

Performance is usually better when stimulus location corresponds with response location, even when stimulus location is irrelevant to the task. This Simon effect is found for different stimulus and response modalities, and across a wide variety of specific stimulus types and task conditions. The Simon effect is typically attributed to activation of the corresponding response code that is produced automatically at stimulus onset. Much evidence suggests that the activation produced by the irrelevant stimulus dimension dissipates over time, at least for visual stimuli that appear in left or right positions. Stimulus location is thought to be coded either as a function of shifting attention or in reference to other objects, and considerable evidence now indicates that the Simon effect can occur relative to multiple frames of reference. When both the relevant and irrelevant stimulus dimensions refer to location, whether the relevant or irrelevant stimulus mode (spatial or verbal) matches the response mode is an important factor in determining the size of the Simon effect. The Simon effect also occurs for dimensions other than location along which stimulus and response sets can overlap. The primary message of the Simon effect for applied purposes is that correspondence or lack of correspondence of displays and controls on irrelevant dimensions can impact performance, and this impact varies systematically as a function of several factors.

5 S-R Compatibility Effects for Multidimensional Stimulus and Response Sets

5.1 INTRODUCTION

When the elements of a display and their controls vary along only one dimension, most people would predict that performance would be best if the controls were mapped to the display elements in a manner that maintains spatial correspondence (Vu & Proctor, 2003). However, for many interfaces, the displays and controls vary along two or more dimensions. Predicting performance with various mappings for these display–control configurations is often more difficult than one might assume. For multidimensional S-R sets, issues of compatibility become more complex because performance may depend on which dimension is defined as relevant for the task, structural relations between the stimulus and response sets play an increasing role, and compatibility may be more important for some dimensions than for others.

5.2 TWO-DIMENSIONAL NONSPATIAL STIMULI MAPPED TO KEYPRESS RESPONSES

Compatibility effects occur when the stimulus and response configurations can be classified as two dimensional, even for situations in which the stimulus dimension bears no resemblance to the response dimension. This point was first illustrated in a study by Fitts and Biederman (1965), following up on a finding reported by Morin, Forrin, and Archer (1961). Morin et al. used a task in which four two-dimensional stimuli (one or two squares or circles) were mapped to four keypresses made by the fingers on the right hand. The left-to-right mapping of stimuli to responses was single circle to index finger, single square to middle finger, two circles to ring finger, and two squares to the pinkie finger (see Figure 5.1). Morin et al. found that RT was 190 ms longer for this four-choice task than for a two-choice task in which a circle was mapped to the right index finger and a square to the right middle finger.

Although this increase in RT as the number of S-R alternatives increased is consistent with the Hick–Hyman law, introduced in Chapter 2, Fitts and Biederman (1965) noted that the conditions examined by Morin et al. (1961) did not seem to be highly compatible with respect to either the set of fingers used for responding or the mapping of stimuli to the fingers. Consequently, they conducted a follow-up

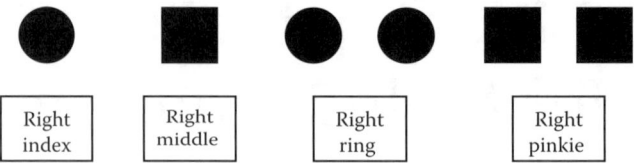

FIGURE 5.1 The two-dimensional stimulus–response assignments used in Morin, Forrin, and Archer's (1961) study.

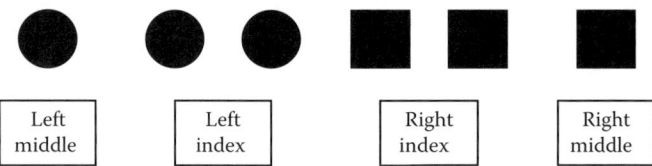

FIGURE 5.2 An optimized version of the two-dimensional stimulus–response assignments identified by Fitts and Biederman (1965).

study to determine whether there was a more compatible S-R configuration by asking subjects to indicate which four fingers of the left and right hands they would prefer to use and how they would map the stimuli to responses. Fitts and Biederman found, "Out of approximately 100 Ss [subjects] so queried, nearly all selected the middle and forefingers of the right and left hands, assigning stimulus *form* to hands, and stimulus *number* to fingers" (p. 409). Thus, subjects judged that it would be most natural to have the two left and two right responses distinguished by hands—rather than by four fingers on one hand—with the form-identity feature differentiating responses on the two hands (see Figure 5.2). In a four-choice–reaction experiment, Fitts and Biederman then demonstrated that RT was more than 100 ms shorter with this response set and mapping than with that used by Morin et al., for which the responses were four right-hand fingers, with the two left and two right responses distinguished by number rather than form identity. This result suggests that the cost in RT for increasing the number of alternatives from two to four is much less than suggested by Morin et al.'s results, as long as high compatibility is maintained for the four-choice task.

Miller (1982) conducted a series of experiments examining which mappings of features to the preferred index and middle finger responses of the left and right hands yielded the best performance. Miller's experiments were follow-ups to his four-choice spatial precuing experiment, described in Chapter 2, which he interpreted as showing that only pairs of finger responses on the same hand could be prepared in advance. In his Experiment 2, the stimuli were consonant-vowel letter pairs BE, BO, ME, and MO. One group of subjects performed with the left-to-right mapping of BE, BO, ME, MO to the left middle, left index, right index, and right middle fingers, for which the consonant distinguishes responses on the left and right hands (see left side of Figure 5.3), and another group with the mapping BE, ME, BO, MO, for which the vowel distinguishes the two hands. Either the consonant or vowel was

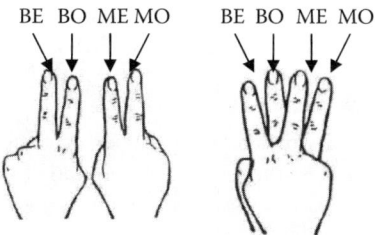

FIGURE 5.3 Illustration of the adjacent- and overlapped-hands placements used in Proctor and Reeve's (1985) study, along with example S-R mappings.

presented first on each trial as a precue, with the other letter appearing 100, 350, or 600 ms after onset of the first. RT was shorter overall for the BEBOMEMO mapping than for the BEMEBOMO mapping (see also Proctor & Reeve, 1985, Experiments 1 and 4), and only the BEBOMEMO mapping showed a difference as a function of whether the vowel or consonant was presented first. For this mapping, there was no difference between presenting the vowel or consonant first at the 100-ms precuing interval, but there was an advantage of about 70 ms for consonant-first trials at the two longer intervals. Miller interpreted this precuing benefit as due to subjects being able to use the consonant information to prepare two responses on the same hand in advance of the complete stimulus information. Because the BEMEBOMO mapping did not show an advantage for the vowel first, he suggested that the system is constrained in such a manner that the consonant must be specified before the vowel.

Proctor and Reeve (1985) showed for the four-choice tasks with two-dimensional stimuli, as for the spatial precuing task, there is nothing special about precuing fingers on the same hand. In their Experiment 1, all subjects performed with the BEBOMEMO mapping used in Miller's (1982) Experiment 2, but half of the subjects performed with the overlapped-hands placement for which the fingers from each hand are alternated on the keyboard (see Figure 5.3, right side). Although responses were slower overall with the overlapped-hands placement than with a normal, adjacent one, a consonant-first advantage of 65 ms was evident for both hand placements. This result indicates that, as for the spatial precuing task, the critical factor for the consonant-first advantage is that the two left or two right locations are precued, rather than that the left and right hands are.

Miller's (1982) Experiment 3 used two-dimensional stimuli that differed in terms of letter identity (S or T in one condition and I or T in another) and font size (small or large). Preliminary two-choice experiments established that the letter-identity discrimination could be performed faster than the size discrimination for the ST set, whereas the size discrimination could be performed faster than the letter-identity discrimination for the IT set. Thus, the idea was that letter identity could serve as an implicit precue for the ST set, and size for the IT set, because it was available before the other dimension, much like the explicit consonant precue in the BEBOMEMO experiment.

In Experiment 3 itself, half the subjects performed with the ST set and half with the IT set. For each set, groups of subjects performed with mappings to the left middle, left index, right middle, and right index fingers that assigned letter to the same hand (e.g., sSᴛT), size to same hand (e.g., sᴛST), or neither (e.g., sTᴛS). For the ST stimulus set, RT was approximately 100 ms faster when letter identity was assigned to the same hand compared with the neither condition, and there was only a slight, but nonsignificant, 20-ms advantage when size identity was assigned to the same hand compared with the neither condition. For the IT stimulus set, letter identity again yielded the largest benefit (75 ms), but this was not reliably different from the 40-ms benefit for the size distinction. Thus, both letter sets showed S-R compatibility effects for which performance was best when letter identity distinguished the responses on the left and right hands, and worst when neither letter identity nor size did. Miller (1982) attributed this S-R compatibility effect to decisions about letter identity typically preceding those of size, with preparation of the responses prior to the size decision possible only when the two responses were on the same hand.

Proctor and Reeve (1985, Experiment 2) replicated the findings for the ST set using a different letter set, OZ, for which a preliminary two-choice task established that the letter-identity discrimination could be made faster than the size discrimination. With an adjacent-hands placement, RT was 85 ms faster with a mapping of the type OozZ, for which letter identity distinguished the two left and two right responses, than for a mapping of the type OzoZ, for which it did not. Unlike the BEBOMEMO experiment described above, an overlapped-hands placement did not produce the same advantage for the OozZ mapping type seen for the normal, adjacent-hands placement. Instead, there was no significant difference between the two mappings. Proctor and Reeve noted that the major difference in results between the two experiments was that responses for the OzoZ mapping were not slowed as much with the overlapped-hands placement as were responses for the other conditions. They interpreted this result as suggesting that subjects were able to benefit from the consistent relation between letter identity and hand in that case. In Proctor and Reeve's Experiment 3, the possibility of using the hand distinction was removed by having all four responses made by the fingers on a single hand. With the single-hand placement, the advantage for the OozZ mapping type was again obtained, consistent with the idea that RT is shorter when the letter identity feature maps onto a salient response feature.

It is important to note that, as for other S-R compatibility effects, the advantage for the OozZ mapping does not disappear with practice. Dutta and Proctor (1992, Experiment 3) had subjects perform with either the OozZ or OzoZ type of mapping for eight sessions of 300 trials each. The advantage for the OozZ mapping was 88 ms in the first session, and this was reduced to 57 ms in the last session. Thus, even after 2,400 trials of practice, a large compatibility effect was still evident.

Proctor and Reeve (1985) explained their results in terms of the salient features coding principle, described previously for spatial precuing tasks: "Both stimuli and responses are coded in terms of the salient features of each, with the translation processes based on the relations between stimulus and response codes" (p. 623). The general idea of this principle is that the stimulus and response sets do not have to be perceptually or conceptually similar to each other. S-R compatibility effects can be obtained solely as a consequence of the mapping of salient structural features of the stimulus set to salient structural features of the response set.

An implication of the salient features coding principle is that performance should benefit from correspondence of the more salient dimensions for any two-dimensional S-R sets. Gordon and Meyer (1984) reported an S-R mapping effect consistent with this implication for a study in which subjects had to respond to an auditory consonant–vowel syllable by saying another one. For the set PUH, BUH, TUH, and DUH, the stimuli can be distinguished according to the phonetic features of voicing (voiced: BUH, DUH; unvoiced: PUH, TUH) and place of articulation (bilabial: BUH, PUH; alveolar: DUH, TUH). RT was faster when the response assigned to a stimulus had the corresponding voicing feature (e.g., respond "DUH" to BUH) than when it did not (e.g., respond "TUH" to BUH), but correspondence for place of articulation had no influence on performance. Gordon (1990) reported similar correspondence effects for nasal/stop consonant and nasal/fricative distinctions. He emphasized that the correspondence effects were found only for the most salient perceptual features of speech.

Proctor, Dutta, Kelly, and Weeks (1994) used the voicing/place consonant–vowel set to determine whether it would yield an S-R compatibility effect when mapped to four horizontal spatial locations. In Experiment 1, the syllables were used as stimuli assigned to keypress responses, whereas in Experiment 2, they were used as responses assigned to spatial location stimuli. In the former case, RT was 33 ms shorter when the salient voicing display feature was mapped to the salient left–right response feature than when it was not, whereas in the latter case, RT was 34 ms shorter when the salient left–right display feature was mapped to the salient voicing response feature than when it was not. In Experiment 3, the two-dimensional visual stimuli used by Proctor and Reeve (1985), large and small sizes of the letters O and Z, were mapped to the speech responses such that the salient letter-identity feature corresponded with the salient voicing feature or did not. Responses were 41 ms faster when the salient features corresponded. Thus, there is considerable evidence that the salient features coding principle captures many aspects of both the spatial precuing results described in Chapter 2 and the mapping effects for two-dimensional stimulus and response sets for which there is no perceptual or conceptual similarity.

5.3 THE RIGHT–LEFT PREVALENCE EFFECT FOR TWO-DIMENSIONAL SPATIAL STIMULI AND RESPONSES

S-R compatibility effects are also obtained for situations in which the stimulus and response sets vary along two spatial dimensions simultaneously. However, the compatibility effect is usually larger along one dimension than the other, and this has been the central focus of much of the research conducted on two-dimensional spatial compatibility effects.

Nicoletti and Umiltà (1984, 1985) were the first to examine S-R compatibility when stimuli and responses varied along vertical and horizontal dimensions simultaneously. For their stimulus arrangement, they mounted two light-emitting diodes in the top-left and bottom-right or top-right and bottom-left corners of a panel, and for the response arrangement they mounted microswitches on two wooden cylinders of different heights along a diagonal corresponding or opposite to that of the stimulus arrangement (Nicoletti & Umiltà, 1984, Experiment 4; 1985, Experiment 1).

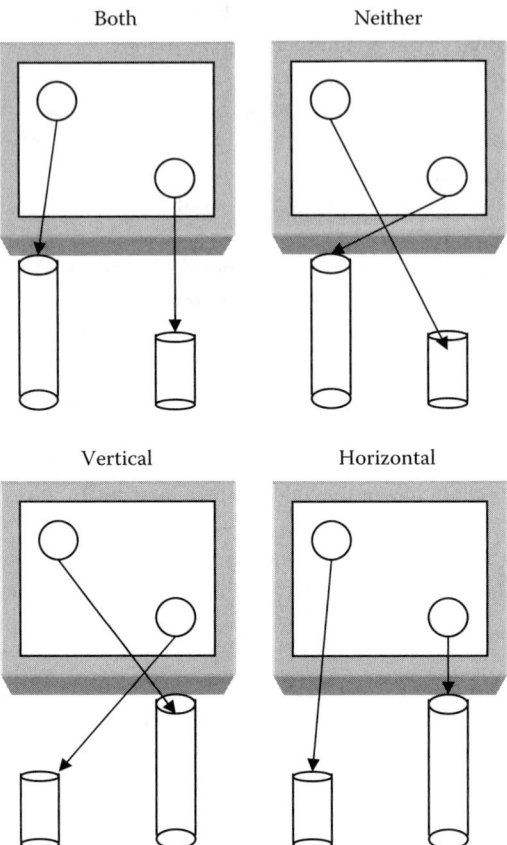

FIGURE 5.4 Illustration of the four mapping conditions used in studies of the right–left prevalence effect. In the first row, the diagonal for both the stimuli and responses is aligned from top-left and bottom-right; both dimensions are mapped compatibly when the top-left stimulus is assigned to the top-left response and bottom-right stimulus to bottom-right response (the both condition), and incompatibly when the opposite assignment is used (the neither condition). In the second row, the diagonal for the stimulus set is aligned from top-left and bottom-right, but the diagonal for the response set is aligned from bottom-left and top-right. In this case, only the vertical dimension is mapped compatibly when the bottom-right stimulus is assigned to the bottom-left response and the top-left stimulus to the top-right response (the vertical condition), and only the horizontal dimension is mapped compatibly with the opposite assignment (the horizontal condition).

Nicoletti and Umiltà (1984, 1985) were able to examine conditions that were compatible on one (horizontal or vertical), both, or neither of the two dimensions by varying the mapping instructions as well as the diagonals in which stimuli and responses were arranged (see Figure 5.4). The compatibility effect for the horizontal dimension was greater than that for the vertical dimension, even though the compatibility effects obtained for the horizontal and vertical dimensions in isolation were of similar magnitude (Vallesi, Mapelli, Schiff, Amodio, & Umiltà, 2005; Vu, Proctor,

& Pick, 2000). This dominance of the horizontal dimension was obtained despite the fact that subjects were instructed in terms of top–bottom locations, rather than right–left locations. In subsequent studies, which used left–right hand and foot responses or auditory stimuli, Nicoletti and Umiltà (1985) and Nicoletti, Umiltà, Tressoldi, and Marzi (1988) found that the compatibility effect for the vertical dimension was virtually eliminated. They called this relative restriction of the S-R compatibility effect to the horizontal dimension the *right–left prevalence* effect and attributed the prevalence effect to an inability to code vertical locations when horizontal ones were presented simultaneously (Nicoletti et al., 1988; Umiltà & Nicoletti, 1990). Right–left prevalence is thus a robust phenomenon, evident when the stimuli are visuospatial or auditory, and responses are made by left and right hands or a hand and a foot.

5.3.1 Accounts for the Right–Left Prevalence Effect

5.3.1.1 Left–Right Effectors

Nicoletti and Umiltà (1984) initially proposed an effector-based account of the right–left prevalence effect in which they attributed it to responses being made with right–left effectors (i.e., right and left hands). An effector-based account can be derived from Heister, Schroeder-Heister, & Ehrenstein's (1990) hierarchical model of spatial compatibility in which the magnitude of the stimulus-response compatibility effect is a function of three factors: spatial coding of response keys' position, spatial coding of effector position, and spatial coding of response effectors. Because the condition with horizontal compatibility benefits from all three factors, but the condition with vertical compatibility benefits from only the first two (the hands are right–left effectors, not top–bottom ones), the advantage of horizontal compatibility over vertical compatibility is assumed to be due to the additional benefit of responding with left–right effectors.

To evaluate the effector-based account, Nicoletti and Umiltà (1985) had subjects respond to the stimuli by using top–bottom effectors (i.e., a hand and a foot). If the benefit for the horizontal dimension was due to the use of right–left effectors, then the prevalence effect should not be obtained when top–bottom effectors were used. However, Nicoletti and Umiltà still obtained a right–left prevalence effect with this procedure. In fact, the vertical compatibility effect present when responses were made by the left and right hands was eliminated completely when responding was made by the hand and foot, leading Nicoletti and Umiltà to reject the effector-based account.

Hommel (1996a) noted that in Nicoletti and Umiltà's (1985) study, the left hand was always paired with the right foot and the right hand with the left foot. Because subjects were still using left and right effectors for responding, the effector-based account could not be ruled out. Hommel evaluated the effector-based account by having subjects respond using one hand by moving a joystick to the response location. With unimanual joystick movements, the right–left prevalence effect was eliminated. This finding led Hommel to conclude that the benefit for the horizontal dimension is due to the use of left–right effectors. Vu and Proctor (2001a) replicated Hommel's finding that the prevalence effect is eliminated with unimanual joystick

FIGURE 5.5 The adjacent (left panel) and top–bottom (right panel) hand placements used by Vu and Proctor (2001a).

movements. Furthermore, Vu and Proctor showed that when the left–right effector distinction was removed by using ipsilateral top–bottom effectors (e.g., left hand and left foot or right hand and right foot), the result was a top–bottom prevalence effect.

Vu and Proctor (2001a) showed that it is not just the use of right–left effectors that leads to the right–left prevalence effect, but rather that the effectors that are used provide a salient frame of reference for horizontal coding. Subjects responded with two fingers on one hand. Because the fingers can be coded in terms of the left–right placement, they are considered to be left–right effectors. For example, Heister et al. (1990) found that when responses were made with two fingers on one hand for left–right stimuli and responses, an S-R compatibility effect is obtained that is of similar magnitude to that obtained when responses are made by two fingers on different hands. Vu and Proctor showed that with two-dimensional stimuli and responses, no right–left prevalence effect was obtained when responses were made with left–right effectors on the same hand. This finding suggests that the reduction/elimination of the vertical compatibility effect relative to the horizontal compatibility effect is not due to the use of right–left effectors. Furthermore, Vu and Proctor obtained the top–bottom prevalence when responses were made with the left and right hands, but the hands were positioned to make the vertical dimension more salient. As shown in Figure 5.5, Vu and Proctor had subjects position their hands on top of one another, so that the index fingers crossed to the contralateral side of the body (e.g., the left finger of the bottom hand is on the bottom-right key and the right finger of the top hand is on the top-left key). With this top–bottom hand placement, the horizontal compatibility effect was reduced compared with the vertical one, yielding a top–bottom prevalence effect.

5.3.1.2 Midline-Barrier Account

Having rejected the effector-based account, Nicoletti and Umiltà (1985) proposed a midline-barrier account. In this account, the reduction/elimination of the vertical compatibility effect relative to the horizontal compatibility effect is due to the stimuli that are compatible only on the horizontal dimension assigned to the effectors on the ipsilateral sides of the body, whereas stimuli that are compatible only on the vertical dimension are assigned to the contralateral effectors. However, Nicoletti and Umiltà rejected this explanation because Nicoletti, Anzola, Luppino, Rizzolatti, & Umiltà (1982) had previously shown that compatibility effects of similar magnitudes are obtained when stimuli and responses are presented to the same side or opposite sides of the body's midline.

5.3.1.3 Attentional Hypothesis

Nicoletti et al. (1988) next proposed an attentional hypothesis to explain the fact that when the vertical compatibility effect is reduced or eliminated for two-dimensional S-R sets, the horizontal compatibility effect remains strong. According to the attentional hypothesis, right–left prevalence is a consequence of attention being directed to the horizontal dimension because it is more difficult to discriminate visual left–right locations than top–bottom ones. Nicoletti et al. attributed this difficulty in discriminating horizontal locations to the fact that people tend to show right–left confusion but not top–bottom confusion. To evaluate the attentional hypothesis, Nicoletti et al. presented subjects with auditory stimuli. Note that it is more difficult to discriminate top–bottom locations than right–left locations in this modality. They reasoned, therefore, that a top–bottom prevalence effect should be obtained if the attentional hypothesis were correct. However, their results showed right–left prevalence, with the vertical compatibility effect being virtually absent.

5.3.1.4 Dimensionally Biased Response Account

The finding of right–left prevalence in Nicoletti and Umiltà's (1984, 1985; Nicoletti et al., 1988) studies was surprising because all their subjects were instructed in terms of top–bottom locations, which should lead to a bias toward the vertical dimension and not the horizontal one. However, Hommel (1996a) proposed that the right–left prevalence effect obtained in Nicoletti and Umiltà's studies was a byproduct of their subjects not following the instructions to respond in terms of the top–bottom locations. He conducted a study in which he instructed one group of subjects to respond only in terms of vertical locations and another group only in terms of horizontal locations. Hommel found both vertical and horizontal compatibility effects, but their magnitudes varied as a function of instructions: The horizontal compatibility effect was larger with horizontal instructions and the vertical compatibility effect was larger with vertical instructions. However, the horizontal compatibility effect under horizontal instructions was larger than the vertical compatibility effect under vertical instructions, indicating an overall right–left prevalence effect. Because the magnitude of the vertical compatibility effect was reduced with horizontal instructions relative to vertical instructions, Hommel attributed the right–left prevalence effect in Nicoletti and Umiltà's studies to subjects ignoring the vertical instructions and adopting a horizontal coding strategy instead. He argued that this strategy of horizontal coding was due to subjects being biased toward the horizontal dimension because they have to discriminate between right and left effectors to execute the response.

Vu and Proctor (2002) noted that although instructions tend to influence the relative magnitudes of the compatibility effects, they mainly do so in situations where the stimulus or response set does not provide a salient frame of reference for horizontal or vertical coding, as in the case of Nicoletti and Umiltà's (1984, 1985; Nicoletti et al., 1988) and Hommel's (1996a) studies. The role of instructions decreases as the frame of reference for the stimulus display, response configuration, or both is made to bias coding of one dimension over the other. That is, when the horizontal dimension is made salient by the stimulus and response set, a right–left advantage is obtained regardless of instructions. Similarly, when the vertical dimension is made salient by the stimulus and response sets, a top–bottom advantage is obtained regardless of instructions. However, when the salient features of the stimulus and response sets do not correspond (i.e., one favors the horizontal dimension and the other the vertical dimension), any prevalence effect is evident only when collapsed across instructions. Thus, although instructions can bias responding by making the frame of reference compatible or incompatible with the response effectors, when the strength of the frame of reference provided by the display–control configuration is strong enough to favor coding along one dimension, an advantage for that dimension will be obtained even when the instructions to respond in that dimension yield an incompatible relation to the effectors.

5.3.1.5 Salient Features Coding Account

Vu et al. (2000) made a distinction between two types of right–left prevalence effects. A weaker form of right–left prevalence occurs when there is a compatibility effect for both the horizontal and vertical dimensions, but the horizontal compatibility effect with horizontal instructions is larger than the vertical compatibility effect with vertical instructions, and right–left prevalence is evident only when collapsed across both instruction conditions. A stronger form occurs when the vertical compatibility effect is eliminated completely, or when the horizontal compatibility effect is larger than the vertical compatibility effect, even when subjects are instructed in terms of the vertical dimension. Vu et al. also showed that the weak form of right–left prevalence was obtained when subjects responded with their hands placed in close proximity on the numeric pad of a keyboard, but the strong form was obtained when subjects responded with their hands held widely apart using a handgrip apparatus.

Vu et al. (2000) suggested that the stronger form of right–left prevalence was evident in the latter situation, but not the former, because the horizontal dimension was more salient when the hands were held apart on the grips than when they were placed in close proximity on the keys. Because the same left–right effectors were used in both cases, the effector-based account cannot explain why the strong form of prevalence occurred with the handgrip apparatus, but not with the keyboard apparatus. To account for this difference in magnitude of the prevalence effect, Vu et al. suggested that the magnitude of prevalence effect depends on the relative salience of the horizontal dimension provided by the response environment.

The salient features coding account can explain several of Vu and Proctor's (2001a) findings described earlier: When responses were made with left–right fingers on the same hand, no right–left prevalence effect was obtained because the horizontal frame of reference was not salient. When the response configuration was made to favor the vertical dimension, by having subjects place their hands on top of each other so that the responding fingers were crossed to the contralateral side of the body, a top–bottom prevalence effect was obtained even when subjects responded with left–right effectors. Thus, Vu and Proctor showed that the use of right–left or top–bottom effectors is not the critical factor for obtaining the prevalence effect; rather, the critical factor is whether these effectors provide a salient frame of reference for horizontal (or vertical) coding.

Consistent with the notion that the prevalence effect is a consequence of the specific effectors used for responding, Vu and Proctor (2002) showed that different types of prevalence effect could be obtained by manipulating the relative salience of the stimulus display. In Vu and Proctor's Experiment 1, the spatial proximity of the stimulus position was manipulated to make the horizontal dimension or vertical dimension more salient, and responses were made with the left and right index fingers on a numeric pad of the keyboard. A control condition was also included in which the spatial proximity along the horizontal and vertical dimensions was equivalent. With the horizontal salient displays, a right–left prevalence effect was obtained, and with the vertical salient displays, a top–bottom prevalence effect was obtained. This finding is in support of the salient features coding account and cannot be explained by the effector-based account.

As mentioned earlier, according to Proctor and Reeve's (1985, 1986) salient features coding principle, S-R translation should be faster when the salient features of the display and response configurations correspond than when they do not. Vu and Proctor (2002) evaluated the salient features coding principle in their Experiments 2 and 3. In Experiment 2, the hands were placed in the top–bottom positions for the vertical-salient response configuration and in adjacent positions for the horizontal-salient response configuration. Consistent with the salient features coding principle, the benefits of horizontal and vertical compatibility were larger when the salient features of the S-R sets corresponded to the specific dimension than when they did not. When the dimension made salient by the stimulus display did not correspond to the dimension made salient by the response configuration, there were no significant differences between the conditions with horizontal and vertical compatibility alone. To generalize the findings, Vu and Proctor's Experiment 3 used contralateral hand–foot responses for the horizontal salient response configuration and ipsilateral hand–foot responses for the vertical salient response configuration. A similar pattern of results was obtained for the hand–foot responses. Thus, the results support the view that the relative benefits for horizontal and vertical compatibility are directly influenced by the coding frame of reference provided by the stimulus display and response configuration.

Rubichi, Nicoletti, Pelosi, and Umiltà (2004, Experiment 1) also found evidence in favor of salient features coding when the stimuli were two-dimensional, but responses were unidimensional. When left–right effectors were used (i.e., left and right hands or feet), only a horizontal S-R compatibility effect was obtained. Similarly, when top–bottom effectors were used (i.e., the ipsilateral hand and foot), only a vertical S-R compatibility effect was obtained. This finding is consistent with the notion that responding to two-dimensional stimuli can be based on either horizontal or vertical codes, and coding of responses along one dimension increases the compatibility effect for that dimension and decreases the effect for the alternative dimension.

However, Rubichi et al. (2004) showed that when the task required processing of both dimensions to select the correct response, a right–left prevalence effect was evident. In their Experiment 2, Rubichi et al. orthogonally mapped stimuli to effectors. For example, when the effectors were hands, subjects responded to both the top-left and top-right stimuli with the right hand and both the bottom-left and bottom-right stimuli with the left hand, or vice versa. A horizontal S-R compatibility effect was present regardless of whether responses were made with horizontal or vertical effectors, but a vertical S-R compatibility effect was evident only when vertical effectors were used. Rubichi et al. suggested that when responses must be based on a conjunction of the horizontal and vertical codes, the right–left prevalence effect may be due to nonenvironmental factors such as the temporal dynamics of horizontal and vertical codes or the number of reference frames that can be used for coding the spatial locations, and that the salient features account should include consideration of intrinsic factors in addition to environmental ones.

In summary, although the different accounts of right–left prevalence are able to explain specific incidences of right–left prevalence, only the salient features coding account currently is able to encompass the different types and magnitudes of prevalence effects obtained (see Rubichi, Vu, Nicoletti, & Proctor, in press). Overall, the results with two-dimensional S-R sets show that both horizontal and vertical dimensions can be coded when they are presented simultaneously, and a compatibility effect is obtained for each dimension. However, because both dimensions give redundant information (i.e., responses can be coded with respect to either dimension), making one dimension more salient than the other through manipulating the relative salience of horizontal or vertical dimension will result in the compatibility effect being enhanced for the salient dimension and reduced for the alternative dimension.

5.3.2 The Role of Preparation

In the studies examining the right–left prevalence effect, the mapping of stimuli to responses along a particular dimension is held constant within a block of trials. Thus, subjects are fully prepared to respond with a single mapping of stimuli to responses. One question that arises is whether the prevalence effect is influenced by the state of preparation for responding based on one dimension. The influence of preparation can be examined by randomly cuing the dimension on which a response is to be based. If the right–left prevalence effect is due to a bias toward horizontal coding that can be overcome to some extent by preparing for the vertical dimension, then the right–left prevalence effect should be larger at short preparation intervals than at longer ones. If, in contrast, the effect is a consequence of representations of the horizontal and vertical codes provided by the task environment, then the prevalence effect should not be influenced by the subjects' preparation state.

The role of preparation in the right–left prevalence effect can be examined by using a precuing paradigm. Meiran (1996; Meiran, Chorev, & Sapir, 2002) has

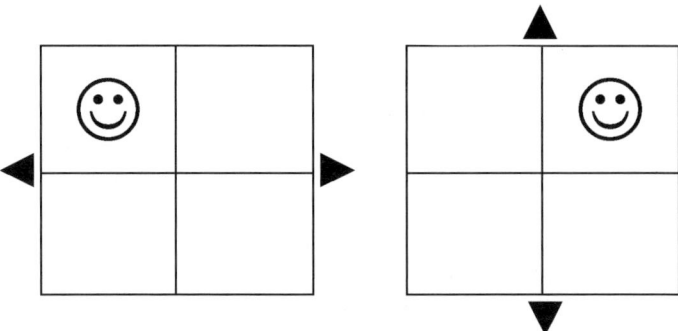

FIGURE 5.6 Illustration of the cuing paradigm used in Meiran's (1996) study. The dark arrowheads indicate whether to respond to the horizontal (left panel) or vertical (right panel) stimulus dimension, and the faces are examples of imperative stimuli.

successfully used a precuing paradigm to study task switching with two-dimensional S-R sets of the type used in studies on the prevalence effect. In his studies, the display consisted of a 2×2 grid, and the target stimulus could appear in one of the four squares (see Figure 5.6). The responses were arrayed diagonally along the positive (1 and 9 keys) or negative (7 and 3 keys) axis of the numeric keypad of a keyboard. Prior to the onset of the target stimulus, two outward-pointing arrowheads were presented either outside the center vertical axis (to signal that the responses should be based on the vertical location of the stimulus) or outside the horizontal axis (to signal that responses should be based on the horizontal location of the display). With this arrangement, the same response was to be made to stimuli presented in locations along the same diagonal as the responses, regardless of which task was cued (rule-congruent trials). In contrast, a different response was to be made to stimuli presented along the opposite diagonal to that of the responses (rule-incongruent trials), depending on which dimension was cued. Meiran (1996, Experiments 1 and 2) found that RT was shorter when the task repeated than when it switched, and that RT was shorter at the long preparation interval compared with the short one. However, rule congruency did not interact with task repetition or preparation interval.

Because Meiran's (1996; Meiran et al., 2002) studies were conducted primarily to examine the effects of task switching, many of his results were not reported in a manner that allows evaluation of the prevalence effect (i.e., the data were collapsed across whether the horizontal or vertical task was cued). However, data from three of his experiments were reported as a function of the cued task (Meiran's Experiment 4; Meiran et al.'s Experiments 2 and 3). Because responses were made with the left and right index fingers placed adjacently on a numeric keypad, a right–left prevalence effect should be evident. In Meiran's Experiment 4, the main effect of task was not significant, indicating no significant right–left prevalence effect. However, mean RT was 10 ms shorter for the right–left task than the top–bottom task, with this difference being largest for rule-incongruent trials on which there was a task switch (34 ms at the short cue–target interval of 132 ms and 19 ms at the long cue–target interval of 1,632 ms). The main effect of task was significant, however, in Meiran et al.'s Experiments 2 and 3, with the right–left prevalence effect being 14 and 30 ms, respectively. In Meiran et al.'s Experiment 2, there was also an interaction of cue dimension with cue–target interval, with the right–left prevalence effect being larger at the short cue–target intervals than at the longer ones (RT differences of 35 and 15 ms, at the shortest and longest cue–target intervals of approximately 100 ms and 2 s, respectively). Thus, across the experiments of Meiran and Meiran et al., it is clear that a right–left prevalence effect was present, but it is unclear whether the effect was influenced by the preparation interval.

Proctor, Koch, and Vu (in press) conducted a study to evaluate whether preparation influences the right–left prevalence effect. In Proctor et al.'s study, the cue was a centered double arrow, pointing to the left/right (to signal the horizontal task) or up/down (to signal the vertical task), and the stimuli were solid disks that could appear in one of the four corners of the screen (see Figure 5.7). In their Experiments 1 and 2, only rule-incongruent trials were examined (i.e., the stimuli appeared along the diagonal opposite to the diagonal in which the responses were arrayed) because Proctor et al. wanted to ensure that subjects were preparing for the cued dimension (i.e., for rule-congruent trials, the same response is made regardless of which dimension is cued). One group of subjects was instructed to respond with a compatible mapping for the cued dimension and the other with an incompatible mapping. Proctor et al.'s Experiment 1 also examined two methods for evaluating the cue–target interval. For the varied response–stimulus interval (RSI), the intertrial interval was held constant within a trial block so that the RSI varied as a function of cue–target interval. For the constant RSI condition, the intertrial interval varied inversely to the cue–target interval so that the RSI was kept at a constant 2 s. The short cue–target interval was 100 ms and long cue–target interval was 800 ms in the varied RSI condition and 900 ms in the constant RSI condition. Proctor et al. found a 25-ms right–left prevalence that did not depend significantly on cue–target interval, mapping condition, or RSI condition.

Proctor et al.'s (in press) Experiment 2 used only a constant RSI of 4 s and increased the length of the cue–target interval to 300 ms for the short interval and 3,000 ms for the long interval. A significant 20-ms right–left prevalence was evident, which did not vary as a function of cue–target interval or mapping condition. In Proctor et al.'s Experiment 3, both rule-congruent and rule-incongruent trials were presented within a test block using a constant RSI of 2 s and a compatible mapping condition. For different groups of subjects, the cue–target intervals of 100 ms and 900 ms were varied in alternating blocks of trials or within a block of trials. Proctor et al. found an overall 13-ms right–left prevalence effect that did not interact with cue–target interval or blocked versus randomized presentation of the intervals. The prevalence effect did, however, interact with rule-congruency: A significant 24-ms right–left prevalence effect was evident on rule-incongruent trials, but no prevalence effect was evident on rule-congruent trials. This finding suggests that the advantage for horizontal location coding occurs primarily when the horizontal and vertical codes do not signal the same response.

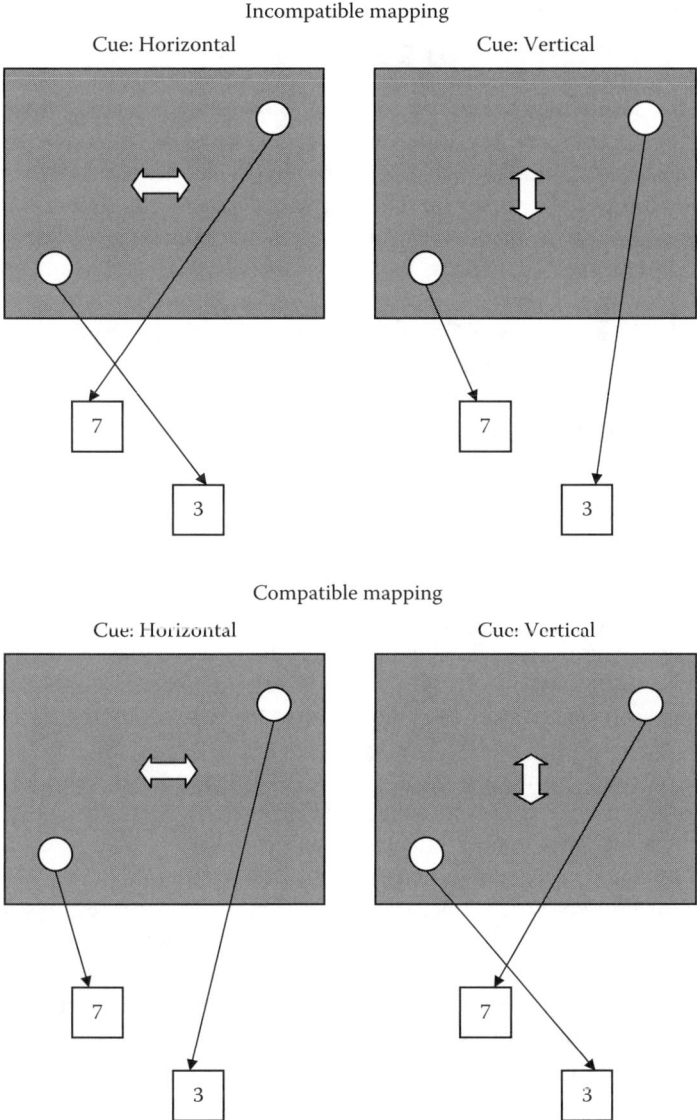

FIGURE 5.7 Illustration of the compatible and incompatible mapping conditions for the horizontal and vertical cued dimensions in Proctor, Koch, and Vu's (in press) study. Copyright 2006, Psychonomic Society. Reprinted with permission.

Overall, the data from Meiran's (1996; Meiran et al., 2002) and Proctor et al.'s (2005) studies suggest that the right–left prevalence effect occurs when the dimension to which a compatible or incompatible mapping should be applied is cued on a trial-to-trial basis. The prevalence effect occurs regardless of whether the preparation time is short or long, and seems to reflect the relative ease with which the stimulus codes can be intentionally translated into response codes.

5.4 SIMON EFFECTS FOR TWO-DIMENSIONAL SPATIAL STIMULI AND RESPONSES

Proctor, Vu, and Nicoletti (2003) conducted a study to examine whether Simon effects would be obtained for the horizontal and vertical dimensions when the S-R sets varied along both dimensions simultaneously. When the stimulus and response locations were equally far apart on the horizontal and vertical dimensions, positive Simon effects of similar magnitude were obtained for both dimensions (12 and 13 ms, respectively). In other words, there was no prevalence effect favoring either dimension. However, when the relative salience for one dimension of the display was increased, by separating the stimuli farther on that dimension than on the other, the Simon effect increased for the salient dimension and decreased for the nonsalient dimension.

Proctor et al. (2003) suggested that the salient features coding account could be applied to the Simon effect for two-dimensional S-R sets. With two-dimensional S-R sets, horizontal and vertical codes provide redundant information, and responding could be based on either dimension. Thus, when salient features of the stimulus display were made to favor one dimension, they provided a strong frame of reference for coding along the salient dimension, leading to stronger activation of location codes for that dimension. It is important to note, however, that Simon effects for both dimensions were obtained, which indicates that spatial codes along both dimensions were produced and processed. Making one dimension more salient than the other influences the degree to which the different location codes are activated or the degree to which subjects utilize coding along the two dimensions, but does not prevent activation of location codes for the nondominant dimension.

The finding of no overall right–left prevalence with Simon stimuli, with neutral S-R sets and across horizontally salient and vertically salient S-R sets, was recently challenged by Rubichi, Nicoletti, and Umiltà (2005), who found a right–left prevalence effect using two-dimensional stimuli and unidimensional responses and with a four-choice version of the task that required responding with both hands and feet. In their Experiments 1 and 2, the stimuli could appear in any corner of an imaginary square and the response dimension was varied along only one dimension. In Experiment 1, both hands or both feet were used as horizontal effectors and the ipsilateral hand and foot as vertical effectors. Only a horizontal Simon effect was evident with horizontal effectors and only a vertical Simon effect was evident with vertical effectors. However, the horizontal Simon effect obtained with horizontal effectors was larger than the vertical Simon effect obtained with vertical effectors (48 ms vs. 16 ms), which Rubichi et al. took as evidence for a right–left prevalence effect.

Vu, Pellicano, and Proctor (2005) noted some problems with Rubichi et al.'s (2005) method for measuring right–left prevalence. Rubichi et al. varied the position of the effectors in that experiment along only one dimension at a time (e.g., when responses were made with the hands, they were positioned in the top-left and top-right locations, but when the responses were made with the feet, they were positioned in the bottom-left and bottom-right locations). Thus, subjects may have coded the response only in terms of a single dimension, rather than along both dimensions. Consistent with this suggestion, Ansorge and Wühr (2004) provided evidence that subjects code responses only along a discriminating dimension (i.e., horizontal dimension when responses varied in left and right locations) and not a nondiscriminating dimension (i.e., vertical dimension because it was held constant within a block: top-left or top-right).

Rubichi et al.'s (2005) Experiment 2 also showed a right–left prevalence effect when contralateral hand and foot responses were used: Only a horizontal Simon effect of 26 ms was evident. Rubichi et al. reasoned that because the contralateral hand and foot are usually considered to be top–bottom effectors and not left–right effectors, the use of contralateral effectors should make the responses equally salient along both dimensions or, if anything, to favor the vertical dimension. However, Vu et al. (2005) argued that the right–left prevalence effect found in Rubichi et al.'s Experiment 2 was due to their use of horizontally salient effectors. Vu and Proctor (2001a, 2002) showed that the contralateral hand and foot tended to be coded as left–right effectors, whereas the ipsilateral hand and foot tended to be coded as vertical effectors. Thus, it would not be surprising to find a horizontal advantage with contralateral hand–foot responses. To provide evidence for an overall right–left prevalence effect, Rubichi et al. would need to show that the advantage for the horizontal dimension with the contralateral hand–foot condition was larger than the vertical advantage with the ipsilateral hand–foot condition. Unfortunately, Rubichi et al. did not include an ipsilateral effector condition in which the response sets were arrayed along both dimensions.

Vu et al. (2005) showed no right–left prevalence effect with a stimulus display that was equally salient along the horizontal and vertical dimensions in two experiments that were conceptually similar to those of Rubichi et al.'s (2005) study. In Vu et al.'s Experiment 2B, subjects responded with the hands placed on horizontally arrayed or vertically arrayed keys, and in Experiment 3, they responded unimanually with a joystick. Thus, Vu et al.'s results indicate that the right–left prevalence effect does not reflect an automatic benefit for horizontal codes, but is due to intentional translation processes that occur when the task environment provides a more salient frame of reference for horizontal coding.

5.5 STATIC AND DYNAMIC DIMENSIONS

5.5.1 COMPATIBILITY EFFECTS FOR DESTINATION AND ORIGIN LOCATIONS

Michaels (1988) investigated another task in which the stimuli can be coded along two dimensions. In her experiment, subjects viewed squares that appeared to start moving from an origin location toward a destination location. Two squares were visible on each trial and, as shown in Figure 5.8, one square expanded around a constant position, to simulate movement toward the destination of the ipsilateral hand, or while shifting toward the contralateral side (but not crossing midline), to simulate movement toward the destination of the contralateral hand. Subjects were to respond to the destination of the moving square in some trial blocks and to the original position of the square in others, with either a compatible or incompatible spatial mapping. Subjects held a joystick with each hand and responded by pushing the appropriate joystick forward (to mimic a catching action) in response to apparent

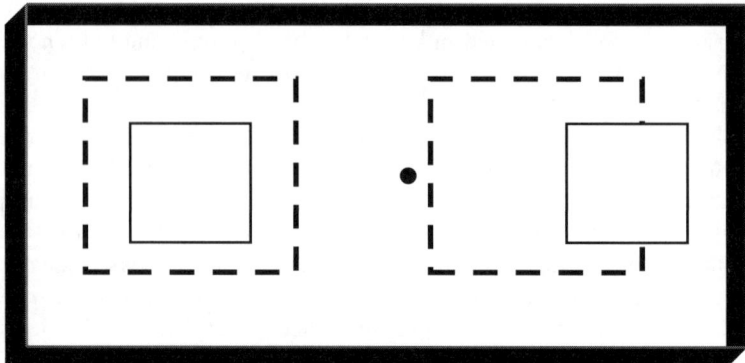

FIGURE 5.8 Illustration of the stimuli used in Michaels' (1988) and Proctor et al.'s (1993) studies. Solid line = initial square; dashed line = final expanded square.

movement of one of two squares on a display screen that appeared to move toward the ipsilateral or contralateral hand. Michaels found typical S-R compatibility effects not only for actual position when it was relevant, but also for destination of movement when it was relevant. Michaels attributed the destination compatibility effect to a catching affordance: The apparent motion afforded a catching action at its apparent destination.

Because the catching affordance account implies that compatibility will be highest when the direction of motion is toward the absolute position of the hand, Michaels (1988, Experiment 2) tested subjects not only with centrally located joystick positions, as in Experiment 1, but also with both joysticks located in the left or right hemispace. As is typical, the advantage for the compatible S-R mapping did not vary with response position when the position of origin was the relevant stimulus dimension, but it did when destination was relevant. The RT advantage for the compatible destination mapping was 98, 183, and 161 ms for the left, center, and right response placements, respectively, which Michaels took as supporting the affordance interpretation: "As the affordance interpretation predicted, *RT is indeed influenced by the hand toward which the square appears to move rather than merely its momentary position or actual motion on the screen*" (pp. 235–236; italics in original).

Note, though, that the S-R compatibility effect at the lateral response positions averaged 130 ms and that the effect was only 22 ms larger at the centered response position than at the right one. Thus, even if one accepts that absolute response position is a factor for destination compatibility, as the catching affordance account implies, it appears to contribute at most a small amount to the effect magnitude. Proctor, Van Zandt, Lu, and Weeks (1993) obtained additional evidence inconsistent with the catching affordance account in their Experiment 1, showing similar S-R compatibility effects for actual position and destination when the responses were left and right keypresses made with the index fingers of each hand, which do not mimic catching actions. In Experiments 2 and 3, Proctor et al. obtained compatibility effects of similar size when the square moved away from the subject (i.e., decreased in size) or shifted only laterally on the screen when it moved toward the subject (i.e., increased in size).

In an attempt to show a compatibility effect that could not be attributed to left–right movement direction, Michaels (1993) added a third type of destination display that had a simulated curvilinear trajectory. With this trajectory, the image of the stimulus shifted toward the center of the screen, as for the linear contralateral condition, but with a pattern intended to produce apparent movement along a circular route to the ipsilateral hand. The intent was to dissociate direction of movement on the screen (e.g., leftward from the right origin location) from the response afforded by the trajectory (e.g., right). Subjects responded compatibly to destination in some trial blocks and incompatibly in others by squeezing a trigger on the front of a left or right joystick. RT was shorter in the compatible blocks than in the incompatible blocks not only for the linear ipsilateral and contralateral trajectories from Michaels's (1988) earlier study, but also for the curvilinear trajectory. This led Michaels (1993) to conclude that this "experiment demonstrates that destination compatibility can be shown even when contradicted by relative left–right motion" (p. 1121).

However, in discussing the results, Michaels (1993) noted that the error rate was extremely high for the "compatible" curvilinear trials (28% in the first trial block) and that RT for this trial type was longer than that for the linear contralateral trials, suggesting that "both the circular and crossing trajectories afforded responses on the contralateral side" (p. 1126). In other words, for the curvilinear trajectory, the initial direction of motion toward the opposite side was more important than the final ipsilateral affordance, in agreement with coding accounts and counter to Michaels's affordance explanation. Moreover, Proctor, Lu, Van Zandt, and Weeks (1994) pointed out that the linear contralateral and curvilinear trajectories actually differed in relative direction for the critical last part of the movement, meaning that the ipsilateral affordance was not dissociated from relative movement direction.

In summary, for the most part, the destination compatibility effect can be attributed to relative direction of motion. There seems to be little, if any, need to evoke the concept of catching affordance to explain the results.

5.5.2 COMPATIBILITY EFFECTS FOR PITCH AND DIRECTION OF PITCH CHANGE

Walker and Ehrenstein (2000) examined S-R compatibility effects for auditory stimuli that varied along two dimensions, one static and one dynamic. In their study, subjects were presented with dynamic tonal stimuli that differed in initial pitch (high or low) and the direction in which the pitch changed (higher or lower). In their first experiment, Walker and Ehrenstein had subjects respond to corresponding or non-corresponding pitch–direction tones with a top (9 key on a numeric keypad) or bottom (6 key) keypress executed with the right-hand middle and index fingers when the arm was bent at a 90 angle. Half the subjects were instructed to respond to the initial tone pitch with a compatible (high tone = top key; low tone = bottom key) or incompatible (high tone = bottom key; low tone = top key) mapping, and half were instructed to respond to the change in direction of the tone pitch with either a compatible or incompatible mapping.

Responses were about 150 ms faster when subjects were responding to the initial tone pitch than to a change in tone direction, even when the time needed to detect the tone change was taken into account. More interestingly, when the pitch and direction of pitch change was the same, responses were 53 ms faster than when they were different. Both onset pitch and pitch change judgments were affected by the irrelevant dimension, but the onset pitch had a larger effect on pitch change judgments than vice versa. However, there was no mapping effect in that, for the relevant dimension, the compatible mapping of high-pitch stimuli mapped to the top key and low-pitch stimuli mapped to the bottom key did not produce any benefit in performance compared with the incompatible mapping. Moreover, mapping did not interact with correspondence of the initial tone pitch and change of direction.

Walker and Ehrenstein (2000) attributed the lack of mapping effect to the possibility that the 9 and 6 keys were not coded as top and bottom but rather as left and right due to the positioning and use of the index and middle fingers of the right hand (see Lippa's, 1996, referential coding hypothesis in Chapter 8). To ensure that the responses were coded in terms of the vertical dimension, Walker and Ehrenstein's Experiment 2 used a vertical response arrangement and had subjects respond by moving their right index finger vertically in the frontal plane. With this vertical response apparatus, they found an initial tone pitch and change of direction correspondence effect, as in their Experiment 1. Furthermore, Walker and Ehrenstein found an effect of compatibility, with RT being 95 ms shorter when the high stimuli were mapped to the top response and low stimuli to the bottom response.

Rosenbaum, Gordon, Stillings, and Feinstein (1987) examined S-R compatibility effects for high- and low-pitch tone stimuli for situations in which the responses were spoken sequences of one, two, or three syllables that differed in whether the vowel sound was high or low pitch (e.g., *gi* or *gu*; *gibi* or *gubu*; *gibidi* or *gubudu*). Responses were initiated more slowly when the tone pitch was mapped incompatibly to vowel pitch than when it was mapped compatibly and as sequence length increased, but these two variables did not interact. In a second experiment, Rosenbaum et al. showed that, for two-syllable sequences, this pitch compatibility effect was attenuated when the choice was between a homogeneous sequence (e.g., *bubu*) and a heterogeneous sequence in which the alternative response differed only in the vowel pitch for the first (e.g., *bibu*) or second (e.g., *bubi*) syllable. As emphasized by Rosenbaum et al., their results are highly similar to those obtained by Inhoff, Rosenbaum, Gordon, and Campbell, (1984) for manual keypress sequences (described in Chapter 4), implying a similar basis for selecting between and programming sequences of responses that are spoken or executed manually.

5.5.3 COMPATIBILITY EFFECTS FOR GESTURES

Stürmer, Aschersleben, and Prinz (2000, Experiment 1) showed subjects illustrations of a model's right hand that performed a grasping movement (fingers moved downward from a neutral position for 500 ms and then upward for 500 ms to return to the neutral position) or a spreading movement (fingers moved upward from a neutral position for 500 ms and then downward for 500 ms to return to the neutral position). Simultaneous with the movement onset, or 400 or 800 ms after it, the color of the hand changed from normal skin color to red or blue, which signaled whether to respond by making a grasping or spreading gesture with the right hand. When the irrelevant stimulus gesture corresponded with the response gesture, an overall 17-ms correspondence (Simon) effect was obtained. However, the effect was only significant at the 0- and 400-ms color change interval, showing a nonsignificant reversal at the 800-ms interval. The finding of a correspondence effect at the 0-ms interval implies that gesture identification can occur shortly after stimulus onset. Stürmer et al. suggested that the nonsignificant reversal of the correspondence effect at the 800-ms interval was due to the reversal in movement direction that occurred after 500 ms, when the hand returned to the neutral position.

Stürmer et al.'s (2000) Experiments 2 and 3 showed that large correspondence effects can be obtained with static "end-state" postures of an open hand (end state of a spreading movement) or fist (end state of a grasping movement), and with movements of the gestures that stopped in a position intermediate to the start and end states. In Experiment 4, a moving bar was used as the stimulus instead of a hand. The bar movement was similar to the hand movement in terms of up/down motion patterns, which lasted 1,000 ms for each phase, and the movement was irrelevant to the task. Again, subjects were to execute a grasping or spreading hand gesture in response to the change in color of the stimulus that occurred 0, 500, 1,000, 1,500, or 2,000 ms after movement onset. In this experiment, initial downward movement of the bar was considered to be corresponding for a grasping gesture, and initial upward movement to be corresponding for a spreading gesture. For color-change intervals of 0, 500, and 1,000 ms, RT was shorter for corresponding trials than noncorresponding trials (16, 14, and 7 ms, respectively), but at intervals of 1,500 and 2,000 ms, RT was shorter for noncorresponding trials, yielding reverse effects of –17 and –27 ms, respectively. This finding supports the notion that movement-based mechanisms contributed to the correspondence effect.

Stürmer et al.'s (2000) Experiments 5 and 6 evaluated the role of movement-based and state-based correspondence effects when they occurred in succession. In Experiment 5, the hand moved to the end-state position, which took 500 ms, and remained frozen in that position for the remainder of the trial instead of returning to the neutral position. When the onset of the relevant color occurred at 0 and 400ms, significant correspondence effects were obtained, supporting the movement-based component. However, at 800 ms, the correspondence effect was absent. In Experiment 6, the end-state posture was presented for 500 ms and then the gesture returned to a neutral position, which involved movement in the opposite direction as the initial gesture movement. Again, large correspondence effects were obtained when the relevant color dimension appeared 0 and 400 ms after the posture was presented and reversed when the color changed occurred at the 800-ms interval. These findings suggest that movement-based mechanisms are dominant when dynamic stimulus information is available, but when the dynamic information is absent, end-state mechanisms are used.

5.6 JUDGMENT AND DECISION MAKING

Although compatibility effects for multidimensional stimulus sets have been studied primarily in the context of choice–reaction tasks, compatibility effects have also been shown to account for several phenomena that are obtained in studies of judgment and decision making. For example, Shafir (1995) stated, "The significance of compatibility between input and output has recently been evoked by cognitive psychologists and by students of judgment and decision making to account for a series of surprising yet systematic patterns of behavior" (p. 248). Two types of compatibility distinguished by Shafir are scale compatibility and strategy compatibility. Both can be conceived of as types of set-level compatibility for which one of two or more stimulus dimensions, or attributes, is weighted more or less heavily in a judgment or choice as a function of the type of decision that is required.

Scale compatibility is that when one stimulus dimension is highly compatible with the scale on which a response must be based, that dimension will have more influence on the decision than it would otherwise. Scale compatibility accounts for the phenomenon of preference reversal, for which subjects prefer one alternative with a particular response scale but another alternative with a different response scale. Preference reversals have been illustrated for situations in which subjects must choose between or evaluate two bets of equal expected value that differ in probability of winning and the amount that can be won on the individual bet. When asked to choose between the two bets, the majority of subjects will pick the one that has a higher chance of winning over the one for which the payoff would be higher. But, when asked to specify the least amount of money for which they would be willing to sell the bet if they owned it, the majority of the same subjects will specify a higher price for the bet with the higher payoff than for the one with the higher chance of winning (e.g., Tversky, Slovic, & Kahneman, 1990). The scale compatibility explanation of this preference reversal is that the prices are relatively more compatible with payoffs than are choices between alternative bets, resulting in the payoffs being weighted more heavily in the pricing decisions.

Strategy compatibility refers to the hypothesis that different response methods, specifically those that are qualitative as opposed to quantitative, will evoke different strategies (Fischer & Hawkins, 1983). Strategy compatibility has been used to explain the prominence effect: After having matched two alternatives to be equally valued, most subjects, when asked to choose between the two, will choose the one that is higher on the dimension that they have previously specified as being more important to them (Slovic, 1975). In other words, the dimension that is judged to be more prominent, or salient, is weighted more heavily in choice than in matching. According to Shafir (1995), "The prominence effect ... can be viewed as a type of compatibility effect. In particular, it seems to arise out of the compatibility between a particular task — choice versus matching — and the strategies that the task invokes" (p. 259). The qualitative aspect of choice tends to cause a strategy of focusing on the most important attribute, whereas the quantitative aspect of matching or pricing tends to cause a strategy based on relative weighting of dimensions.

5.7 CHAPTER SUMMARY

Compatibility effects occur for S-R sets that vary along more than one dimension. From a theoretical point of view, this research allows examination of the nature of how different codes are formed and interact. From a practical point of view, these studies allow prediction of performance when more than one dimension is involved. Studies with two-dimensional, nonspatial stimuli mapped to keypress responses indicate that compatibility effects can arise when structural features of the stimulus set can be mapped to structural features of the response set, even when there is no conceptual or perceptual overlap between the sets. The importance of structural correspondence will be elaborated in Chapter 8, which discusses S-R compatibility effects for orthogonal spatial dimensions.

Studies of two-dimensional spatial compatibility demonstrate the flexibility with which people can utilize the spatial codes available for selecting an action. When stimuli vary along two dimensions simultaneously, spatial codes are formed for both dimensions. When the codes are redundant, responding can be based solely on the processing of the spatial code for one dimension. In that case, a right–left prevalence effect often occurs, in which horizontal codes exert a greater influence on responding than vertical codes. This dominance of the horizontal dimension is usually a consequence of the right and left hands providing a salient frame of reference for horizontal coding. However, the benefit for horizontal coding is not a consequence of automatically coding horizontal locations over vertical locations because there is no overall prevalence effect for the Simon task. If the vertical dimension is made more salient than the horizontal dimension, a top–bottom prevalence effect can occur.

Stimuli that differ along both a dimension of static initial value and a dimension of dynamic change will tend to be coded according to both features. Mapping effects are typically obtained for either dimension when it is defined as relevant for the task and, in some cases, Simon-type effects occur for the irrelevant dimension as well. Findings of this type have been obtained for origin location and movement direction of a visual stimulus, initial tone pitch and direction of pitch change of an auditory stimulus, and the static and dynamic components of a hand gesture. In judgment and decision-making tasks, where speed of responding is not a factor and probabilities of outcomes must be considered as well as their values, compatibility effects influence which alternative will receive the most favorable judgment.

6 Reversing the Simon Effect for Irrelevant Location

6.1 INTRODUCTION

As described in Chapter 4, the Simon effect is a ubiquitous phenomenon that is obtained across a variety of relevant stimulus dimensions, modalities of irrelevant stimulus information and responses, and task variations. It is almost a certainty that performance will be better when a spatially corresponding response, rather than a noncorresponding one, is to be made to a stimulus, even when stimulus location is nominally irrelevant. Yet, several exceptions have been found for which RT is shorter for the spatially noncorresponding response than for the corresponding response.

In Chapter 4, we described examples in which the Simon effect was determined by correspondence with the locations of action goals instead of the locations at which the physical responses were executed. In the present chapter, we describe four different types of reversals of the Simon effect that do not directly involve the location goals assigned to the relevant stimulus dimension: (a) the Hedge and Marsh reversal that occurs when the relevant S-R mapping is incompatible; (b) a relative frequency reversal found when noncorresponding trials are more frequent than corresponding trials; (c) a mixed-task reversal that occurs when location-irrelevant trials are intermixed with incompatibly mapped location-relevant trials; and (d) a reversal that occurs after practicing with an incompatible spatial mapping. These reversals are important because of their implications for human information processing and the nature of the Simon effect, as well as for their design implications about situations in which the expected benefit of spatial correspondence may not occur.

6.2 THE HEDGE AND MARSH REVERSAL

The most well-known reversal of the Simon effect is known as the Hedge and Marsh reversal. Hedge and Marsh (1975) had subjects respond to the color (red or green) of a stimulus, which could occur in a left or right location, by moving a hand from a home key to a red or green key located to the left or right (see Figure 6.1). With a compatible mapping of stimulus color to response color, a standard Simon effect was obtained for which RT was 23 ms shorter when the stimulus and response locations (and colors) corresponded than when they did not. However, with an incompatible mapping of stimulus color to key color (i.e., red stimulus mapped to

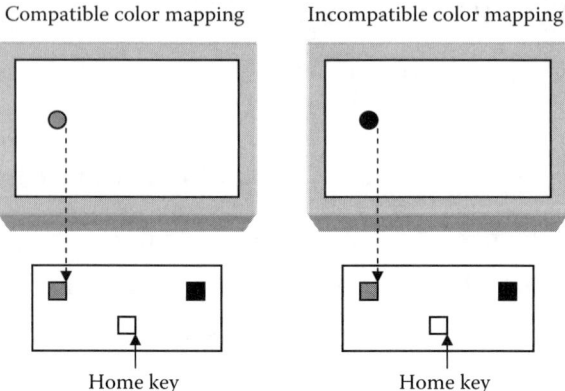

FIGURE 6.1 Illustration of the apparatus used by Hedge and Marsh (1975). One response key was red and the other green. The hand was to be moved from the home key to the corresponding color key for a compatible color mapping and to the noncorresponding color key for an incompatible color mapping.

green key and green stimulus to red key), a reverse Simon effect of –55 ms was evident for which responses were faster when the stimulus location was on the side opposite the response key rather than on the same side. Positive and reverse Simon effects of 67 and –40 ms, respectively, were obtained when the apparatus was rotated so that both the stimulus and response locations varied along the vertical dimension.

As with the Simon effect, the Hedge and Marsh reversal occurs across several task variations. Arend and Wandmacher (1987, Experiment 1) demonstrated the reversal using left- and right-pointing arrows, appearing in irrelevant left and right locations, to which a left or right keypress was to be made (with the middle and index fingers of the left hand in some trial blocks and of the right hand in others). Positive Simon effects of 30 ms were obtained with a compatible mapping of arrow direction to response locations, but negative Simon effects of about the same size were obtained with an incompatible mapping. Arend and Wandmacher found similar results in their Experiment 2 when the relevant stimulus dimension was the word LEFT or RIGHT. In this case, though, the reverse Simon effect for the incompatible mapping (–55 ms) was larger than the positive Simon effect for the compatible mapping (22 ms). Finally, Arend and Wandmacher obtained positive and reverse Simon effects of 15 and –20 ms, respectively, when a square and triangle were assigned to left and right keypresses, with a dot in the center of the stimulus on half of the trials designating an "incompatible" mapping of the stimulus form to the response keys.

Ragot and Guiard (1992) demonstrated the Hedge and Marsh reversal with auditory stimuli. Subjects made a left or right keypress to the spoken word "left" or "right" presented to the left or right ear. At the start of each trial, a green or red light came on to indicate whether the word–response mapping for the trial was compatible or incompatible. RTs were 115 ms shorter for trials on which the mapping was compatible than for trials on which it was not. More important, the results showed a 40-ms Simon effect when the mapping was compatible, but a –52-ms reverse Simon effect when it was incompatible. Ragot and Fiori (1994) replicated this result pattern, obtaining normal and reverse Simon effects of 31 ms and –58 ms, respectively, for compatible and incompatible mappings.

Although the Hedge and Marsh reversal of the Simon effect is obtained for auditory and visual presentation of stimuli for which the relevant information is color, arrow direction, or location word, it does not occur when stimulus location is designated as relevant. Brebner (1979) reversed the instructions for Hedge and Marsh's (1975) version of the task, making stimulus location relevant and stimulus color irrelevant. In this case, he observed a 55-ms benefit when the relevant S-R mapping was spatially compatible compared with when it was incompatible, but there was no effect of correspondence of the irrelevant stimulus color feature with the response-key color. Arend and Wandmacher (1987) included conditions in their Experiment 2, part of which was described above, in which subjects were instructed to respond to left–right physical location as the relevant stimulus dimension and to ignore the word (LEFT or RIGHT) that occurred in that position. They also found no Simon effect for the compatible mapping or Hedge and Marsh reversal for the incompatible mapping as a function of word meaning, even though word meaning overlaps conceptually with left–right keypresses (see Section 4.3, Verbal and Non verbal S-R Modes, in Chapter 4).

Simon, Acosta, and Mewaldt (1975) reported another failure to obtain the Hedge and Marsh reversal, in this case for warning tone location. They examined a condition in which an irrelevant 200 Hz warning tone was presented to the left ear, right ear, or both ears, followed 200 or 400 ms later by a 500 Hz target tone in the left or right ear. No Simon effect as a function of position of the warning tone was obtained when the S-R mapping for the target tone was compatible, and the Simon effect was positive when the mapping was incompatible. However, Proctor and Pick (1998) were unable to replicate these findings in an experiment similar to Simon et al.'s. Instead, they found results consistent with what most other studies would lead one to expect: a positive Simon effect of 57 ms with a compatible mapping and a reverse Simon effect of –42 ms with an incompatible mapping. Proctor and Pick obtained similar, though smaller, effects for a task in which the warning tone preceded a relevant left or right visual stimulus. Thus, at present, Simon et al.'s (1975) results should be regarded as an anomaly.

Several explanations have been offered for the Hedge and Marsh reversal of the Simon effect. We describe and evaluate each of these explanations in turn.

6.2.1 LOGICAL RECODING ACCOUNT

Because Hedge and Marsh (1975) found a positive Simon effect when the S-R color mapping was compatible and a reverse Simon effect when it was incompatible, they attributed both effects to a strategy of misapplying the response rule appropriate for the relevant color–response dimension to the irrelevant spatial dimension. This explanation is called the *logical recoding account*. In Hedge and Marsh's words,

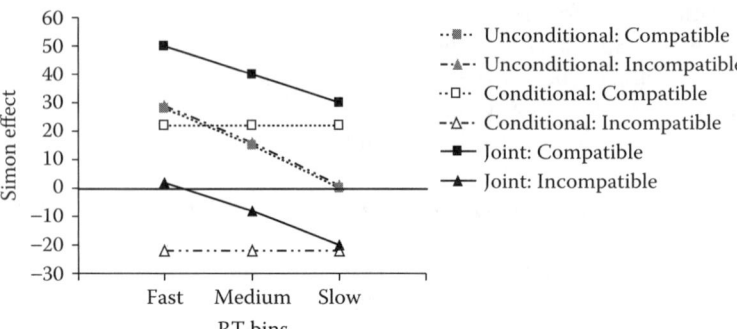

FIGURE 6.2 The time-course of the contributions of the conditional and unconditional components to the Simon effect (Noncorresponding–Corresponding) for De Jong et al.'s (1994) model, along with the predicted Simon effect and its reversal (the solid lines marked as joint). From "Conditional and unconditional automaticity: A dual-process model of effects of spatial stimulus–response correspondence," by R. De Jong, C.-C. Liang, and E. Lauber, 1994, *Journal of Experimental Psychology: Human Perception and Performance, 20*, p. 733. Copyright 1994, American Psychological Association. Reprinted with permission.

> The logical character of the recoding which would relate either of these attributes [colour and position] to the attributes of the response might be either 'identity' (same colour, or same position) or 'reversal' (alternate colour, or alternate position) .... *For a given logical recoding* (identity or reversal) *of the relevant attribute* (colour), *responding was faster for trials in which the recoding of the irrelevant attribute* (position) *was of the same logical type as that of the relevant attribute, than for trials in which the logical recoding of the irrelevant attribute was opposite in type* (p. 435, italics in original).

For a stimulus presented to the left, the identity rule yields a "left" spatial response code, whereas the respond opposite rule yields a "right" code. Responding is facilitated if the assigned response is the same as the location code generated by the rule but is slowed if the assigned response is different from the generated location code. This yields an advantage for corresponding responses when the mapping is compatible and for noncorresponding responses when the mapping is incompatible.

De Jong, Liang, and Lauber (1994) extended the logical recoding account to explain the RT distribution functions for compatibly and incompatibly mapped trials. They proposed a model according to which two processes with different time courses contribute to the Simon effect and its reversal (see Figure 6.2). According to the model, a stimulus automatically activates its spatially corresponding response shortly after onset, and this activation then decays. The second component is a task-defined S-R translation (identity or reversal) that is applied to the relevant stimulus feature based on the mapping instructions. This conditional component, which generalizes to the irrelevant stimulus location dimension, is not time dependent. For a compatible S-R mapping of the relevant dimension, both components produce activation of the corresponding response, which leads to a large Simon effect that decreases across the RT distribution. However, for an incompatible mapping, the two components counter each other immediately after stimulus presentation and the conditional, reversed component comes to dominate across time as the unconditional, automatic activation of the corresponding response decays.

De Jong et al. (1994) conducted several experiments to test this model. Their Experiment 1 included conditions in which the assignment of colors to the response keys varied randomly from trial to trial, as indicated by color labels on the screen or, in another condition, by color-word labels. In a third condition, the mapping of stimulus colors to response colors was varied from trial to trial, while the color of the keys remained constant. All three conditions yielded a positive Simon effect averaging 18 ms when the mapping was compatible and a reverse Simon effect of –47 ms when it was incompatible. RT distribution analyses showed that although the Simon effect decreased across the distribution, the reverse Simon effect increased, as predicted by the dual-process model. The primary deviation from predictions was that the reverse Simon effect was larger than the normal Simon effect, as is often the case. Additionally, lateralized readiness potentials measured in Experiment 4 showed an initial deflection toward the corresponding response for all trials, regardless of whether the relevant color mapping was compatible or incompatible. However, for the incompatible color mapping, this activation then shifted to the noncorresponding response and, if that response was incorrect, then back to the corresponding response. Thus, both the RT distribution analyses and the lateralized readiness potentials were in general agreement with the dual-process account.

Lu and Proctor (1994) obtained mean RT results consistent with De Jong et al.'s (1994) two-process account for several conditions. The primary purpose of their study was to determine whether the patterns of results obtained with compatible and incompatible mappings of the relevant stimulus dimension to responses depended on the conceptual similarity of that dimension to the irrelevant stimulus-location dimension. Conceptual similarity did not seem to affect performance much, but across seven conditions in three experiments that differed in whether the relevant stimulus dimension was arrow direction, location word, or color, a strong correlation was found between mean RT with the compatible relevant mapping and the relative sizes of the Simon effect and the Hedge and Marsh reversal. For conditions that yielded fast responding, the Simon effect tended to be slightly larger than the reversal. In contrast, for the three slowest conditions, the Hedge and Marsh reversal was 26 ms larger than the positive Simon effect. In Lu and Proctor's Experiment 4, the relevant color discrimination was easy (red or green) or difficult (dark green or light green). For the easy condition, which yielded mean RTs more than 100 ms shorter than the hard condition, the Simon effect and reverse Simon effects were 12 and –12 ms, respectively. However, for the hard condition, the mean Simon effect decreased to 7 ms, as often occurs when responding is slower (see Chapter 4), but the reverse Simon effect increased to –27 ms.

The timecourse of the Hedge and Marsh reversal provides some of the strongest evidence for the hypothesis that the reversal is due to logical recoding. Results from other studies described in the next subsection are less favorable toward the logical recoding hypothesis.

Compatible S-R mapping (same-color condition)

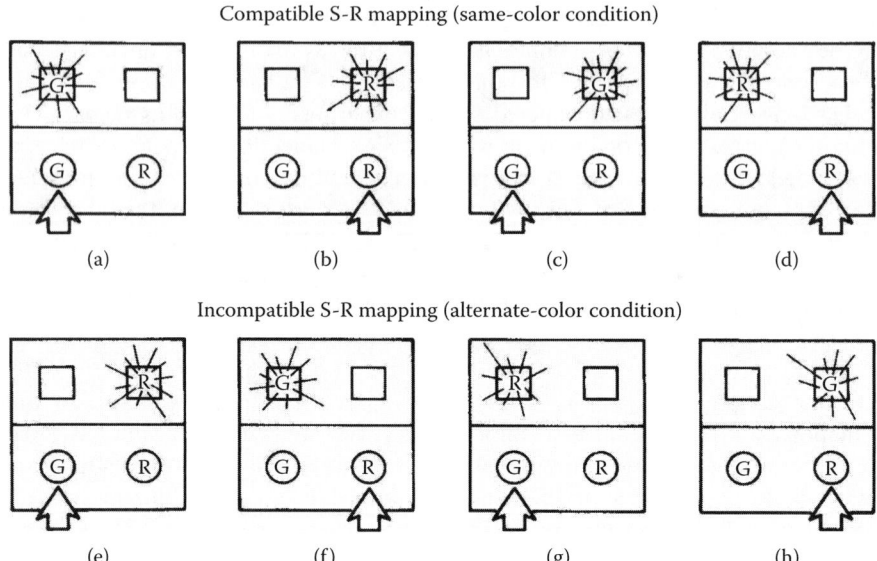

Incompatible S-R mapping (alternate-color condition)

FIGURE 6.3 Representation of the stimulus–response relations for the compatible and incompatible mappings in Hedge and Marsh's (1975) and Simon, Sly, and Vilapakkam's (1981) studies. The arrows indicate the correct responses. From "The effects of an irrelevant directional cue on human information processing," by J. R. Simon, 1990. In R. W. Proctor and T. G. Reeve (Eds.), *Stimulus–response compatibility: An integrated perspective,* p. 65. Copyright 1990, Elsevier Science Publishers. Reprinted with permission.

6.2.2 DISPLAY–CONTROL ARRANGEMENT CORRESPONDENCE

Simon, Sly, and Vilapakkam (1981) proposed display–control arrangement correspondence as an alternative to the logical recoding account of the Hedge and Marsh reversal. According to Simon et al., the reversal is a consequence of correspondence between the location of the color stimulus and the location of the colored response key. The general idea is illustrated in Figure 6.3. For a compatible mapping of stimulus color to response color, the two locations correspond when the stimulus is on the same side as the designated response but not when it is on the opposite side. In contrast, for an incompatible mapping, the two locations correspond for trials on which the designated response is on the opposite side of the display.

Simon et al. (1981) conducted three experiments to test their display–control arrangement correspondence hypothesis. Experiment 1 was similar to Hedge and Marsh's (1975) experiment, with the left key always colored green and the right key red. However, rather than varying correspondence and noncorrespondence of the stimulus and key colors randomly from trial to trial, Simon et al. blocked the trial types so that stimulus and response locations corresponded on all trials in one block (green stimulus in left location and red stimulus in right location) and did not on all trials in the other block (green stimulus in right location and red stimulus in left location). Even with blocked presentation, a reverse Simon effect of –64 ms was found with an incompatible color mapping, although this was smaller than the positive Simon effect of 113 ms found with a compatible color mapping. Note that the reversed effect with the incompatible color mapping can be conceived of as an advantage for the condition with display–control arrangement correspondence.

Simon et al.'s (1981) crucial test was their Experiment 3, in which the red or green stimulus was always presented in a centered location and a tone occurred simultaneously at either the left or right ear. The compatible S-R color mapping condition showed a positive Simon effect of more than 50 ms, but there was no Simon effect, positive or negative, for the incompatible mapping condition. The absence of a Hedge and Marsh reversal in the accessory stimulus version of the task is predicted by the display–control arrangement correspondence hypothesis because there was neither correspondence nor noncorrespondence of the centered visual stimulus with the designated response key on any trial. In contrast, the logical recoding account would seem to predict a reversal because there is no a priori reason why the opposite mapping rule for S-R color would not be applied to irrelevant auditory location if it is applied to irrelevant visual location.

Because of the importance of Simon et al.'s (1981) findings, Proctor and Pick (2003) evaluated whether they were replicable. In their Experiment 1, Proctor and Pick obtained results with color stimuli that were similar to those reported by Simon et al., differing only in that the compatible S-R color mapping, as well as the incompatible mapping, did not show a significant Simon effect. Proctor and Pick obtained additional results supportive of the display–control arrangement correspondence account in three other experiments. Most important were the results of their Experiment 2. In that experiment, a condition that was similar to Hedge and Marsh's (1975) original experiment, except that the left and right index fingers were placed on the color-labeled response keys, yielded a positive Simon effect even when the mapping of stimulus color to response color was incompatible. In contrast, when visible labels were placed at the bottom of the display screen, a standard Hedge and Marsh reversal of 53 ms was evident with the incompatible mapping. There is no reason why logical recoding should depend on visible color labels, whereas display–control arrangement correspondence would seem to require visible response labels that can be matched to the colored stimuli.

Without labeled keys, the reversal of the Simon effect seems to occur primarily when the relevant and irrelevant dimensions share physical location codes. Proctor and Pick (2003) found that when the irrelevant stimulus dimension was the location of a tone, a compatible mapping of a visual relevant stimulus dimension yielded a Simon effect, regardless of whether the relevant attribute was physical location or color. However, with an incompatible mapping, the Hedge and Marsh reversal of the Simon effect was obtained only when the visual relevant dimension was physical location and not when it was color. The Hedge and Marsh reversal with colored stimuli apparently depends on the response keys being visibly labeled, and the reversal does not occur when the irrelevant location information is conveyed by tones; this points toward display–control arrangement correspondence as a critical factor in the Hedge and Marsh reversal.

6.2.3   STIMULUS-IDENTIFICATION ACCOUNT

In an article titled "Stimulus–Response Compatibility and the Simon Effect: Toward a Conceptual Clarification," Hasbroucq and Guiard (1991) claimed that the Simon effect and the Hedge and Marsh reversal are not S-R compatibility effects at all but stimulus–stimulus (S-S) correspondence effects that influence the duration of stimulus identification. Stoffels, Van der Molen, and Keuss (1989) made a similar proposal for the auditory accessory stimulus Simon effect based on an additive factors analysis, showing that the effect of irrelevant tone location did not interact with the number of S-R alternatives. Hasbroucq and Guiard's account is straightforward for tasks in which the relevant stimulus dimension conveys spatial information (e.g., the word LEFT or RIGHT, or a left- or right-pointing arrow). In such tasks, the relevant stimulus and response dimensions overlap conceptually on the location dimension, and the relevant stimulus value can be congruent or incongruent with the irrelevant stimulus location. Consequently, the effects could be due to slower identification of the relevant stimulus dimension when it is incongruent with the irrelevant stimulus dimension than when the two dimensions are congruent. With this account, the Hedge and Marsh reversal obtained with an incompatible S-R mapping is not really a reverse Simon effect because performance is best when the relevant and irrelevant stimulus dimensions correspond.

For the standard Simon task and Hedge and Marsh's (1975) variations, however, there is no overlap between the relevant color dimension for the stimulus and its irrelevant location dimension. So, how can the Simon effect and its reversal be attributed to S-S correspondence? According to Hasbroucq and Guiard (1991), the instructions that define the mapping of relevant stimulus dimension to responses "amount to the definition of a correspondence code linking the two dimensions" (p. 250). With respect to Simon tasks that require a left or right response to a low- or high-pitch tone, Hasbroucq and Guiard stated,

> Specifically, instructions ensure the tonal coding of lateral position .... The use of long-standing linguistic signs (words) or new events (tones) as relevant signifiers of lateral position does not alter the logic of the paradigm. Because the S [stimulus] event amounts to the presentation of two simultaneous left–right messages, it must be either intrinsically congruent or intrinsically incongruent. (p. 250)

Hasbroucq and Guiard (1991) interpreted the relation between the relevant S-R color dimension and the irrelevant location dimension in Hedge and Marsh's (1975) task in a slightly different manner, saying

> Notice, however, that the Hedge and Marsh paradigm makes the color and the position of the subject's R [response] systematically covary. Because of the permanent color labeling of the R buttons, selection of the green R always amounts to selection of the left R, whereas selection of the red R always amounts to selection of the right R. In other words, here it is the output side of the paradigm that defines a one-to-one correspondence between color and position. (p. 250)

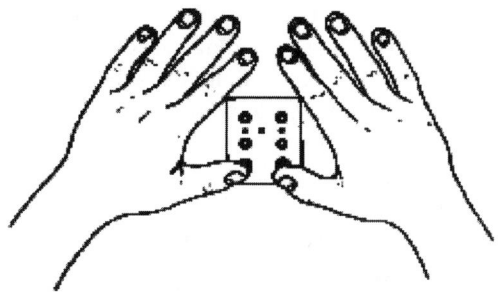

FIGURE 6.4 Depiction of the hand-held apparatus used by Hasbroucq and Guiard (1991). From "Stimulus–response compatibility and the Simon effect: Toward a conceptual clarification," by T. Hasbroucq and Y. Guiard, 1991, *Journal of Experimental Psychology: Human Perception and Performance, 17,* p. 253. Copyright 1991, American Psychological Association. Reprinted with permission.

O'Leary, Barber, and Simon (1994) pointed out that this definition of stimulus congruence in terms of response attributes for the Hedge and Marsh task involves a subtle shift from the definition in terms of the instructed assignment of relevant stimulus dimension to responses that Hasbroucq and Guiard (1991) provided for the standard Simon task. (See Guiard, Hasbroucq, & Possamai, 1994, and Barber, O'Leary, and Simon, 1994, for further discussion of this issue.)

Hasbroucq and Guiard's (1991) stimulus-identification account predicts that the Simon effect obtained with a compatible S-R mapping and the Hedge and Marsh reversal obtained with an incompatible mapping should be of similar magnitude because the effects arise in the stimulus-identification stage, prior to the response–selection stage on which mapping exerts its effect. They reported results consistent with this prediction in their Experiment 1, which replicated Hedge and Marsh's experiment using a small hand-held apparatus (see Figure 6.4) that included two pairs of response buttons (pressed by the thumbs), over which two permanently lit light-emitting diodes (LEDs, one red and one green) designated the key colors. Just above the key diodes was a warning row of small yellow LEDs, and above this row, a pair of bicolor LEDs that could be either red or green, to display the target stimulus. With this apparatus, Hasbroucq and Guiard found a Simon effect of 35 ms with the compatible color-to-color mapping and a reverse effect of –29 ms with the incompatible color-to-color mapping. This finding of similar magnitudes for the Simon effect and the Hedge and Marsh reversal does not generalize to several other studies summarized earlier in this chapter.

Another prediction of Hasbroucq and Guiard's (1991) stimulus-identification account is that when the colors of the response keys vary randomly from trial to trial, there should be no Simon effect. This prediction is made because in that situation there is no systematic covariation of response key color and key location, as there is when the key color is held constant. Consequently, color cannot signify location when the assignment of colors to response keys varies from trial to trial. To test this prediction, Hasbroucq and Guiard used the hand-held apparatus in their

Experiment 2, but randomly varied the assignment of the red and green colors to the two response keys. The two LEDs providing the color labels onset simultaneously with the stimulus LED on each trial, meaning that subjects could not know prior to stimulus onset which response key was to be pressed to which stimulus color. With this procedure, neither the compatible mapping of stimulus color to response color nor the incompatible mapping showed Simon effects, positive or negative.

De Jong et al. (1994) showed that the Simon effect and the Hedge and Marsh reversal can be obtained under at least some conditions in which the assignment of response colors to response locations varies from trial to trial. In a study described earlier, they used a procedure in which subjects responded on a keyboard, with the keys labeled at the bottom of the display screen. Similar patterns of positive Simon effects for compatible mappings and reverse Simon effects for incompatible mappings were obtained when the response color labels varied randomly from trial to trial and when they were held constant. However, because De Jong et al. presented the response-key labels 1,500 ms prior to the onset of the target stimulus, one could argue that subjects had sufficient time to establish temporary color-location associations prior to stimulus onset that would allow S-S correspondence effects to occur.

Hommel (1995) provided even stronger evidence that the Simon effect can be obtained reliably when the response color-key assignment varies randomly in an article titled "Stimulus–Response Compatibility and the Simon Effect: Toward an Empirical Clarification," which was directed at Hasbroucq and Guiard's (1991) findings and conclusions. His Experiment 1 was intended to allow response selection to be completed before the signal to respond appeared. In that experiment, left- or right-pointing arrows, centered on the display screen, designated the left or right response, respectively, to be made on that trial. However, the response was to be withheld until a go (green) signal to execute it or no go (red) signal not to do so appeared to the left or right approximately 1 s later. The results showed a 42-ms Simon effect as a function of go-signal location, even though the color of the go signal was not correlated with the response that was to be made (i.e., the signal was green regardless of whether the correct response was left or right). Hommel's Experiments 2 and 3 used displays similar to those of Hasbroucq and Guiard, but with all of the display elements appearing on a computer display screen and the responses made with index fingers on two keys of the computer's keyboard. In Experiment 2, the stimuli were much larger than the response labels in one condition and much smaller in another condition. In Experiment 3, the warning row located between the stimulus and the labels disappeared 200 ms before stimulus presentation or remained on until a response was made. Both experiments showed substantial Simon effects that did not vary as a function of the manipulations.

Hommel (1995, Experiment 4) was able to replicate the absence of a Simon effect when he used a hand-held display–response apparatus designed to be virtually identical to that used by Hasbroucq and Guiard (1991). He noted that this apparatus tended to elicit a perceived line connecting the stimulus to its correct response, regardless of whether the S-R locations were corresponding or noncorresponding, which Hommel concluded likely led to the lack of difference in RT for corresponding and noncorresponding trials. As support for this perceptual grouping hypothesis, he modified the apparatus by removing the center light of the row of three warning

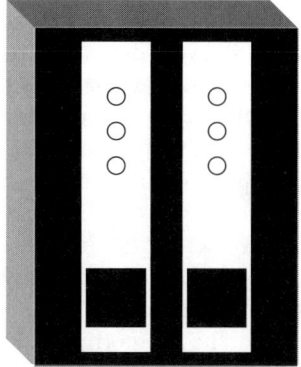

FIGURE 6.5 Diagram of Hommel's (1995) apparatus.

lights and covering parts of the apparatus so that perceptual grouping produced two distinct columns for the left and right sides (see Figure 6.5). These modifications, which removed the diagonal "lines" for the noncorresponding trials, yielded a significant Simon effect of 18 ms. Thus, Hasbroucq and Guiard's failure to obtain correspondence effects when the assignment of colors to response keys varied from trial to trial apparently was an artifact of the specific apparatus that they used.

Hasbroucq and Guiard's (1991) conceptual analysis highlights some of the difficulties involved in interpreting the results from variations of the Simon task and attributing them unambiguously to a particular processing stage. However, the relevant results currently do not conform well to the predictions of their stimulus-identification account once artifacts engendered by their specific apparatus are removed. The logical recoding and display–control arrangement correspondence hypotheses generally fare better than the stimulus-identification account, but an unequivocal case for any of the hypotheses cannot be made at present.

6.3 MANIPULATING PERCENTAGES OF CORRESPONDING AND NONCORRESPONDING TRIALS

In a typical Simon task, stimulus location corresponds with response location on 50% of the trials and does not correspond on the other 50%. Consequently, stimulus location is completely uninformative as to which response is correct on any given trial. By varying the relative percentages of corresponding and noncorresponding trials, correspondence between stimulus and response locations can be made more likely than noncorrespondence, and vice versa. This may allow expectancies to be developed that weight stimulus location into the decision process even though it is nominally irrelevant.

6.3.1 BASIC FINDINGS

Logan and Zbrodoff (1979) varied the relative percentages of corresponding and noncorresponding trials in spatial versions of the Stroop task. For these tasks, stimuli were the words ABOVE and BELOW presented above or below a fixation cross, and a left or right keypress was to be made to the stimulus dimension defined as relevant. For the spatial task, the relevant dimension was whether the stimulus was located above or below the fixation cross, whereas for the word task, it was whether the word was ABOVE or BELOW. Experiment 1 varied the percentages of corresponding and noncorresponding trials, with those for corresponding trials being 20% or 80%, and those for noncorresponding trials being the opposite, in two blocks of trials. The spatial task showed no effects of the percentage of corresponding trials and, indeed, no correspondence effect for the irrelevant location word, most likely because the relation between physical location and keypresses is more natural than that between location words and keypresses.

The word task, however, did show a correspondence effect for the irrelevant stimulus location, and this was dependent on the percentage of corresponding trials. When corresponding trials predominated, a positive correspondence effect of 81 ms was obtained, with RT shorter for corresponding than noncorresponding trials. However, when noncorresponding trials predominated, a reverse correspondence effect of –42 ms was obtained, with RT longer for corresponding than noncorresponding trials. In their other experiments, Logan and Zbrodoff (1979) used several different relative percentages and showed that as corresponding trials decrease systematically from 90% to 10%, the correspondence effect decreases monotonically from a large positive effect (159 ms in Experiment 3) to a substantial, but not as large, reverse effect (–89 ms).

Hommel (1994) obtained similar results for the Simon task itself in which dimensional overlap exists between the irrelevant stimulus dimension and the response dimension, and not the relevant and irrelevant stimulus dimensions, as in Logan and Zbrodoff's (1979) study. Subjects responded to the red or blue color of a stimulus presented to the left or right by pressing a left or right key, and the ratio of corresponding to noncorresponding trials was 50–50 in one trial block and 25–75 in another. The Simon effect was 28 ms larger when the percentages of corresponding and noncorresponding trials were equal, and it decreased to 14 ms when noncorresponding trials predominated. Stürmer, Leuthold, Soetens, Schröter, and Sommer (2002) also found the Simon effect to decrease as the percentage of corresponding trials decreases. They used vertically arrayed stimuli and responses, and had subjects respond to a square or diamond stimulus while ignoring whether it appeared in a top or bottom location. When 80% of the trials were corresponding, an 80-ms Simon effect was evident. This effect decreased to 33 ms when 50% of the trials were corresponding trials and reversed to 17 ms when only 20% were.

Toth, Levine, Stuss, Oh, Winocur, and Meiran (1995) used left- and right-pointing arrows mapped to left and right response buttons as the relevant task. The arrows were presented on the left, center, or right of a display screen, with stimulus location as the relevant dimension. With this procedure, conceptual overlap exists between the two stimulus dimensions and the response dimension. The percentage of corresponding to noncorresponding trials within a block was varied, being 83–17%, 50–50%, and 17–83%. The Simon effect was 39 ms for the 50–50% condition, and it shifted to 83 ms when corresponding trials predominated and –22 ms when noncorresponding trials did.

6.3.2 ACCOUNTS OF RELATIVE FREQUENCY EFFECTS

Hommel (1994) and Logan and Zbrodoff (1979) attributed the influence of relative frequency of corresponding and noncorresponding trials on Simon and Simon-type effects to a strategy of attending to the nominally irrelevant location dimension. Hommel indicated that his finding of a reduced Simon effect when noncorresponding trials predominated "demonstrates strategic preparation of stimulus processing and/or response selection based on irrelevant location information" (p. 261). Similarly, Logan and Zbrodoff summarized their view as, "The advantage for conflicting stimuli when they were frequent was taken as evidence for a strategy involving dividing attention between reported [relevant] and unreported [irrelevant] dimensions" (p. 166).

Toth et al. (1995), however, attributed their results to an implicitly learned component of automatic processing. Specifically, they proposed that as subjects perform a block of trials in which the percentages of corresponding and noncorresponding trials are unequal, associations between the stimulus locations and their most likely responses are acquired. In addition, a nonassociative component produces activation of the corresponding response, regardless of the relative percentages. When corresponding trials predominate, both automatic components activate the response corresponding to the stimulus location, resulting in a large Simon effect. When noncorresponding trials predominate, the associative component activates the noncorresponding response and the nonassociative component the corresponding response, leading to a reduced or reverse Simon effect. Because the two components are counter to each other when noncorresponding trials predominate, the reverse Simon effect will not be as large as the positive Simon effect obtained when corresponding trials predominate.

Stürmer et al. (2002) also attributed the change in magnitude of the Simon effect as the percentage of corresponding trials varied to practice with the predominant relation. However, they assigned this learning to the formation of short-term memory links in the conditional, or intentional, response–selection route. That is, according to Stürmer et al., when a certain trial type predominates, subjects receive more practice with that trial type and can induce a response bias in favor of it through short-term S-R associations.

Additionally, Stürmer et al. (2002) documented that strong sequential effects occur in Simon tasks, as several other studies have also shown (e.g., Hommel, Proctor, & Vu, 2004; Notebaert, Soetens, & Melis, 2001). The occurrence of such effects suggests that the composition of the trial set could influence the size, and possibly direction, of the Simon effect through altering the relative proportions of the different sequential relations. Two types of sequential analyses have been performed to examine the trial-to-trial changes in performance on the Simon task: correspondence versus noncorrespondence on the preceding trial and complete versus partial repetition. Both of these analyses examine performance on the current trial as a function of the preceding trial but each emphasizes different aspects of the trial sequences.

The typical finding for the correspondence versus noncorrespondence analysis is that the Simon effect is evident primarily on those trials that immediately follow trials for which the location relation was corresponding and not following trials for which it was noncorresponding (e.g., Hommel et al., 2004; Mordkoff, 1998; Stürmer et al., 2002; Valle-Inclán, Hackley, & de Labra, 2002). Stürmer et al. and others have interpreted this pattern of results as indicating trial-to-trial modulation (or gating) of the direct, or automatic, response–selection route. This route is suppressed after a noncorresponding trial because the response code activated by stimulus location was incorrect. For the opposite reason, the suppression is released following a corresponding trial. Consequently, automatic activation of the corresponding response location occurs primarily on those trials that immediately follow ones on which the stimulus and response locations corresponded.

According to the suppression hypothesis, when corresponding trials predominate, the automatic route should not be suppressed on most trials, which will speed responding on corresponding trials and slow responding on noncorresponding trials. In contrast, when noncorresponding trials predominate, the automatic route should be suppressed on most trials and the correspondence relation on the current trial should thus have little influence on performance. The suppression account predicts a reduction of the Simon effect when noncorresponding trials predominate but not a reversal of the Simon effect. To account for the reversal, Stürmer et al. (2002) proposed that the mechanism involved in learning of relative frequencies in the intentional response–selection route, described above, produces a separate influence.

In contrast to the corresponding or noncorresponding repetition analysis, the complete versus partial repetition analysis categorizes the trials according to whether they are a complete repetition of the previous trial (same relevant stimulus feature and response, and same irrelevant location), a complete change (different relevant stimulus feature and different response, and different irrelevant location), or a partial repetition (same relevant feature and response, but different irrelevant location or different relevant feature and response, but same irrelevant location). RT is shorter when there is complete repetition of the previous trial or complete change than when there is partial repetition (Hommel et al., 2004; Notebaert et al., 2001).

Hommel et al. (2004) proposed a feature integration account of the sequential effects for complete repetition/change and partial repetition based on the assumption that stimulus features and responses are bound in an "event file" (Hommel, 1998b). Unlike the suppression hypothesis, the feature integration account does not assume suppression and release of the automatic response–selection route on a trial-to-trial basis. Instead, the overall Simon effect is due to activation of the corresponding response on all trials, with complete repetition or alternation adding in a separate beneficial effect because the stimulus unambiguously signals whether the response should be the same or different from the one made on the previous trial. No such benefit exists on partial repetition trials because one feature signals that the response should be the same as on the previous trial and the other signals that it should be different.

The feature integration account explains the effects of the percentage manipulations as follows: When one type of trial predominates, the majority of trials in that category will be either complete repetitions or complete changes. This is evident by considering a Simon task, with green assigned to a left response and red to a right response, for which stimulus location is always corresponding. If the stimulus is green on one trial, it will be in the left location and require the left response. On the next trial, if the stimulus is green, it also has to be in the left location to maintain spatial correspondence, which is a complete repetition of the previous trial. If the stimulus is red, then it has to be in the right location to maintain correspondence and requires a right response, which is a complete change from the previous trial. Likewise, when 90% of the trials are corresponding, most of the corresponding trials will follow other corresponding trials and thus will be complete repetitions or complete changes. In contrast, most noncorresponding trials will follow correspond- ing trials and thus be partial repetitions. As a consequence, when corresponding trials predominate, performance on those trials will benefit from being primarily complete repetitions or complete changes, whereas performance on noncorrespond- ing trials will suffer from being primarily partial repetitions. When 90% of the trials are noncorresponding, everything described above reverses. This example illustrates that the feature integration account predicts a reversal of the Simon effect that the suppression account does not. In this case, though, a separate process is assumed to produce the overall Simon effect to account for why the positive Simon effect is larger than the reverse effect.

6.3.3 Tests of the Suppression and Feature Integration Accounts

Wühr (2004, 2005; Wühr & Ansorge, 2005) recently reported several studies com- paring the suppression and feature integration accounts. Wühr (2004) used a variant of the Hedge and Marsh task in which participants wore a red glove on one hand and a green glove on the other. With a compatible S-R color mapping, the typical pattern of modulation of the Simon effect was obtained: The effect was present following spatially corresponding trials and absent following noncorresponding tri- als. However, with an incompatible S-R color mapping, the reverse Simon effect occurred after noncorresponding trials but not after corresponding trials. This out- come indicates that if trial-to-trial modulation of the direct response–selection route occurs, noncorrespondence between the response activated by the stimulus color and that activated by response position is the critical factor that triggers suppression. Wühr pointed out that the feature integration account could also explain the quali- tative patterns of results obtained with both S-R color mappings. Based on the assumption that the feature integration hypothesis does not predict a complete absence of the Simon effect following a noncorresponding trial or of the reverse Simon effect following a corresponding trial, he concluded that the suppression model fit the data "somewhat better" (p. 105).

Wühr and Ansorge (2005) incorporated neutral trials (centered colored stimuli) into their experiments and varied the interval between onsets of successive stimuli. The Simon effect was largest following a corresponding trial, intermediate following a neutral trial, and smallest following a noncorresponding trial. Although the sequen- tial modulations were reduced when the interval between stimulus onsets was 6 s compared with when it was 1.5 s, the pattern was qualitatively similar. Wühr and Ansorge noted that both the suppression account and feature integration account could explain this result by assuming that the suppression or binding weakened across the longer interval. They concluded that although neither account fits all aspects of the data, a combination "seems to provide a rather comprehensive account" (pp. 728–729).

Wühr (2005) evaluated whether sequential changes in the Simon effect could occur when the amount of feature overlap in the different conditions was the same. The stimuli were red or green forms that could occur in lower left, upper left, lower right, and upper right positions on a display screen, and responses were made using two vertically arrayed response buttons. Wühr performed separate analyses for partial repetition trials, in which either the stimulus position or the color (and response) repeated and complete alternation trials. In both cases, a Simon effect was obtained that was larger after corresponding trials than after noncorresponding trials. Because the amount of feature overlap was the same within each of these trial-types, Wühr reckoned that these results cannot be explained by feature integration processes. He concluded,

> Together, the results reported by Hommel et al. (2004) and those of the present study suggest that at least two different mechanisms can cause sequential modulations of the Simon effect. One can be seen in the feature-integration processes that affect performance in Simon tasks, in addition to spatial S-R correspondence. A second cause for sequential modulations of the Simon effect may be a mechanism that gates direct response activation in the Simon task. (p. 286)

6.4   EFFECTS OF INTERMIXED LOCATION-RELEVANT MAPPINGS

When trials with an incompatible location-relevant mapping are intermixed with Simon trials for which stimulus location is irrelevant, the mapping has a substantial influence on direction and magnitude of the Simon effect. In the most basic case, stimuli for all trials occur in left and right locations, with stimulus location being relevant on half of the trials and irrelevant on the other half. As discussed in more detail in the next chapter, the mapping in effect for a given trial can be indicated by a mapping signal. De Jong (1995) used a high- or low-pitch tone to signal whether stimulus color or direction of arrow tilt was relevant, whereas Proctor and colleagues have signaled location as relevant with white stimuli and color as relevant with a red or green stimulus (Marble & Proctor, 2000; Proctor, Marble, & Vu, 2000; Proctor & Vu, 2002a, 2002b; Proctor, Vu, & Marble, 2003; see Figure 6.6). With physical location stimuli, Proctor and Vu (2002b) showed that when the location-relevant mapping was compatible, a positive Simon effect of 42 ms was obtained, but when the location-relevant mapping was incompatible, the Simon effect reversed to −44 ms. Similar results were reported by Marble and Proctor and Proctor, Vu, & Marble (2003), who also showed that, compared with the Simon effect obtained in pure blocks of location-irrelevant trials, the Simon effect is enhanced when the location-relevant mapping is compatibly mapped. It is important to note, though, that the reverse Simon effect when the location-relevant mapping is incompatible is as large as the positive effect when the mapping is compatible, indicating that the mapping in effect for the intermixed location-relevant task is the sole determinant of the Simon effect.

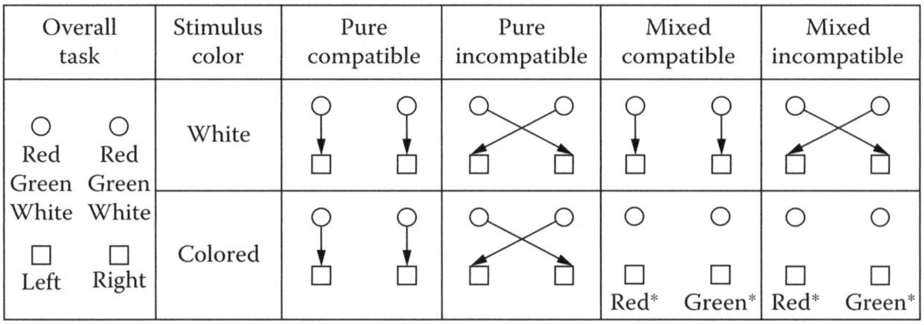

Overall task	Stimulus color	Pure compatible	Pure incompatible	Mixed compatible	Mixed incompatible
○ ○ Red Red Green Green White White ☐ ☐ Left Right	White				
	Colored				Red* Green*	Red* Green*

FIGURE 6.6 Conditions from the mixed location-relevant and location-irrelevant tasks studied by Proctor and colleagues. *The response keys were not colored and the labels indicate the keys to be pressed to particular stimulus colors with one of the two possible color–response mappings.

6.4.1 SYMBOLIC AND VERBAL STIMULI

The finding that the Simon effect reverses when the location-relevant mapping is incompatible also generalizes to left- and right-pointing arrow stimuli and, in some cases, to location-word stimuli. With arrow stimuli, reverse Simon effects were consistently obtained when the location-relevant mapping of arrow direction to response was incompatible (Proctor et al., 2000; Proctor & Vu, 2002b). However, unlike the symmetric effects obtained with physical location stimuli, the reverse Simon effect when the location-relevant mapping was incompatible was smaller than the positive Simon effect when the location-relevant mapping was compatible. With location-word stimuli, a small reverse Simon effect was usually obtained when the location-relevant mapping was incompatible, but this effect was relatively unstable. Proctor et al. found reverse Simon effects for three experiments in which location words were mapped incompatibly to keypress responses, but the reverse effect was significant only in the RT data for one of those experiments. Proctor and Vu conducted a similar study, but they did not obtain a reverse Simon effect for incompatibly mapped location words. Proctor et al. showed that words tend to yield a bimodal distribution, with subjects showing either a large positive or large negative S-R compatibility effect. They suggested that the absence of a reverse effect in their study was due to the fact that, by chance, a majority of the subjects in their sample were taken from the positive end of the distribution.

When the responses are the spoken words "left" or "right," the Simon effect reverses for physical location and arrow stimuli when the location-relevant mapping is incompatible (e.g., say "left" to a right stimulus and "right" to a left stimulus) and is eliminated for location-word stimuli. For physical locations and arrows, the reverse Simon effect is smaller in magnitude than the positive Simon effect obtained with the compatible location-relevant mapping. Thus, the symmetry in effect size for positive and reverse Simon effects with physical location stimuli mapped to keypresses disappears when the response modality is vocal. This suggests that the benefit of the corresponding response is due, at least in part, to subjects naming the stimuli prior to making the verbal response.

6.4.2 Accounts of the Simon Effect with Mixed Presentation

Several accounts have been proposed to explain the reduction, elimination, and reversal of the Simon effect when location-relevant and location-irrelevant trial types are mixed. These include explanations based on bias and uncertainty, intertrial variability, and dual response–selection routes.

6.4.2.1 Bias and Uncertainty

When location-relevant trials occur 50% of the time and Simon trials 50%, uncertainty exists about which dimension will be relevant for a trial. When the location mapping is compatible, the correct response is the corresponding response on 75% of the trials (all location-relevant trials and half of the location-irrelevant trials). Likewise, when the mapping is incompatible, the noncorresponding response is correct on 75% of the trials. As noted, the Simon effect tends to increase in size when corresponding trials predominate and reverse when noncorresponding trials predominate. Thus, it is possible that the effect of mixing could be due to the alteration of overall percentages of corresponding and noncorresponding trials.

Marble and Proctor (2000) evaluated this biasing hypothesis for physical location stimuli by manipulating the frequency of corresponding and noncorresponding trials (75–25, 50–50, or 25–75) for pure blocks of location-irrelevant trials and comparing the magnitudes of the Simon effects with the mixed conditions. With pure presentation, the Simon effect was 15 ms for the unbiased condition. The effect increased to 58 ms when the corresponding trials predominated and reversed to –36 ms when the noncorresponding trials predominated. Thus, the proportion of corresponding and noncorresponding trials affected performance. This asymmetric pattern of positive and reverse Simon effects also replicates the findings of Toth et al. (1995) for arrow stimuli. However, for the mixed conditions, Marble and Proctor found that the reverse Simon effect (–42 ms) with the incompatible location-relevant mapping was not smaller than the positive Simon effect (26 ms) with the compatible mapping. Because the proportion of corresponding and noncorresponding trials is the same in the pure and mixed conditions, but the former condition shows a greater benefit for the compatible mapping than the latter condition, biasing is not the sole factor contributing to the size and direction of the Simon effect although it may be a contributing factor. As noted previously, though, the pattern of the reverse Simon effect is at least as large as the positive Simon effect when location-relevant and location-irrelevant mappings are mixed; this seems to hold only for physical location stimuli (Marble & Proctor, 2000; Proctor & Vu, 2002b). Both arrows and words show the asymmetric pattern of positive and reverse Simon effects under pure and mixed conditions when the proportion of trials favors corresponding and noncorresponding trials, respectively.

Data relevant to uncertainty accounts can be obtained by cuing the trial type in advance because precuing allows subjects to prepare for the appropriate mapping. Although precuing reduces the effects of mixing when compatible and incompatible mappings are mixed, it does not seem to reduce the influence of the location-relevant mapping on location-irrelevant trials when both trial types are mixed. Precuing effects were examined in Marble and Proctor's (2000) study, which mixed incompatibly mapped location-relevant and location-irrelevant trials. In their study, the trial type was precued by the word COLOR or SPACE 150, 300, 600, 1,200, or 2,400 ms prior to the onset of the imperative stimulus. RT was faster at the longest precuing interval than at the shortest interval for both trial types, but the reverse Simon effect for location-irrelevant trials was still evident at all precuing intervals. Thus, incompatible location-relevant mappings still affected performance when subjects were informed 2,400 ms in advanced that the trial was location-irrelevant. This result is consistent with the findings of Proctor and Lu (1999) and Tagliabue, Zorzi, Umiltà, and Bassignani (2000), described in Section 6.5, in which the location-relevant mapping continues to influence performance even though it is no longer relevant to the task, and subjects know that the trial is based on color and not location. Overall, these results indicate that the Simon effect is not a result of automatic activation of the corresponding response that is hard-wired or overlearned, but that responding can be mediated by S-R associations defined for the task.

6.4.2.2 Intertrial Variability

The intertrial variability account attributes the overall pattern of positive and reverse Simon effects to the pattern of corresponding and noncorresponding trials in a sequence. Marble and Proctor (2000) conducted a repetition analysis to examine if performance changes as a function of whether the previous trial was location-relevant (task nonrepetition or switch trials) or location-irrelevant (task repetition trials). Overall, there was a large cost for switching the task. When the task switched, the reverse Simon effect for the incompatible location-relevant mapping was twice as large as when the task was repeated. Marble and Proctor showed that there was a task repetition benefit, but it did not affect the pattern of positive and reverse Simon effects.

When the repetition effects were analyzed as a function of whether the previous trial was corresponding or noncorresponding, Marble and Proctor (2000) showed that for pure and mixed blocks of location-irrelevant trials, the Simon effect was positive when the previous trial was corresponding and negative when the previous trial was noncorresponding, with the effects being stronger for the mixed than for the pure conditions. This pattern of repetition effects is similar to that reported by Mordkoff (1998) and Valle-Inclán et al. (2002) for pure Simon trials, which, as noted earlier in the chapter, some authors (e.g., Mordkoff) have attributed to trial-to-trial changes in suppression of the direct or automatic response–selection route.

Proctor and Vu (2002a) conducted a repetition analysis of Marble and Proctor's (2000) Experiment 1 that partitioned trials into conditions in which stimulus color (and response) and stimulus position were repeated or changed on consecutive trials. In the pure location-irrelevant condition, performance was better when both the stimulus color and position repeated or changed compared with conditions in which one dimension repeated and the other changed. In the mixed location-relevant and -irrelevant condition, repetition of the location-irrelevant task benefited only when both the stimulus color and position were repeated, but not when they both changed.

Furthermore, when the location-relevant mapping was compatible, the Simon effect was larger when both stimulus color and position were repeated or changed than when one dimension was repeated and the other changed. Similarly, when the location-relevant mapping was incompatible, the reverse Simon effect was larger when both stimulus color and position were changed or repeated than when one dimension was repeated and the other changed. This pattern of repetition effects can be accounted for with Hommel's (1998b) feature integration account.

As described earlier, the feature integration account assumes that the stimulus features on a trial and the response made to them are integrated into an event file, and responding is facilitated when all stimulus features are repeated or changed, and slowed when at least one stimulus feature repeats and another changes. Consistent with the feature integration account, Notebaert et al. (2001) and Hommel, Proctor, and Vu (2004) found that for pure blocks of Simon trials, the Simon effect is larger when the stimulus location and response location repeat or change than when only one feature changes and the other repeats. However, if the combination of stimulus location and response location repetition are collapsed together as a function of whether the previous trial is corresponding or noncorresponding, then the collapsed data show that the Simon effect occurs only after corresponding trials. Thus, the event-file hypothesis can account for the pattern of repetition effects without assuming that the suppression of the direct route is released on repetition trials (Hommel et al., 2004; Notebaert et al., 2001; but see Wühr's studies in Section 6.3.3).

6.4.2.3 Transient Activation Account

The transient activation hypothesis has been proposed to account for the Simon effect being reduced in size as a function of increasing reaction time. The hypothesis is based on the notion that automatic activation of the corresponding response occurs quickly and then dissipates (e.g., Hommel, 1993b). Thus, when response times are short, a large Simon effect is obtained because the corresponding response is highly activated, and when response times are slowed, the activation contributes less to performance, thus decreasing the magnitude of the Simon effect. Marble and Proctor (2000) conducted an RT distribution analysis for the location-irrelevant trials in their Experiment 1. Unlike the prediction of the transient activation account that the Simon effect would decrease across RT bins, Marble and Proctor did not find any interaction of RT bin and Simon effect for the mixed condition.

6.5 TRANSFER FROM A PRIOR SPATIAL MAPPING TO THE SIMON TASK

The Simon effect is influenced not only by an incompatible spatial mapping that is currently in effect, as when location-relevant and location-irrelevant trials are mixed, but also by an incompatible spatial mapping that was previously in effect. Proctor and Lu (1999, Experiment 2) had subjects perform 930 trials with a spatially compatible or incompatible mapping over 3 days and transferred them to a Simon task on the fourth day. For the Simon task, subjects were to respond to letter identity (S or H) and ignore its location. Subjects who performed with the spatially compatible mapping in the practice session showed a standard Simon effect of 21 ms. However, subjects who performed with the spatially incompatible mapping showed a reverse Simon effect of –14 ms. That is, after practicing with a spatially incompatible mapping, performance was better when the location of the letter did not correspond to the location of the response than when it did. Proctor and Lu replicated this latter finding in their Experiment 3 and showed that the effect generalized even when the stimuli used in the practice session (colored circles) were different from those used in the transfer session (letters). These findings indicate that the incompatible mapping from the previous spatially incompatible task continued to exert an influence on response selection for the current Simon task.

Tagliabue et al. (2000) showed that practice with an incompatible mapping of only 72 trials is sufficient to eliminate the Simon effect when a Simon task was performed immediately after the practice session. The most likely reason the effect reversed in Proctor and Lu's (1999) study, but was only eliminated in Tagliabue et al.'s study, is that the subjects received more practice in the former study than in the latter. However, Tagliabue et al. showed that when subjects were transferred to the Simon task 1 week after practicing with the spatially incompatible mapping, a reverse Simon effect of 22 ms was evident. They attributed this reversal to consolidation of the short-term S-R associations over the week delay.

Tagliabue, Zorzi, and Umiltà (2002) showed in a follow-up study that the effects of practice could generalize across stimulus modality. Tagliabue et al. had subjects practice with a spatially incompatible mapping of left and right tones to left and right keypresses. They were then transferred to a visual Simon task after delays of 5 minutes, 1 day, or 1 week. The Simon effect was eliminated in all conditions. However, unlike their results with the visual stimuli, practice with an incompatible mapping of auditory stimuli did not produce a reverse Simon effect, even in the 1-week-delay condition. Tagliabue et al. conducted a between-study analysis of the week-delayed Simon data as a function of practice with the visual or auditory incompatible mapping and found no significant difference in the Simon effect between the two conditions. Based on this nonsignificant interaction, Tagliabue et al. concluded that the elimination and reversal of the Simon effect is not due to context-specific coding.

However, this conclusion was not supported by a subsequent study conducted by Vu, Proctor, and Urcuioli (2003). Vu et al. replicated the week-delay conditions of Tagliabue et al.'s (2000, 2002) studies but included a full factorial design of stimulus modalities for the practice and Simon-transfer tasks. When subjects practiced with an incompatible mapping of auditory or visual stimuli and transferred to a visual Simon task, there was a significant difference in the Simon effects obtained. The Simon effect reversed when the practice task was visual, but it was only eliminated when the practice task was auditory. These results replicated the findings of Tagliabue et al., but the significant difference in Simon effects as a function of the stimulus modality in the practice session does not support Tagliabue et al.'s conclusion about context-specific coding. Furthermore, when transferred to an auditory Simon task after practice with an incompatible mapping, positive Simon effects were obtained regardless of whether the practice stimuli were auditory or visual. In fact, the magnitude of the Simon effect was not significantly different from that of a control group who received only the Simon task. These results suggest that generalization of previous task associations is dependent on the stimulus practice–transfer modalities.

Vu (2005) examined the nature of the location codes acquired during practice with a spatially incompatible mapping and how those codes are transferred to the subsequent Simon task. Subjects performed with S-R sets that were arrayed along either the horizontal or vertical dimension. Subjects completed 72 or 600 trials with a spatially incompatible mapping of horizontal or vertical stimulus–response sets during the practice session. Half of the subjects in each practice condition were transferred to a Simon task in which the stimulus–response set was arrayed along either the horizontal or vertical dimension. This yielded four combinations of practice–transfer combinations for the stimulus and response arrangement: horizontal–horizontal, horizontal–vertical, vertical–horizontal, or vertical–vertical. If the elimination of the Simon effect in the transfer session were evident only after practicing with an incompatible mapping of stimuli and responses arrayed in the same dimension as the transfer task, this would suggest that subjects acquired associations between specific stimuli and their assigned noncorresponding responses, and these associations remain active in the transfer task. If elimination of the Simon effect in the transfer session is independent of the dimension along which the stimulus and response arrangements were arrayed in the practice and transfer conditions, it would suggest that subjects are acquiring a more abstract "respond opposite" procedure during the practice task that is then transferred to subsequent conditions in which the procedure can be applied.

Vu (2005) found that when the number of practice trials with the incompatible spatial mapping was small (i.e., 72 trials), the transfer effect was evident only for the horizontal–horizontal practice–transfer condition. This suggests that the associations that subjects acquired were not rule-based procedures. One possible reason the transfer effect was not evident in the vertical–vertical condition is that top–bottom keys were operated by left and right hands and not top and bottom effectors. When the number of practice trials with the spatially incompatible mapping was increased to 600, the Simon effect reversed for the horizontal–horizontal condition and was eliminated for the remaining combinations. This finding suggests that when the number of practice trials is large, subjects begin to develop and proceduralize an "opposite" association that can be transferred automatically to different situations.

Most accounts of the Simon effect emphasize the role of long-term, automatic S-R associations that are innate or acquired through years of experience. However, the reversal and elimination of the Simon effect through relatively little practice with an incompatible spatial mapping prior to performing the Simon task, or when Simon trials are intermixed with incompatibly mapped location-relevant trials, indicate that short-term, task-defined associations can produce automatic activation of their associated responses as well. In fact, De Houwer (2004) demonstrated that Simon effects for location-irrelevant trials can arise through activation of short-term associations that have been defined for an intermixed location-relevant task.

De Houwer (2004) had subjects respond to the word "left" or a left-pointing arrow by saying "cale" and to the word "right" or a right-pointing arrow by saying "cole," or vice versa. Because the words "cale" and "cole" have no prior associations with the concepts of left and right, the S-R associations formed between them must be established through short-term linkages set up by the task instructions. After performing a practice block with the word and arrow stimuli mapped to "cale" and "cole," subjects performed another practice block of trials in which blue and green squares were presented to the left or right side of the screen. Subjects were instructed to respond to the color of the stimulus and ignore its location. Blue was mapped to the vocal response "cale" and green to "cole," or vice versa.

For the experimental trials, all three stimulus types were intermixed. For the location-irrelevant trials, performance for trials on which the position of the square (e.g., left) corresponded to the location with which the vocal response was associated (e.g., correct response of "cale" to the color of the stimulus was associated with left arrows and words) was compared with that for trials on which it did not correspond. Results showed a Simon effect in that performance was better on corresponding than noncorresponding trials. De Houwer, Beckers, Vandorpe, and Custers (2005) showed that, for location-irrelevant trials, the occurrence of a Simon effect through short-term associations of location information with unrelated spoken responses can even be obtained when subjects are instructed to be prepared to respond to location-relevant stimuli, but none are ever actually presented.

6.6 CHAPTER SUMMARY

In this chapter, we described four situations in which the Simon effect is reversed so that performance is better when stimulus and response locations do not correspond than when they do. One such situation is when the relevant stimulus dimension is mapped incompatibly to a relevant response dimension. This Hedge and Marsh reversal may be caused by a tendency to misapply the relevant mapping rule to the irrelevant location dimension, although some evidence suggests that it may be due in part to correspondence relations involving the colored stimulus and colored response keys. The Simon effect also reverses when noncorresponding trials are much more frequent than corresponding trials. Contributing factors to this reversal may include learning or attending to the predominant relation between stimulus and response locations and a change in the sequential dependencies across successive trials.

Reversals of the Simon effect also occur when the incompatibly mapped trials for which location is relevant are intermixed with trials for which stimulus location is not relevant. This reversal seems to be due at least in part to the task-defined associations between stimulus and response locations countering, and in some cases even overriding, the tendency for a stimulus to activate its corresponding response code. Finally, a spatially incompatible mapping that was in effect but is no longer can influence performance when stimulus location is irrelevant. Essentially, the task-defined associations persist for a period of at least 1 week and continue to produce activation of the noncorresponding response code. These reversals illustrate that several factors, some implicit and some explicit, can enter into play to determine whether there will be a benefit for spatial correspondence in any particular situation and how large this benefit (or cost) of correspondence will be.

7 Consequences of Mixed Mappings and Tasks

7.1 INTRODUCTION

When the number of stimuli and responses, or display elements and controls, is small, each stimulus can be mapped compatibly to its spatially corresponding response to ensure fast and accurate performance. However, in many complex display–control configurations, it is impossible to have a unique mapping of each stimulus to a spatially corresponding response. Instead, because of the limited space available, many stimuli may be presented in several fixed locations on the screen and mapped to a limited number of response actuators on the control panel. For these complex display–control configurations, the operator may be required to maintain several concurrent mappings of stimuli to responses, including compatible spatial mappings, incompatible spatial mappings, or mappings of symbols to different responses. For example, a navigation device may present a green arrow pointing to the left or right to signal a left or right turn, respectively, which is a spatially compatible mapping. This device can also be used to signal where traffic accidents or traffic jams are currently located, by placing a red "X" in those locations. In this case, the driver wants to avoid the red Xs, going to the left when the X is presented to the driver's right and going to the right when the X is to the driver's left, which is an incompatible spatial mapping. The driver must remember to respond compatibly to one set of events (green arrows) and incompatibly to the other set (red Xs).

Because of the complexity of our interactions with different systems and interfaces, it is important to determine how performance is affected by the requirement to maintain several S-R mappings concurrently and to switch between them. In Chapter 6, we discussed situations in which the Simon effect was influenced by the mapping that was in effect for intermixed trials on which stimulus location was relevant. In the present chapter, our focus is on consequences of mixing two or more S-R mappings or tasks on S-R compatibility proper.

7.2 MIXING COMPATIBLE AND INCOMPATIBLE MAPPINGS IN SPATIAL CHOICE TASKS

Over the past decade, considerable interest has developed in how people perform when they must switch periodically between two or more tasks. When the task that is to be performed varies from trial to trial, performance is worse than when only a single task is performed on all trials. Four phenomena are typically evident in task-switching experiments (Monsell, 2003). First, there is a switch cost, or task repetition benefit, for which RT is longer for trials on which the task or mapping switches from the previous trial than for trials on which it does not. Second, if the forthcoming task is precued sufficiently far in advance to allow the participant to prepare for it, the switch cost is reduced. Third, a residual switch cost usually remains even when sufficient time to prepare for the task switch is given. Fourth, a cost of mixing the tasks is obtained over and above the slowing that is due to the switch cost. Several mechanisms have been proposed to explain these effects, including reconfiguration of the task set from the previous trial (Rogers & Monsell, 1995), task-set inertia in which inhibition of the task set that was inappropriate for the previous trial is carried over to the current trial (Allport, Styles, & Hsieh, 1994), retrieval of prior instances of task sets associated with the current stimulus (Allport & Wylie, 2000), and preparation to respond (De Jong, 1995).

When two or more S-R mappings are mixed, there is an overall cost in performance compared with "pure" tasks of only single mappings that is due in part to having to switch between the two mappings (Los, 1996; Shaffer, 1965; Vu & Proctor, 2004). Moreover, as described below, this slowing is generally greater for the compatible mapping than for the incompatible mapping, which often leads to the elimination of the S-R compatibility effect (Vu & Proctor). Table 7.1 provides a summary of mean RTs for compatible and incompatible mapping conditions, under blocked presentation of a single mapping or mixed presentation of both mappings, for many of the studies described in the chapter.

7.2.1 Two-Choice Tasks

Shaffer (1965) was the first to examine the effects of mixing compatible and incompatible spatial mappings for two-choice tasks within a block of trials. Subjects in a pure condition performed with a compatible spatial mapping in one trial block and an incompatible mapping in the other. For subjects in a mixed condition, both mappings were in effect, and a horizontal or vertical line presented simultaneously with the imperative stimulus (left or right light) signaled whether the mapping for the trial was compatible or incompatible (see Figure 7.1). The S-R compatibility effect was 54 ms in the pure condition but a nonsignificant –8 ms in the mixed condition.

Heister and Schroeder-Heister (1994) used a method similar to Shaffer's (1965), but with the color of the imperative stimulus serving as the mapping signal for the mixed condition (see Figure 7.2a). They obtained a 60-ms S-R compatibility effect with pure compatible and incompatible spatial mappings, and a nonsignificant 2-ms effect with mixed mappings. Using a similar procedure, Vu and Proctor (2004) found compatibility effects of 71 ms and 4 ms for pure and mixed mappings, respectively. Thus, there are now several demonstrations that, in two-choice tasks, the S-R compatibility effect is eliminated when mappings are mixed.

The elimination of S-R compatibility effects in two-choice tasks with mixed mappings also extends to arrow stimuli. De Jong (1995, Experiment 1) used a method similar to Shaffer's (1965), with a high- or low-pitch tone as the signal to indicate whether the mapping of an upright arrow, tilted to the left or right, was compatible or incompatible for the trial. When the tone was presented 100 ms prior to the imperative stimulus, an 18-ms S-R compatibility effect was obtained. This effect was significantly smaller than the 45-ms S-R compatibility effect evident with pure

TABLE 7.1
Mean Reaction Time for Pure and Mixed Blocks of Compatible and Incompatible Mappings

	Stimuli	Responses	Pure Blocks			Mixed Blocks		
			Comp	Incomp	SRC Effect	Comp	Incomp	SRC Effect
Two-Choice Studies								
Shaffer (1965)	Locations	Keypresses	387	441	54	706	698	−8
Heister and Schroeder-Heister (1994)	Locations	Keypresses	310	370	60	521	523	2
Vu and Proctor (2004)	Locations	Keypresses	318	389	71	587	591	4
Vu and Proctor (2004)	Arrow	Keypresses	370	450	80	650	653	3
Vu and Proctor (2004)	Words	Keypresses	455	490	35	688	805	117
Vu and Proctor (2004)	Locations	Vocal	434	458	24	730	748	18
Vu and Proctor (2004)	Arrow	Vocal	484	526	42	735	761	26
Vu and Proctor (2004)	Words	Vocal	502	632	130	688	763	75
De Jong (1995)	Arrow-tilt	Keypresses	n/a	n/a	45	n/a	n/a	18
Four-Choice Studies								
Duncan (1978)	Locations (mixed inner–outer)	Keypresses	432	559	127	592	605	13
Duncan (1978)	Locations (mixed left–right)	Keypresses	502	662	160	634	733	99
Stoffels (1996b)	Locations (mixed inner–outer)	Keypresses	398	462	64	494	477	−17
Stoffels (1996b)	Locations (mixed left–right)	Keypresses	400	469	69	487	516	29
Stoffels (1996a)	Arrows (mixed inner–outer)	Keypresses	508	591	83	563	585	22
Ehrenstein and Proctor (1998)	Locations (mixed left–right)	Keypresses	n/a	n/a	n/a	568	624	56
Ehrenstein and Proctor (1998)	Locations (mixed inner–outer)	Keypresses	n/a	n/a	n/a	571	621	50
Ehrenstein and Proctor (1998)	Locations (mixed alternating)	Keypresses	n/a	n/a	n/a	666	749	83
Van Duren and Sanders (1988)	Digits	Vocal	465*	555*	90	520*	545*	25

Comp = Compatible mapping
Incomp = Incompatible mapping
SRC = S-R compatibility
* Denotes estimated values from figures

Mapping signal

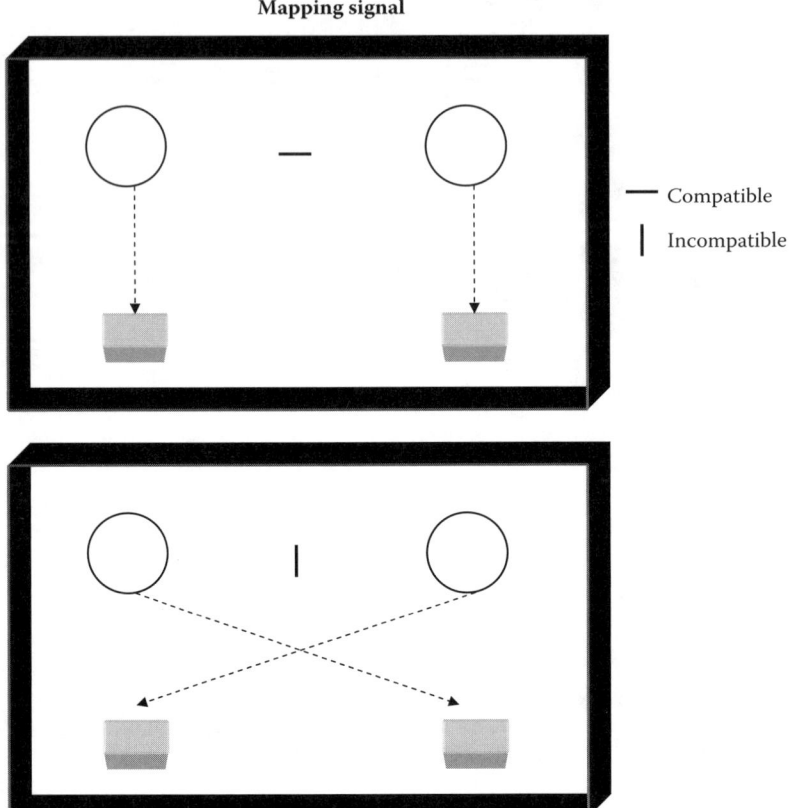

FIGURE 7.1   Illustration of the mixed compatible and incompatible mapping procedure used by Shaffer (1965). The line orientation signaled which mapping was in effect for the current trial.

presentation of the two mappings. Using stimulus color as the mapping signal (see Figure 7.2b), Vu and Proctor (2004) found an 80-ms S-R compatibility effect for left- and right-pointing arrows in the pure condition that was reduced to 3 ms in the mixed condition.

Although S-R compatibility effects disappear when compatible and incompatible mappings of spatial locations or arrow directions to keypresses are mixed, this result does not generalize to location words. Vu and Proctor (2004) found that when the stimuli were the location words LEFT and RIGHT and the responses were keypresses (see Figure 7.2c), the S-R compatibility effect was in fact much larger when the two mappings were mixed (117 ms) than when they were presented in pure blocks of trials (35 ms).

When responses are the spoken words "left" and "right" instead of keypresses, the S-R compatibility effect for mixed mappings is only reduced significantly for location–word stimuli (see Figure 7.2f): Vu and Proctor (2004) found that with location words, the S-R compatibility effect was 130 ms in pure conditions and 75

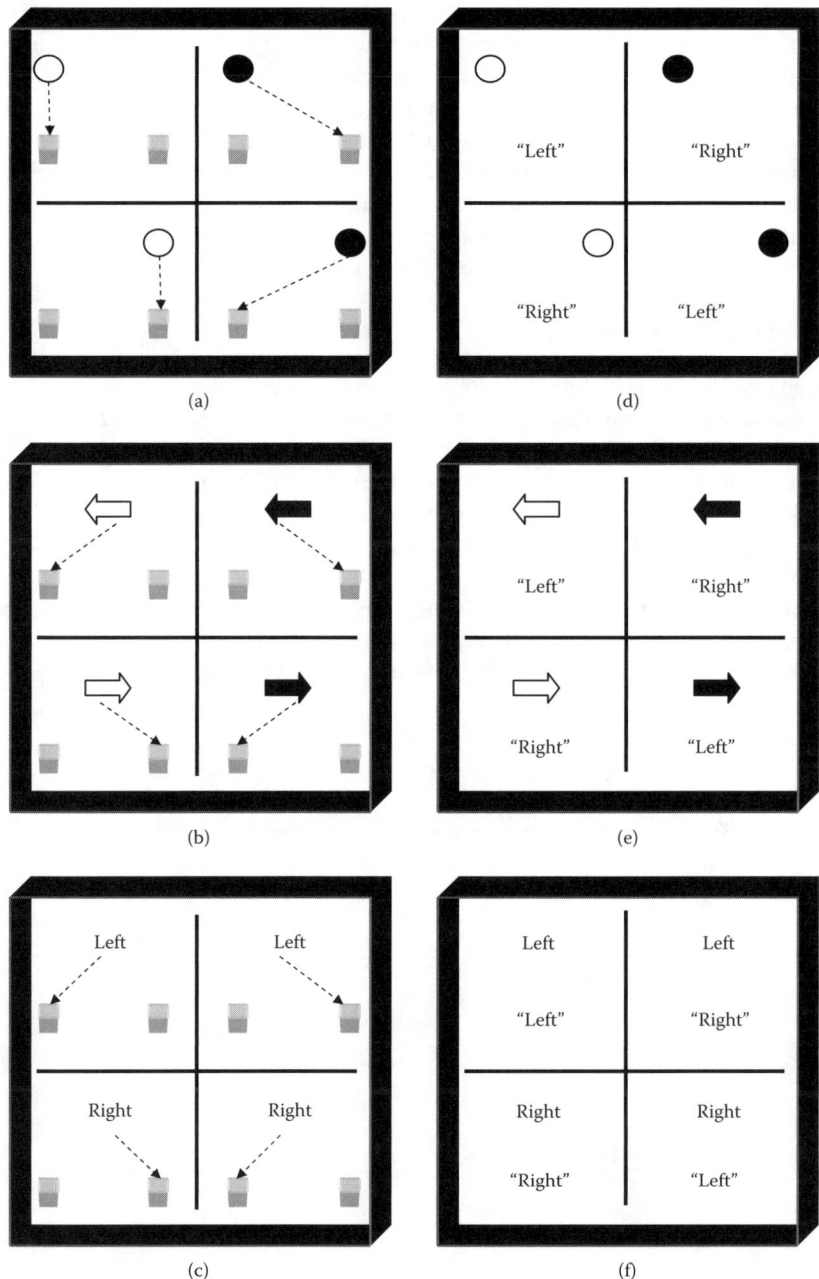

FIGURE 7.2 Illustration of the mixed compatible and incompatible mapping procedure for a two-choice task with three stimulus modes (physical locations, arrow directions, and location words) and two response modalities (manual and vocal). Stimulus color indicated which mapping was in effect for the current trial. The S-R sets illustrated in panels a, b, and f have higher set-level compatibility than those illustrated in panels c, d, and e.

ms in the mixed condition. For physical locations or arrow directions (see Figures 7.2d and 7.2e), there was a nonsignificant tendency for the compatibility effects to be smaller with mixed mappings (18 ms for physical locations and 26 ms for arrows) than with pure mappings (24 ms for physical locations and 42 ms for arrows). Although mixing tended to reduce the benefit for the compatible mapping, for all stimulus sets, the S-R compatibility effects remained significant, unlike the results obtained with keypress responses. These residual compatibility effects for mixed mappings when the responses are the spoken words "left" and "right" may be due to a tendency to name the stimulus even under mixed conditions.

7.2.2 Four-Choice Tasks

Duncan (1977b) devised a method for studying the effects of mixed mappings using four stimuli and four responses instead of two. In his study, the four stimulus locations were arranged in a row and mapped to four response keys that were also arranged in a row. The index and middle fingers of each hand rested on these keys, with the appropriate finger used to execute a specific keypress response. For the mixed mapping condition, subjects responded to the two inner stimulus locations with a compatible mapping and to the two outer locations with an incompatible, crossed mapping, or vice versa (see Figure 7.3). Duncan initially reported that mixing slowed RT equally for both mappings. However, subsequent studies using the four-choice procedure (e.g., Duncan, 1978; Stoffels, 1996a) showed that mixing slows responding more for the compatible than incompatible mapping, essentially eliminating the S-R

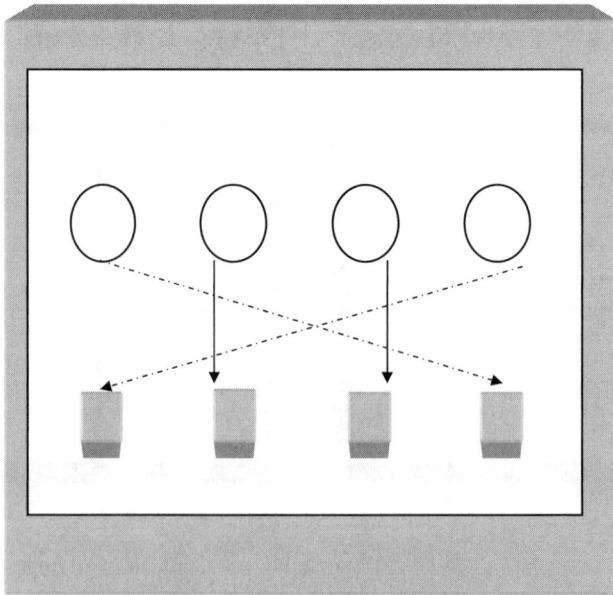

FIGURE 7.3 Illustration of mixed compatible and incompatible mapping procedure for a four-choice task. In this example, the inner pair of stimuli is mapped compatibly to responses and the outer pair is mapped incompatibly.

compatibility effect (13 ms in Duncan's study and −17 ms in Stoffels's study), as in the two-choice tasks.

In other versions of the four-choice task, each mapping is assigned to the two left or two right stimulus locations or to pairs of alternating stimulus locations. For these versions, the S-R compatibility effect is reduced by approximately 50 ms compared with pure mapping conditions but not eliminated entirely (e.g., Duncan, 1978; Stoffels, 1996b). Although the studies by Duncan and Stoffels found that manipulating the mapping between inner/outer pairs and left/right pairs yielded different patterns of results, Ehrenstein and Proctor (1998) found significant S-R compatibility effects of about 50 ms with both types of mixing procedures [compared with an S-R compatibility effect of about 120 ms that is obtained when blocks of pure compatible or pure incompatible (crossed) mappings are performed for the four-choice task; e.g., Vu & Proctor, 2003].

As with the two-choice task version, the reduction of the S-R compatibility effect in the four-choice task with mixed mappings is not limited to spatial location stimuli. Stoffels (1996a) obtained a reduction under mixed mapping conditions for a four-choice task in which the stimuli were centered left-/right-pointing arrows of two different colors mapped onto a row of four response keys. The color of the arrow determined whether the mapping for the current trial was compatible or incompatible and whether the response keys were the inner or outer pairs (see Figure 7.4). For example, if white arrows mapped to the outer response keys signaled the compatible mapping, then the left-pointing arrow would be mapped to the outer left response and right arrow to the outer right response. Similarly, if black arrows mapped to the inner response keys signaled the incompatible mapping, then the left-pointing arrow would be mapped to the inner right response and right-pointing arrow to the inner left response. Results showed that the S-R compatibility effect was approximately 83 ms for pure blocks of compatible and incompatible mappings, but it was reduced to 22 ms with mixed presentation of compatible and incompatible mappings. Thus, unlike Stoffels's finding of a reverse S-R compatibility effect with physical locations, arrows showed only a reduced effect.

7.2.3 MIXED STIMULUS MODES

Proctor and Vu (2002c) evaluated whether presenting the stimuli for the compatible and incompatible trials in distinct modes (spatial and verbal) would reduce the cost of mixing the two mappings. Because, as noted previously, mixing compatible and incompatible mappings eliminated the compatibility effect for physical locations and enhanced it for location words, Proctor and Vu used one stimulus mode for trials with a compatible mapping and the other for trials with an incompatible mapping. Although presenting the compatible and incompatible mappings in distinct modes (in addition to colors) reduced the overall mixing cost, the S-R compatibility effect was still eliminated for physical location stimuli (8 ms) but enhanced for location words (141 ms). This pattern of S-R compatibility effects is similar to that obtained when there is no mode distinction, which suggests that the S-R associations for the compatible and incompatible mappings are not mode specific (see also De Houwer, 2004).

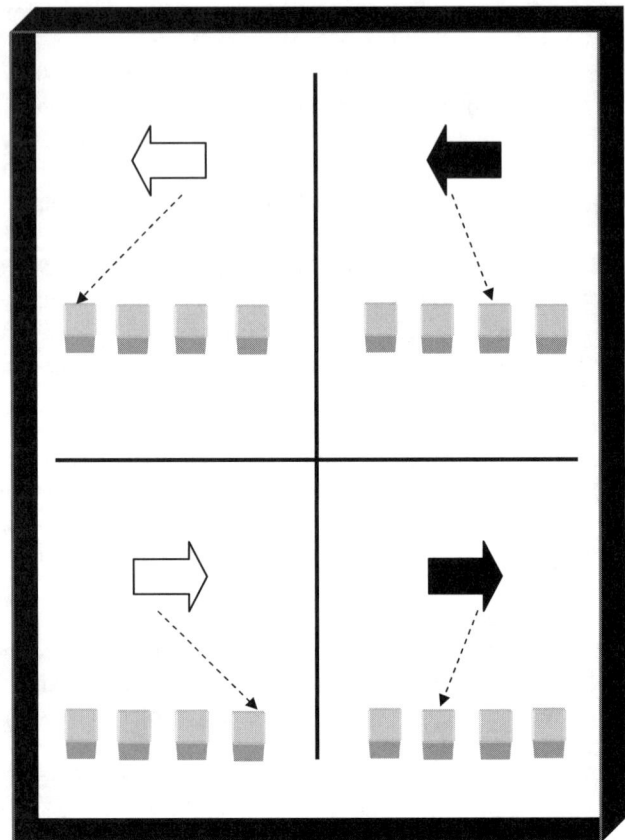

FIGURE 7.4 Example of stimuli used by Stoffels (1996a). Color of the arrow indicated whether the mapping on the trial was compatible or incompatible, and whether the response was to be an inner or outer location. In the example, white arrows are mapped compatibly to outer responses and black arrows are mapped incompatibly to inner responses.

7.3 MIXING NONSPATIAL S-R MAPPINGS

The S-R compatibility effect for nonspatial stimulus information is also reduced when mappings are mixed. Van Duren and Sanders (1988) conducted a study using vocal numeral responses to digits. Subjects were to respond with the corresponding names to the digits 2 and 3 and with the name of the opposite member of the pair to the digits 4 and 5. With a pure presentation of compatible or incompatible mappings, a 90-ms S-R compatibility effect was obtained. However, the S-R compatibility effect was reduced to 25 ms with mixed presentation. Similar to previous studies, the reduction/elimination of the S-R compatibility effect with mixed mappings was a consequence of the compatible responses being slowed more by mixing than the incompatible ones.

Related studies by Morin and Forrin (1962; Forrin & Morin, 1967) suggest that the greater effect of mixing on a compatible S-R mapping occurs even when the "incompatible" mapping is an arbitrary one of a stimulus set that has no dimensional overlap with the response set. In Morin and Forrin's study, subjects responded with spoken digit names to digit stimuli or shapes (e.g., square, triangle). The correct response to a digit stimulus was the digit name, whereas the correct response to a shape stimulus was an arbitrarily assigned digit name. Thus, the former task can be characterized as a compatible mapping and the latter as an incompatible mapping.

Morin and Forrin (1962) examined performance in two- and four-choice tasks with a single stimulus type, and in a four-choice task in which two stimuli were digits and two were shapes. Results showed that responding to the digit stimuli did not take longer when the number of alternatives increased from two to four digits, but when the extra two stimuli were shapes, RT to the digits increased by about 90 ms. However, for shape stimuli, increasing the number of alternatives from two to four increased RT by about 120 ms, regardless of whether the additional stimuli were digits or shapes. In other words, mixing the mappings had no effect for the incompatible mapping beyond that due to the increase in set size but slowed responding for the compatible mapping.

In a Hedge and Marsh version of the Simon task, for which the responses were presses of keys that were labeled with colors, De Jong et al. (1994) mixed a compatible mapping of stimulus color and key color with an incompatible one. A mapping signal ("SAME" or "OPPO") was presented 1.4 s prior to the imperative stimulus. Responses were 109 ms faster with the compatible S-R color mapping than with the incompatible mapping. The finding of a sizeable benefit for the compatible color mapping is likely because the 1,400-ms precuing interval allowed subjects time to prepare the appropriate mapping prior to the onset of the imperative stimulus. The effects of irrelevant location conformed to those from Hedge and Marsh's (1975) study (see Chapter 6). For a compatible color mapping, responses were 15 ms faster when the stimulus location corresponded to the response location than when it did not. However, for an incompatible color mapping, responses were 50 ms slower when the stimulus location corresponded to the response location than when it did not. The finding that the reverse Simon effect was larger than the positive Simon effect suggests that automatic activation of the corresponding response did not contribute to performance. If it had, this activation should have added to the positive effect when the mapping was compatible and subtracted from the reverse effect when the mapping was incompatible.

7.4 ACCOUNTS FOR ELIMINATION OF THE S-R COMPATIBILITY EFFECT WITH MIXED MAPPINGS

7.4.1 Uncertainty

The reduction and elimination of the S-R compatibility effect under mixed mapping conditions is often attributed to uncertainty about whether the appropriate mapping for the trial will be compatible or incompatible. Shaffer (1965) proposed a translation efficiency account to explain the elimination of the S-R compatibility effect for two-choice mixed mapping conditions. According to this account, when the assigned mapping is not known in advance, S-R translation is of similar difficulty for the compatible and incompatible mappings, and the subject has to choose the appropriate transformation. Shaffer attributed the benefit for the compatible mapping with pure presentation to a special "null transformation" that is easy to compute when the mapping is the same for all trials.

Evidence for the uncertainty hypothesis comes from studies in which the mapping for the trial is precued, such as that of Shaffer (1965), which included a condition in which the mapping was precued 333 ms before onset of the imperative stimulus. In contrast to the absence of S-R compatibility effect when the mapping signal was presented simultaneously with the imperative stimulus, the precued condition showed an S-R compatibility effect of similar magnitude to that obtained with pure presentation of compatible and incompatible mappings. Jennings, Van der Molen, Van der Veen, and Debski (2002) reported similar results when the mapping was indicated by a change in background screen color (yellow for compatible and blue for incompatible) either 50 ms or 500 ms prior to the imperative stimulus. The S-R compatibility effect was –4 ms at the short precuing interval and 24 ms at the long one.

Shaffer's (1965) translation efficiency account has not received much attention, most likely due to the fact that he did not develop the account in much detail and only devoted a few lines to it in his papers. In addition, the translation efficiency account cannot explain why errors in four-choice tasks tend to be the response that would be correct if the alternative mapping were in effect on the current trial (Duncan, 1977a, 1977b, 1978; Ehrenstein & Proctor, 1998; Stoffels, 1996b) or why the compatible mapping benefits more from repetition than the incompatible mapping under mixed conditions (see, e.g., Proctor & Vu, 2002a).

Duncan (1977a, 1977b, 1978) proposed a rule-based translation account for the reduction of the S-R compatibility effect when mappings are mixed for the four-choice task. According to Duncan, when compatible and incompatible mappings are mixed within a block of trials, response selection occurs in two steps. First, the appropriate mapping rule for the current trial is determined, and, second, that mapping rule (respond same or opposite) is applied to select the correct response. Thus, this two-step model predicts that the effect of mixing should be additive with that of mapping because the mapping-selection step is added when compatible and incompatible mappings are mixed. This prediction is consistent with Duncan's (1977b) atypical finding of additive effects in his initial study, but this additivity has not been replicated in later studies (e.g., Ehrenstein & Proctor, 1998). However, the two-step model can be modified to explain the reduction of the S-R compatibility effect for the four-choice task by assuming that when the mapping is determined, the four-choice task is reduced to a two-choice task, and this reduction benefits the incompatible mapping more than the compatible one.

The strongest support for the two-step model comes from analyses of error patterns showing that, when a subject makes an error, it is likely to be the appropriate response if the alternative mapping rule were in effect (Duncan, 1977a, 1978; Ehrenstein & Proctor, 1998; Stoffels, 1996b). Additional evidence for the two-step model comes from precuing studies, in which the precues reduce the four S-R alternatives to two (Ehrenstein & Proctor; Stoffels). With this procedure, the two cued alternatives can be from the compatible subset, the incompatible subset, or a mixed subset (one from the compatible and one from the incompatible subset). Overall, the S-R compatibility effect increases in magnitude because the precue is more beneficial when the mapping for the precued pair is compatible than when it is incompatible, and when the mapping for the precued subset is mixed, for the compatible member than for the incompatible one. Ehrenstein and Proctor showed that precuing S-R pairs with the same mapping is more beneficial than precuing S-R pairs with mixed mappings. However, regardless of whether the precue designated S-R pairs with the same or mixed mappings, the compatible mapping benefited more than the incompatible mapping at shorter precuing intervals, and there was no benefit for precuing S-R pairs with an incompatible mapping at long intervals. These findings suggest that precues improved performance by reducing the number of alternatives from four to two.

Like Shaffer's (1965) translation efficiency account, the two-step model alone cannot explain several findings. For example, the model does not account for the fact that a precue is more beneficial for the compatible than incompatible mapping because the reduction in uncertainty to two alternatives at the second step that is more beneficial for the incompatible mapping than the compatible mapping should contribute to performance when the mapping rule is precued as well as when it is not. Moreover, the model seems more applicable to the four-choice task rather than for the two-choice task because it assumes that the larger mixing costs for the compatible than incompatible mapping is due to the incompatible trials benefiting from a reduction in number of possible response alternatives from four to two.

A preparation hypothesis proposed by De Jong (1995) assumes that when compatible and incompatible mappings are mixed, participants intentionally adopt a strategy of preparing for the incompatible mapping to minimize the detrimental effect of uncertainty. This preparation strategy eliminates the benefit for the compatible mapping. Consistent with this preparation hypothesis, De Jong found that the reduction of the S-R compatibility effect when compatible and incompatible mappings are mixed was smaller when the percentage of compatible trials was 67% than when it was 33%. However, De Jong did not view this outcome as favoring the preparation hypothesis, because positive S-R compatibility effects were obtained even when incompatible trials predominated, indicating that the elimination of the S-R compatibility effect is not due solely to a tendency to be biased toward making the incompatible response. Furthermore, Ehrenstein and Proctor's (1998) analysis of the error types that were committed indicated that participants were more likely to make the compatible response when the mapping was incompatible than vice versa, counter to the hypothesis that preparation is biased toward the incompatible mapping.

Although the results of De Jong (1995) and Ehrenstein and Proctor (1998) indicate that the reduction/elimination of the S-R compatibility effects with mixed mappings is not solely due to a response bias for the incompatible mapping, Vu and Proctor (2006) showed that manipulating the ratio of compatible to incompatible trials does affect the sign and magnitude of the S-R compatibility effect. They used spatial locations, arrow directions, or words to convey location information in a two-choice mixed mapping procedure and varied the percentage of compatible and incompatible trials within a block (see discussion in Chapter 6). Three conditions were examined: compatibly biased (75% compatible and 25% incompatible), incompatibly biased (25% compatible and 75% incompatible), and unbiased (50% compatible and 50% incompatible). With spatial locations and arrows, the S-R compatibility effect was a nonsignificant 6 ms in the unbiased condition, replicating previous studies (e.g., Shaffer, 1965; Vu & Proctor, 2004). However, the compatibility effect averaged 104 ms when compatible trials predominated and –90 ms when incompatible trials predominated. Thus, relative to the unbiased condition, the influence of bias was of similar magnitude for compatible (98 ms) and incompatible (96 ms) mappings.

When the location information was conveyed by words, a large 164-ms S-R compatibility effect was obtained in the unbiased condition, again replicating the findings of Vu and Proctor (2004). When compatible trials predominated, the S-R compatibility effect was enhanced to 207 ms, and when incompatible trials predominated, it reversed to a nonsignificant –4 ms. For words, the effect of bias was smaller for the compatible (43 ms) than incompatible (168 ms) mapping, but the smaller effect for the compatible mapping could be due to a ceiling effect.

The results of Vu and Proctor's (2006) study suggest that the S-R compatibility effect is reduced or eliminated when mappings are mixed because subjects are uncertain about which mapping will be in effect for the current trial. When compatible and incompatible mappings are mixed and presented on an equal number of trials, the elimination of the S-R compatibility effect may be due not to a tendency to be biased toward making incompatible responses, but rather to waiting for the mapping signal to appear before selecting the response based on the mapping. Vu and Proctor's experiment also showed that biasing occurs for physical location, arrow, and word stimuli (see also De Jong, 1995), and that the positive and reverse effects were symmetrical for physical locations and arrows. However, De Jong found asymmetric effects for arrows, suggesting that the prior associations of stimuli to responses were still playing a role in response selection. The difference between De Jong's study and Vu and Proctor's study is likely due to the different percentages of compatible and incompatible trials used in the two studies (33% vs. 66% in De Jong's study and 25% vs. 75% in Vu and Proctor's study) or to the fact that Vu and Proctor used left-/right-pointing arrows and De Jong used upward-pointing arrows that were slightly tilted to the left or right. Regardless, both studies show that there is an effect of response preparation.

7.4.2 Intertrial Variability

Intertrial variability created by factors associated with task switching may also cause the reduction and elimination of the S-R compatibility effect with mixed presentation (Los, 1996). The effects of intertrial variability can be examined by performing repetition analyses. Many robust repetition effects have been reported in mixing studies (e.g., Proctor & Vu, 2002a; Shaffer, 1965, 1966; Stoffels, 1996b). When compatible and incompatible mappings are mixed, responding is faster when the mapping from the previous trial is repeated than when it is not, with the benefit of repetition being larger for the compatible mapping than the incompatible mapping. Consistent with the outcome that the compatible mapping benefits more from repetition, De Jong (1995) found that when compatible and incompatible mappings were mixed, an S-R compatibility effect for arrow-tilt stimuli was obtained on mapping-repetition trials but not on nonrepetition trials. Stoffels obtained similar patterns of repetition effects for the four-choice task version.

Shaffer (1965, 1966) found a benefit for repeating the mapping from the previous trial and a benefit for repeating the same stimulus position from the previous trial. However, he did not examine the pattern of mapping repetition benefits as a function of whether stimulus position was also repeated. Proctor and Vu (2002a) showed that the effects of mapping repetition varied as a function of whether the stimulus position also repeated. That is, when the stimulus position repeated, there was a benefit for the compatible mapping over the incompatible mapping for mapping-repetition trials. However, when the stimulus position changed, there was a cost for the compatible mapping on mapping nonrepetition trials. This pattern of mapping- and stimulus-position repetition effects can be interpreted in terms of Hommel's (1998b; Hommel, Proctor, & Vu, 2004) feature integration account, described in Chapter 6, which emphasizes that the stimulus features and the response on a trial are integrated into an event file. When all stimulus features are repeated, the response with which they were integrated on the previous trial is reactivated, facilitating responding. Similarly, when all stimulus features change, they produce no reactivation of the previous response and signal that the alternative response is to be made. Response selection is more difficult on trials for which at least one stimulus feature repeats and another changes because one stimulus feature produces reactivation of the previous response and the other signals the alternative response. The conflict on these trials slows selection of the correct response.

Although robust repetition effects are obtained in various studies, Proctor and Vu (2002a) noted that these effects are independent of the overall mean effects. For example, in Vu and Proctor's (2004) study, the S-R compatibility effect for physical locations and arrow directions was eliminated when the stimuli were mapped to keypress responses, but a significant S-R compatibility effect was still evident when the stimuli were mapped to vocal "left"–"right" responses. However, both conditions yielded similar patterns of repetition effects. In addition, Proctor and Vu (2002c) showed that when the compatible and incompatible mappings were signaled by a mode distinction (e.g., physical locations when the mapping was compatible and location words when the mapping was incompatible, or vice versa), and responses were keypresses, the S-R compatibility effect was eliminated when the location information was conveyed by physical locations and enhanced when it was conveyed by location words, similar to when there was no mode distinction. However, the pattern of repetition effects in the mixed mode condition was not similar to that in the same mode condition.

Although intertrial variability and uncertainty are not mutually exclusive accounts for the effects of mixing, most researchers have tended to favor an uncertainty explanation for the elimination of the compatibility effects (De Jong, 1995; Proctor & Vu, 2002a; Shaffer, 1965). Uncertainty and intertrial variability are general accounts for the elimination and reduction of the S-R compatibility effect. At least two other specific explanations have been proposed to explain the results obtained for mixed mappings: transient activation and alternative routes accounts.

7.4.3 Transient Activation

The transient activation hypothesis is based on the notion that when a stimulus is presented, it automatically activates its corresponding response. However, this activation is assumed to occur quickly and then dissipate (e.g., Hommel, 1993b). As in Kornblum, Hasbroucq, and Osman's (1990) dimensional overlap model, the stimulus is presumed to activate its corresponding response directly, facilitating responses with a compatible mapping and inhibiting responses with an incompatible mapping in pure blocks of trials. Because mixing slows responding, the response activation produced by the direct route dissipates before response selection is completed, thus eliminating the benefit for the compatible mapping.

Vu and Proctor (2004) evaluated the transient activation hypothesis by performing RT distribution analyses. If the elimination of the S-R compatibility effect were due to transient activation, then the S-R compatibility effect should be large at the shorter RTs and decrease progressively at longer RTs. However, in Vu and Proctor's study, the RT distributions showed little evidence of transient activation. When compatible and incompatible mappings were presented in pure blocks, the S-R compatibility effect was an increasing function of RT bin for all stimulus and response modes, a pattern opposite to that which is predicted by the transient activation account. This finding replicates results obtained by Roswarski and Proctor (1996) for a pure mapping S-R compatibility task in which location was also relevant. For the mixed conditions in which the S-R compatibility effect was reduced or eliminated, the magnitude of the effect decreased across RT bins, which is a pattern suggestive of transient activation. However, Vu and Proctor noted that this pattern was due to the fact that repetition trials, for which the S-R compatibility effect is evident, predominate in the short RT bins and nonrepetition trials, for which it is not, predominate in the long RT bins. Thus, there is little support for transient activation decaying as the explanation for the reduction and elimination of the S-R compatibility effects.

7.4.4 Alternative Routes Model

Another explanation for the elimination and reduction of the S-R compatibility effect with mixed presentation is the alternative routes model. The alternative routes model is a dual-route model of the type proposed by Kornblum et al. (1990) in which response selection occurs by way of the direct (automatic) and indirect (intentional) routes. Response selection occurs via the direct route when the mapping is compatible. When the mapping is incompatible, or compatible and incompatible trials are mixed, this direct route is suppressed because it would lead to the wrong response on many trials, and an indirect route is used for responding. Responding via the indirect route leads to slower responses because the assigned response must be actively retrieved. Thus, the major difference between the alternative routes model and Kornblum et al.'s dimensional overlap model is that in the latter, direct activation occurs when dimensional overlap is present regardless of mapping or task set, whereas the former allows for the direct activation to be suppressed under certain situations.

The alternative routes model is the most widely accepted account for the reduction and elimination of the S-R compatibility effect (De Jong, 1995; Proctor & Vu,

2002a; Stoffels, 1996a, 1996b; Van Duren & Sanders, 1988). Like the uncertainty accounts, the alternative routes model can explain the precuing effects obtained when compatible and incompatible mappings are mixed. That is, the S-R compatibility effect is reinstated because the precue allows the direct route to be used for the compatible mapping, leading to a benefit for it compared with the incompatible mapping. In addition, the alternative routes model can accommodate the pattern of repetition effects. De Jong interpreted the finding of an S-R compatibility effect for repetition trials as indicating that suppression of the direct route is released on repetition trials because the preparatory state for the mapping of the previous trial carries over automatically and counteracts the suppression (see Chapter 6 for discussion of the suppression/release account for sequential effects in the Simon task). Stoffels (1996b) also attributed this finding of an S-R compatibility effect on mapping repetition trials for the four-choice task to release of the suppression of the direct route on repetition trials. However, as with the Simon task (see Chapter 6), Hommel, Proctor, and Vu (2004) noted that the pattern of repetition effects can be accounted for in terms of Hommel's (1998b) feature integration account, which does not assume suppression release of the direct route and attributes the repetition effects to a separate sequential comparison process.

Furthermore, the alternative routes account cannot easily explain two findings. First, with mixed mappings, Vu and Proctor (2004) found that, although the S-R compatibility effect is eliminated or reduced for S-R sets with high set-level compatibility (e.g., physical locations and arrows mapped to keypresses or location words mapped to vocal responses), it is not for S-R sets with lower set-level compatibility (e.g., physical locations and arrows mapped to vocal responses or location words mapped to keypresses). Because the direct route should be suppressed when compatible and incompatible mappings are mixed regardless of the degree of set-level compatibility, the alternative routes model predicts that the effects should be reduced or eliminated in all cases.

Second, the alternative routes account cannot explain why the benefit for the compatible mapping was reduced in the study of Morin and Forrin (1962), which mixed arbitrarily mapped shapes to digit names with compatibly mapped digit stimuli. In this case, the correct response to a digit is always its corresponding name, and the direct route should not be suppressed. Furthermore, Forrin (1975) showed that an advantage for digit naming over letter naming was eliminated when the compatibly mapped digits were mixed with compatibly mapped letters. In this case, activation of the corresponding digit or letter name via the direct route should yield the correct responses for all stimuli. In contrast, Forrin's results are in complete agreement with the view that all responses involve the indirect route and, consequently, that RT is an increasing function of set size. RT would be shorter for digits presented in a pure list than for letters because the set size for digits (10) is smaller than that for letters (26); however, mixing would be more deleterious for digits because the increase in set size to 36 would be larger than that for letters. Although the mixing effects in Forrin's study were small, they suggest that mixing stimulus categories is sufficient to prolong response selection more for the easier task than the more difficult one, even when both tasks require highly compatible responses.

7.5 MIXING LOCATION-RELEVANT AND LOCATION-IRRELEVANT TRIALS: THE S-R COMPATIBILITY EFFECT

Another variant of a mixing procedure is when a task for which location information is relevant on some trials is mixed with a Simon task for which the location information is irrelevant on other trials. As described in Chapter 6, the Simon effect is enhanced when such Simon trials are mixed with trials for which stimulus location has a compatible S-R mapping and reversed when they are mixed with trials for which stimulus location has an incompatible mapping. In this section, we examine the effects of the Simon trials on performance of the location-relevant task.

7.5.1 Spatial Locations

The spatial S-R compatibility effect is reduced when location-relevant trials are mixed with Simon-type, location-irrelevant trials, much as it is when compatible and incompatible spatial mappings are mixed. Marble and Proctor (2000) conducted experiments with mixed location-relevant and location-irrelevant trials in which the location information for both trial types was conveyed by physical locations (see also Proctor & Vu, 2002a, 2002b; Proctor, Vu, & Marble, 2003). The mapping (compatible or incompatible) for location-relevant trials was varied between subjects, and subjects were to respond on the basis of the stimulus's color for the location-irrelevant trials. Marble and Proctor found that the S-R compatibility effect was eliminated for the location-relevant trials. Proctor and Vu (2002b, Experiment 1) replicated Marble and Proctor's experiment but included control conditions in which all trials were location-relevant, with the mapping being compatible or incompatible. The S-R compatibility effect for location-relevant mappings of 77 ms with pure presentation was a nonsignificant –16 ms with mixed presentation.

7.5.2 Symbolic and Verbal Modes

Proctor and Vu (2002b) showed that elimination of the S-R compatibility effect when location-relevant and location-irrelevant trials are mixed does not occur when the location information is conveyed symbolically or verbally. As in their previous study, stimulus color indicated whether to respond with a location mapping or color-to-response mapping. The Simon effect data for location-irrelevant trials were discussed in Chapter 6; the data below are for the location-relevant trials. When left- and right-pointing arrows conveyed the location information and responses were keypresses, the S-R compatibility effect of 42 ms obtained with mixed presentation did not differ statistically from the 32-ms effect obtained with pure presentation. This result is similar to that obtained by De Jong (1995) with arrow-tilt stimuli. Furthermore, when the words LEFT and RIGHT conveyed the location information and responses were keypresses, the S-R compatibility effect was larger with mixed presentation of location-relevant and location-irrelevant trials than with pure presentation (172 and 21 ms, respectively). Moreover, with vocal "left"–"right" responses, the S-R compatibility effect was larger with mixed than with pure presentation for all stimulus modes (physical locations: 65 and 21 ms; arrows: 67 and 34 ms; location words: 182 and 115 ms). Thus, the elimination of the S-R compatibility effect with mixed location-relevant and location-irrelevant trials tends to be limited to physical location stimuli mapped to manual responses.

7.5.3 MIXED-MODES PRESENTATION

To evaluate whether the mode distinction reduces the mixing effect for the location-relevant trials, Proctor and Vu (2002a) presented the location-relevant information through left–right physical locations or the location words LEFT and RIGHT, with the location-irrelevant information presented in the alternative mode. The S-R compatibility effect in these mixed-mode conditions was intermediate in size compared with the large effect obtained for location words and the absence of effect obtained for physical locations when the relevant and irrelevant modes were either both words or both physical locations. These results indicate that the mode distinction reduces the influence of mixing on the spatial compatibility effect, suggesting that the associations between location information and responses defined for the tasks are to a considerable degree specific to the stimulus mode. These findings are counter to those of Proctor and Vu (2002c), noted previously, when compatible and incompatible location-relevant mappings were mixed. In that case, the mode distinction did not influence the mixing effects that were obtained.

Although Proctor and Vu's (2002a) data suggest that the task-defined associations are mode specific when two different tasks are mixed, there is some evidence to suggest that mode-independent associations can be formed in such situations. As described in Chapter 6, De Houwer (2004) and De Houwer, Beckers, Vandorpe, and Custers (2005) established short-term, task-defined associations for a location-relevant task by associating unrelated spoken-word responses to the concepts *left* and *right*. For trials on which the location information was relevant, the stimulus was either a left- or right-pointing arrow or the word LEFT or RIGHT. When intermixed with location-irrelevant trials, for which the location information was left or right position and the responses were the words associated with those concepts, a correspondence effect was obtained for which responding was faster when the position corresponded with the location associated with the response. Proctor and Vu's (2002a) and De Houwer's studies differed in at least two potentially important respects. First, Proctor and Vu used only one stimulus mode to convey the information on location-relevant trials, whereas De Houwer used two modes, which might encourage development of associations that are more abstract. Second, Proctor and Vu used visual–spatial stimuli for one task and verbal stimuli for the other, but De Houwer used visual–spatial stimuli for both the location-relevant task (the arrows) and the location-irrelevant task (physical positions), which means that both of his tasks had some trials that were within the visual–spatial mode.

7.6 ACCOUNTS FOR THE ELIMINATION OF S-R COMPATIBILITY EFFECTS WITH MIXED LOCATION-RELEVANT AND LOCATION-IRRELEVANT TRIALS

7.6.1 UNCERTAINTY

When location-relevant and -irrelevant trials are mixed, subjects must keep both types of associations activated and then select the set appropriate for the particular trial. The elimination of the S-R compatibility effect with mixed location-relevant and -irrelevant trial types may be due to uncertainty in determining on which stimulus dimension to base responding. Precuing the trial type in advance can reduce this uncertainty. De Jong's (1995) Experiment 3 used a high- or low-pitch tone presented 10 or 800 ms prior to the imperative stimulus to indicate whether responses were to be based on the tilt or color of an arrow. The S-R compatibility effect for the trials on which arrow tilt was relevant was 33 ms at the short preparation interval and 38 ms at the long one. Because the effect was not significantly smaller at the short precuing interval, as it is for physical location stimuli, the results suggest that uncertainty regarding whether the trial is location-relevant or -irrelevant is not a major factor when the stimuli are arrows.

7.6.2 Intertrial Variability

As with studies of mixed compatible and incompatible mappings, the influence of intertrial variability can be examined through repetition analyses. Several studies in which compatible and incompatible location-relevant trials were mixed (described earlier in the chapter) reported repetition analyses for pure blocks of compatible and incompatible mappings. For pure blocks, repetition benefits the incompatible mapping more than the compatible mapping. This benefit for the incompatible mapping is usually attributed to the fact that the incompatible mapping has more to gain from a shortcut based on repetition because response selection takes longer for incompatible than compatible mappings (e.g., Soetens, 1998). In contrast, for mixed presentation, the compatible trials benefited more from repetition than did the incompatible trials (De Jong, 1995; Stoffels, 1996b).

Only Proctor and Vu (2002a) discussed repetition effects for compatible and incompatible mappings when intermixed with location-irrelevant trials. They presented an overview of the repetition effects obtained for the data from several of their studies (e.g., Proctor & Vu, 2002b; Proctor, Vu, & Marble, 2003) in which location-relevant and -irrelevant trials were mixed. Proctor and Vu (2002a) partitioned repetition trials for the location-relevant task into three categories: complete task repetition, task repetition of the alternative S-R pair, and task switch. Overall, responses were faster for complete repetition trials than for task repetition of the alternative S-R pair, which was faster than task switch trials. Unlike the findings of mixed compatible and incompatible mappings where the compatible mapping benefits more from repetition, when location-relevant and -irrelevant trial types were mixed, task repetition was more beneficial for the incompatible mapping than for the compatible mapping, regardless of whether the specific S-R pair was repeated. The larger repetition benefits for incompatible than compatible mappings occurred primarily for conditions in which there was a verbal component to at least one of the two intermixed tasks.

7.6.3 Transient Activation

De Jong (1995) evaluated the transient activation hypothesis for the elimination of the S-R compatibility effect for mixed mappings by mixing location-relevant and -irrelevant trial types. He reasoned that if the elimination of the S-R compatibility effect with mixed presentation was due to the activation of the corresponding response decaying before response selection because responding is delayed, then the S-R compatibility effect should be eliminated or reduced significantly when the preparation interval is short compared to when it is long. However, as noted, De Jong found a 33-ms S-R compatibility effect at the short preparation interval that did not differ significantly from the 38-ms S-R compatibility effect at the longer interval, indicating that there was no decay of automatic activation.

Most evidence for the transient activation hypothesis comes from RT distribution analyses. If automatic activation dissipates when responding is delayed, then a large S-R compatibility effect should be obtained at short RT bins, and the effect should be reduced or eliminated at longer bins. Although studies of mixed location-relevant and -irrelevant trials have reported RT distribution analyses for the location-irrelevant trials (e.g., Marble & Proctor, 2000), these analyses have not been reported for location-relevant trials. However, several studies with pure location-relevant trials have found little evidence for automatic activation of the corresponding response that dissipates with time (e.g., Roswarski & Proctor, 1996; Vu & Proctor, 2004), making the transient activation hypothesis an unlikely candidate for the elimination of the S-R compatibility effect.

7.6.4 Alternative Routes Model

As with mixed compatible and incompatible mappings, the alternative routes model has been proposed to explain the elimination of the S-R compatibility effect with mixed location-relevant and -irrelevant trials. The alternative routes model can account for the elimination of the S-R compatibility effect by assuming that the direct response–selection route is suppressed when location-relevant and -irrelevant trials are mixed. Suppression of the direct route occurs because responding based on it would lead to the incorrect response on a large percentage of trials. Although the alternative routes model can account for the elimination of the S-R compatibility effect with physical locations, it cannot explain why the S-R compatibility effect was unaffected for arrows and enhanced for words.

As described, De Jong's (1995) finding that the S-R compatibility effect is of similar magnitude regardless of whether the preparation interval is short or long when location-relevant and -irrelevant trials are mixed does not support the transient activation account. However, he claimed that it does support the suppression hypothesis of the alternative routes model. According to De Jong, when location-relevant and -irrelevant trials are mixed, the interference from the location-relevant mapping is reduced because there is only one mapping in effect for any given block of trials. This reduction of interference will reduce the need to suppress the direct route, yielding an S-R compatibility effect at the short and long preparation intervals. However, although Proctor and Vu (2002b) replicated De Jong's findings with left- and right-pointing arrows instead of arrow-tilt stimuli, they found that when the location information was conveyed by physical locations, the S-R compatibility effect was eliminated. The elimination of the S-R compatibility effect with physical location stimuli does not conform to De Jong's logic that mixing trial types reduces the need for suppression of the direct route.

7.6.5 Task Switching

A possible reason why mixing tasks eliminates or reduces the spatial compatibility effect is that there is a cost of switching tasks. Research on task switching suggests that there are two components associated with changing task set from the previous trial (e.g., Allport et al., 1994; Rogers & Monsell, 1995), one that is under the subject's voluntary control and another that is not. The automatic component is attributed to residual activation of the response that was made to the previous task and can facilitate performance if the task is repeated and interfere with performance if the task changes. The automatic component can be examined at short response–stimulus intervals because short intervals do not allow subjects to prepare for the next task or to reconfigure their task set. At long response–stimulus intervals, the role of subjective strategies can be examined and changes can be evaluated as a function of whether the task is repeated or changed. Analyses of task-repetition and task-switch trials allow examination of whether the elimination of the S-R compatibility effect with mixed location-relevant and -irrelevant trials can be attributed to switching between two tasks.

Proctor and Vu (2002a) conducted repetition analyses to determine the pattern of S-R compatibility effects as a function of task repetition or task switch. The slowest responses for the location-relevant trials occurred when the task for the previous trial was location-irrelevant. This finding is consistent with many results in the literature and has been interpreted as reflecting a need to adopt a new task set when the task on the current trial is different from the one on the preceding trial (e.g., Rogers & Monsell, 1995). Responses were also faster when the specific S-R pair was repeated than when the task repeated but the S-R pair switched. For physical locations and arrows mapped to keypresses, the task-switch cost was of similar magnitude for the compatible and incompatible mappings, indicating that the elimination of the S-R compatibility effect observed for physical locations mapped to keypresses is not due solely to switching costs. However, the large increase in the S-R compatibility effect for location words was primarily evident on the task-switch trials, in which the cost of a task switch was much larger with the incompatible mapping than with the compatible mapping.

Proctor, Vu, and Marble's (2003) study also showed that the elimination of the S-R compatibility effect for physical locations when location-relevant and -irrelevant trials are mixed was not due to the requirement to switch between tasks but to their sharing location codes. In Proctor et al.'s Experiment 2, the location-relevant and location-irrelevant trials were mixed, but the location-irrelevant stimuli were presented in the center of the screen rather than in the same left–right locations as the location-relevant stimuli (see Figure 7.5, top row). With centered location-irrelevant stimuli, an S-R compatibility effect was obtained for location-relevant trials of similar magnitude as that obtained with pure presentations of compatible and incompatible mappings. In their Experiment 3, the stimuli for both trial types were presented in left–right locations, but location-relevant stimuli were presented above fixation and location-irrelevant stimuli below fixation, or vice versa (see Figure 7.5, middle row). With this presentation of stimuli, no S-R compatibility effect was evident. In Proctor et al.'s Experiment 4, location-relevant stimuli were presented in

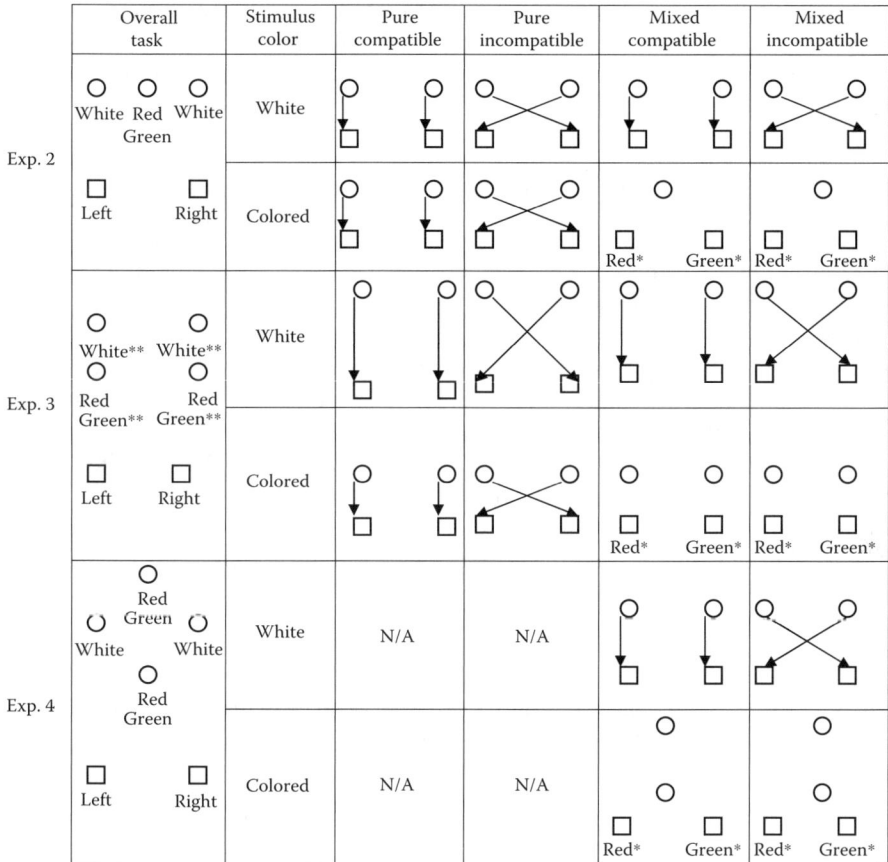

FIGURE 7.5 Stimulus and response arrangements and mappings used by Proctor, Vu, and Marble (2003). * Participants are responding to color of stimuli with spatial location irrelevant. As shown in the figure, participants press the left key if the circle is red and the right key if it is green. Although not illustrated, half the subjects responded to the red stimuli by pressing the right key and green stimuli by pressing the left key. ** For half the participants, white circles appeared in the top row and colored ones in the bottom row, as illustrated. The other half received the colored circles in the top row and white ones in the bottom row.

left–right locations and location-irrelevant stimuli in top–bottom locations (see Figure 7.5, bottom row). A significant S-R compatibility effect was obtained with this method of presentation that was similar in magnitude to that obtained with pure presentation.

The results of Proctor, Vu, and Marble's (2003) study indicate that switching between location-relevant and -irrelevant tasks is not sufficient to eliminate the S-R compatibility effect. That is, when the location-irrelevant trials were presented in the center of the screen or in central top–bottom locations, the S-R compatibility effect for location-relevant trials was unaffected. The elimination of the S-R compatibility effect when the location-relevant and -irrelevant stimuli appeared in left–right locations, but in separate rows, indicates that the elimination is not due to presenting the two trial types in the same locations. Rather, the results suggest that the elimination of the S-R compatibility effect is due to the two trial types sharing left–right codes.

7.7   COMPARISON OF MIXED MAPPINGS AND TRIAL TYPES

The effects of mixing compatible and incompatible mappings can be compared with those of mixing location-relevant and -irrelevant trials by evaluating the pattern of results in Vu and Proctor's (2004) and Proctor and Vu's (2002b) Experiments 1 and 2 because these studies factorially manipulated the same stimulus and response modes for each mixing procedure. Although the methods used in the two studies were comparable, it is important to note that in Proctor and Vu's study, compatibly or incompatibly mapped location-relevant trials were mixed with location-irrelevant trials for which color was the relevant stimulus dimension. The mapping effects, therefore, involved comparing RT across participants who received the different mappings. In contrast, in Vu and Proctor's study, all trials were location-relevant, the mapping (compatible or incompatible) was varied on a trial-to-trial basis, and an S-R compatibility effect was obtained for each subject.

7.7.1   KEYPRESS RESPONSES

For physical locations and location words, but not arrows, the cost of mixing was larger when location-relevant and -irrelevant trials were mixed (Ms = 305 ms for physical locations and 350 ms for words) than when compatible and incompatible S-R mappings of locations were mixed (Ms = 236 ms for physical locations and 274 ms for words), indicating that there is a greater cost for switching between tasks than mappings. When the location information is conveyed by physical locations, the S-R compatibility effect is eliminated for both mixing paradigms. Similarly, when the location information is conveyed by words, the S-R compatibility effect is enhanced for both mixing paradigms. However, for arrow stimuli, mixing the trial types did not alter the magnitude of the S-R compatibility effect, but mixing mappings eliminated it. The fact that physical locations and location words yield similar results for both mixing procedures, but arrows do not, can be attributed to the fact that arrows tend to show results intermediate to those of physical locations and words (Proctor & Wang, 1997a). Arrows show a tendency to automatically activate their corresponding responses (Eimer, 1998), as physical locations show, but because they are symbolic stimuli, responding may rely on some verbal mediation, as location words show.

7.7.2   VOCAL RESPONSES

With vocal responses, the reduction of the S-R compatibility effect for location words when compatible and incompatible mappings were mixed (Vu & Proctor, 2004) is the opposite of the enhanced compatibility effect obtained when location-relevant and -irrelevant trials were mixed (Proctor & Vu, 2002b). For physical locations and arrows, there was a nonsignificant reduction in the magnitude of the

S-R compatibility effect with the mixed mapping paradigm, but both stimulus types showed an enhanced S-R compatibility effect with the mixed trial type paradigm.

7.7.3 EVALUATION

The fact that different patterns of results are obtained for the two mixing paradigms suggests that the elimination, reduction, and enhancement of the S-R compatibility effect in each are not due entirely to common processes. Rather processes unique to the different task requirements seem to play a role as well. Currently, the suppression hypothesis of the alternative routes model is the most widely accepted account for explaining the elimination of the S-R compatibility effect for both mixing paradigms. However, Vu and Proctor (2004) pointed out that, for mixed mappings, the suppression account seems to be limited to situations in which the set-level compatibility is high (i.e., the S-R sets have both perceptual and conceptual similarity), and not those in which it is lower (i.e., the S-R sets have only conceptual similarity). Similarly, Proctor and Vu (2002) noted that, for mixed trial types, the suppression hypothesis can only account for the results obtained with physical location stimuli. Furthermore, the suppression hypothesis cannot predict the enhanced S-R compatibility effect in either mixing paradigm because suppression of the direct route should always result in smaller S-R compatibility effects with mixed presentation compared with pure presentation. Proctor, Vu, and Marble (2003) showed that with mixed location-relevant and -irrelevant trial types, the elimination with physical locations seems to be due to the relevant and irrelevant dimensions sharing the same left–right codes.

7.8   A NEGATIVE CORRESPONDENCE EFFECT FOR MASKED STIMULI

When a prime stimulus conveying location information is presented prior to the target stimulus to which the subject is to respond, the prime can be thought of as an intermixed location-irrelevant trial for which no response is required. Variations of the priming procedure have been used to investigate a phenomenon called the negative correspondence effect for masked primes, first demonstrated by Eimer and Schlaghecken (1998). They examined a situation in which a left- or right-pointing double arrow (<< or >>) was presented for 16 ms as a prime stimulus, followed immediately by a mask composed of superimposed double arrows for 100 ms. At mask offset, another left- or right-pointing arrow was displayed for 100 ms, to which a compatible keypress with the left or right index finger was to be made. Even though the prime stimulus could not be identified at greater than chance levels, responses were 53 ms slower when the direction in which the prime pointed corresponded with that of the target stimulus (and response) than when it did not. Moreover, the lateralized readiness potential showed initial activation of the corresponding response 250 ms after prime onset that switched to activation of the contralateral response beginning approximately 280 ms following prime onset. This pattern of results led Eimer and Schlaghecken to conclude that the masked prime initially produced activation of the corresponding response, which was then inhibited, resulting in relatively greater activation of the noncorresponding response at the time that the response to the target stimulus was being selected, producing the negative correspondence effect.

The conditions under which the negative correspondence effect occurs have been clearly delineated (see Eimer & Schlaghecken, 2003, for a review), and the results have generally supported the interpretation that Eimer and Schlaghecken (1998) provided initially. The effect is found for a variety of response sets, including presses of the index and middle fingers of a single hand or movements of a single finger from a home key to one of two response keys (Eimer & Schlaghecken, 1998; Klapp & Hinckley, 2002), as well as with left–right foot movements, left–right saccadic eye movements, and vocal "left"–"right" responses (Eimer & Schlaghecken, 2001; Eimer, Schubö, & Schlaghecken, 2002).

When stimulus or response modes are intermixed, the negative correspondence effect occurs across modes. Eimer et al. (2002) found that when trials requiring a left–right hand response or a left–right foot response were intermixed, with color of the target arrow designating whether to respond with the hand or foot, negative correspondence effects occurred for both response modes. Similarly, Klapp and Hinckley (2002) found effects for both stimulus modes when the masked prime arrows (which pointed up or down) were paired with up- and down-pointing target arrows on some trials and with high (up) or low (down) pitch tones on others, to which up–down responses were made. However, the left- and right-pointing arrows must be assigned to the responses required to the targets for the negative correspondence effect to occur. Eimer and Schlaghecken (1998) found no effect of the masked arrow primes when all target stimuli were the letters L (left) and R (right). Likewise, when Eimer et al. signaled the hand–foot distinction by whether the target was an arrow or an X presented to the left or right of fixation, whichever response mode was assigned to the arrow stimuli showed the negative correspondence effect but the other did not.

Results of the type described in the previous paragraph provide evidence that the processes producing the negative correspondence effect are at neither a perceptual nor a semantic, conceptual level (Eimer & Schlaghecken, 2003). That is, the effect can be obtained when the stimulus set is physically dissimilar to the masked primes, but it does not generalize to a conceptually similar response set assigned to different stimuli. An additional finding of importance is that the negative correspondence effect does not occur when the prime stimuli are above threshold and can be perceived, with a positive correspondence effect evident instead (Eimer & Schlaghecken, 2002; Klapp & Hinckley, 2002). Findings such as these led Eimer and Schlaghecken (2003) to conclude that the inhibition that produces the negative correspondence effect is exogenous and not under the subject's control.

One of the more interesting findings concerning the negative correspondence effect is that it occurs for free-choice trials on which there is no correct response. Klapp and Hinckley (2002, Experiment 5) included trials on which the target stimulus was an ambiguous diamond symbol, intermixed with ones in which the target was a left- or right-pointing arrow. On the ambiguous symbol trials, participants were to respond with either the left or right response. For those trials, 61% of the responses selected did not correspond with the prime, compared with only 39% that did. Schlaghecken and Eimer (2004) and Klapp and Haas (2005) replicated this result, showing that it does not occur unless the free response trials are intermixed with normal reaction trials. Moreover, Klapp and Haas found that the effect was also absent when the intermixed reaction trials used the letters L and R to designate the response, or plus signs in left and right locations, and not arrows. They demonstrated as well that the response modality (manual or vocal) had to be the same for the reaction and free-choice trials. The free-choice trials showed a negative correspondence effect for vocal "left"–"right" responses as well as for manual left–right responses, but only when the response modality was the same for the intermixed left–right reaction trials. In sum, the negative correspondence effect seems to indicate that inhibitory mechanisms can influence response selection under conditions in which the person is not aware of the activation that is inhibited nor of the stimulus that produced that activation.

7.9 CHAPTER SUMMARY

Compatibility effects are altered systematically by mixing compatible and incompatible mappings with each other or with Simon-type trials for which stimulus location is irrelevant. When set-level compatibility is high, the S-R compatibility effect obtained for blocks of pure compatible or incompatible mappings is reduced, and often eliminated, when mappings are mixed within the trial blocks. When set-level compatibility is low, the S-R compatibility effect is either reduced only slightly (for physical locations or arrows mapped to vocal responses) or increased drastically (for location words mapped to keypresses). When a location mapping is mixed with color-to-response mappings, the spatial compatibility effect is eliminated for visuospatial–manual S-R sets but is enhanced when either the stimulus or response set is verbal in nature. Studies of mixed mappings using nonspatial stimuli also show that the benefit for the easier task is eliminated when that task is mixed with a more difficult task. Although several factors may contribute to these mixing effects, dual-route accounts are at present the most widely advocated explanations. Finally, the negative correspondence effect that occurs when masked arrow primes precede a target stimulus suggests response selection can be influenced by processes operating on response activation produced by a stimulus that is outside of awareness.

8 Compatibility Effects for Orthogonal Dimensions

8.1 INTRODUCTION

For many years, it was assumed that spatial S-R compatibility effects would only arise when the stimulus and response sets shared the same spatial dimensions, or were spatially congruent. Because orthogonally oriented S-R sets, such as up and down stimulus positions mapped to left and right responses, provide no basis for spatial correspondence, there was little reason to think that one mapping might be preferred to another. For example, Bertelson (1963) included conditions in which subjects made left and right keypresses to two vertically aligned stimulus locations, but he averaged across the mappings for a single "perpendicular" condition that he compared with direct and crossed mappings of horizontally aligned stimulus locations. However, studies of direction-of-motion stereotypes provided some suggestion of preferred mappings for orthogonal stimulus and response sets (see Chapter 9). For instance, Vince and Mitchell (1946) found that subjects preferred a rightward movement of a lever to move an indicator up and a leftward movement to move the indicator down. Research beginning in the latter half of the 1970s has confirmed that systematic, sometimes quite strong, S-R compatibility effects occur when stimuli and responses vary along orthogonal dimensions. Considerable effort in recent years has been devoted to explaining why these orthogonal S-R compatibility effects occur and how they relate to the more typical spatial correspondence effects discussed in earlier chapters.

Orthogonal compatibility effects are not only of theoretical interest but of applied interest as well. Andre, Haskell, and Wickens (1991) expressed this point, noting, "Orthogonal S-R compatibility is an equally important issue from an applied design perspective" (p. 1546). They pointed out that displays and controls are often located orthogonally to each other in aircraft cockpits and interfaces for industrial equipment. The example of the four-burner stove, with which we began this book, for which two pairs of horizontally arrayed controls are mapped to pairs of burners arrayed vertically on the corresponding sides of the stove surface, is an example of an orthogonal S-R relation.

Cotton, Tzeng, and Hardyck (1977) were the first to examine S-R compatibility effects with orthogonally oriented stimulus and response sets. In Cotton et al.'s Experiment 1, the stimuli were four lights at the corners of an imaginary rectangle (see Figure 8.1), similar to the displays used for studies of the right–left prevalence effect described in Chapter 5. Responses were movements of an index finger from a start key to one of two response keys located to the left or right of the start key. Half of the subjects were instructed to press the right key to a stimulus above fixation

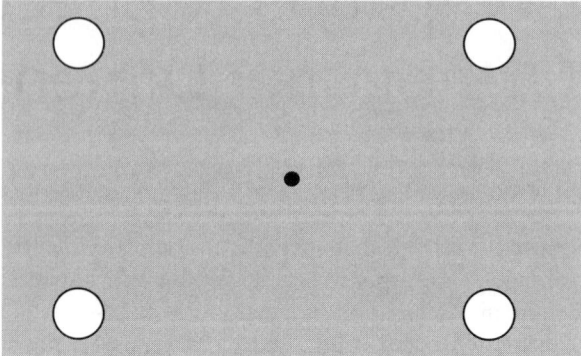

FIGURE 8.1 Illustration of the display used by Cotton, Tzeng, and Hardyck (1977), which had four possible locations arrayed rectangularly about a fixation point.

and the left key to one below fixation, and half were given the opposite instructions. All subjects used their left hand in one block of trials and right hand in another. The only significant effect was that RT was approximately 80 ms shorter when the left or right stimulus position corresponded with that of the signaled response. Note that this phenomenon is simply a Simon effect, that is, a correspondence effect for irrelevant stimulus location and response location on the horizontal dimension. Also, though nonsignificant, RT was approximately 14 ms shorter for the mapping of the upper stimulus locations to the right response and the lower stimulus locations to the left response than for the other mapping. This lack of statistical significance for this effect possibly was due to the small sample size of eight subjects for each mapping.

Bauer and Miller (1982) provided the first demonstrations of statistically significant orthogonal S-R compatibility effects. In their Experiment 1, subjects responded to left and right stimuli by making aimed-movement responses from a home key to one of two keys aligned vertically on a tabletop. The index finger of the responding hand was placed initially on the home key, located on the sagittal midline; in response to onset of a stimulus, the finger was to be moved to the target key located away from the body (the "up" key) or toward the body (the "down" key). When responding with the left hand, RT was 116 ms shorter and percentage error 1.9% lower for the mapping left–up/right–down (i.e., the left stimulus mapped to the up response and the right stimulus to the down response) than for the alternative mapping. However, when responding with the right hand, RT was 2 ms longer and percentage error 4.1% higher for the left–up/right–down mapping than for the other mapping. Bauer and Miller obtained similar results in their Experiment 2, for which the stimuli were centered arrows that pointed to the left or right.

Bauer and Miller's (1982) Experiment 3 was like their Experiment 1, except that subjects made right- or left-aimed movement responses to target keys in response to whether the stimulus was above or below a central fixation point. It should be noted that the mapping of vertical stimulus dimension to horizontal response dimension was the same as that in Cotton et al.'s (1977) Experiment 1, with the major methodological difference being that there were only two stimulus positions, aligned vertically with the fixation point, rather than four positions. The results showed a mapping x hand interaction, which Bauer and Miller interpreted as indicating different mapping preferences for the left and right hands: up-right/down-left (the up stimulus to the right response and down stimulus to the left response) for the left hand and up-left/down-right for the right hand. However, as noted by Weeks and Proctor (1990), both the left and right hands showed an up-right/down-left advantage in RT (64 and 20 ms, respectively) and percentage error (1.9% and 0.9%, respectively), although this advantage was smaller for the right hand. As in Cotton et al.'s (1977) experiment, a likely reason the overall advantage for the up-right/down-left mapping was not significant is that mapping was a between-subjects variable.

Bauer and Miller's (1982) findings that the orthogonal S-R compatibility effects are influenced by the hand used for responding seem on the surface difficult to explain with an account in terms of correspondence of codes in response selection. Consequently, Bauer and Miller concluded:

> The complementary effects observed in the two hands indicate clearly that the structure of the motor system plays a role in producing compatibility effects. Whereas Wallace (1971) found that compatibility effects depended on the response in the external world rather than the responding limb, we have found that compatibility effects may also depend on the responding limb. This finding limits the generality of Wallace's argument that compatibility effects arise in the process of matching internal spatial codes for stimulus and response. (p. 378)

However, the overall up-right/down-left advantage evident when the stimulus dimension is vertical and the response dimension horizontal seems difficult to reconcile with an account in terms of structure of the motor system. Moreover, Ladavas (1987) obtained an up-right/down-left advantage of 28 ms for right-handed individuals (to which most orthogonal S-R compatibility studies are restricted) who responded to up and down stimuli by pressing pushbuttons on cylinders held by the left and right hands (as well as an up-left/down-right advantage of similar magnitude for left-handed individuals, a finding of which we are not aware of having been replicated). Ladavas noted, "Since the response did not require any displacement of the hand but consisted simply in the depression of a key, it is difficult to account for the results in terms of asymmetries in motor preferences between dominant and non-dominant hands, as proposed by Bauer and Miller (1982)" (p. 19). The issue of whether orthogonal S-R compatibility effects are due to the structure of the motor system or to the codes used in response selection of the sort to which most compatibility effects are attributed has been fundamental in most research on the topic conducted to date. Our conclusion is that the effects are primarily correspondence phenomena attributable to response selection.

Most studies of orthogonal S-R compatibility effects subsequent to that of Bauer and Miller (1982) have used vertically oriented stimuli mapped to horizontally oriented responses, as Ladavas (1987) did. The two aspects illustrated by the results of Bauer and Miller's Experiment 3, that there is (a) an overall advantage for the up-right/down-left mapping but (b) an effect of factors such as the hand used for responding, have been the focus of most subsequent research. We distinguish these two aspects here, first considering the up-right/down-left advantage and then the effects that vary as a function of hand and hand placement. We make this distinction mainly for expository reasons, although some authors consider these two aspects of orthogonal S-R compatibility to require fundamentally different explanations (e.g., Lippa & Adam, 2001).

8.2   UP-RIGHT/DOWN-LEFT MAPPING ADVANTAGE

Weeks and Proctor (1990) were the first to emphasize the up-right/down-left advantage. Their Experiments 1A and 1B used left-/right-aimed movement responses but showed no interaction of mapping with responding hand. Instead, across experiments, mean RT was approximately 15 ms shorter for the up-right/down-left mapping than for the alternative mapping. Weeks and Proctor noted, as previously mentioned, that this trend toward an overall up-right/down-left advantage was also evident in Bauer and Miller's Experiment 3, as well as in an experiment by Michaels (1989), described in more detail later, that included a condition in which unimanual responses (left–right toggle switch movements) were made at body midline. Across these experiments, the up-right/down-left advantage averaged approximately 20 ms.

8.2.1   SALIENT FEATURES CODING ACCOUNT

Weeks and Proctor (1990) proposed an account of the up-right/down-left advantage in terms of Proctor and Reeve's (1985, 1986) salient features coding principle, described in Chapters 2 and 5. The general idea behind this principle is that response selection benefits from correspondence of salient structural features of the stimulus and response sets, even when these sets have no perceptual or conceptual similarity. For orthogonal S-R sets, Weeks and Proctor noted that evidence from the word–picture verification literature indicates that relative position on the vertical and horizontal dimensions is coded asymmetrically (e.g., Chase & Clark, 1971; Olson & Laxar, 1973; Seymour, 1974). Specifically, above (or up) and right are coded as salient, or positive polarity, and below (or down) as nonsalient, or negative polarity. Weeks and Proctor proposed that in two-choice tasks, the up stimulus position and right response are coded as salient and that the advantage for the up-right/down-left mapping occurs because each stimulus is mapped to the response of corresponding salience, or polarity. According to Weeks and Proctor, this correspondence of relative salience leads to faster translation of stimuli to responses than when the salient features of the stimulus and response sets do not correspond.

Weeks and Proctor (1990) tested implications of their salient features coding account in their remaining experiments. Experiment 2 used keypresses made with the left and right index fingers, with the mean RT data showing a 26-ms up-right/down-left advantage for a normal, uncrossed-hands placement that increased to 56 ms for a crossed-hands placement. Because Weeks and Proctor's subjects were right-handed, this outcome is in agreement with that of Ladavas (1987), described earlier. Similarly, Proctor, Wang, and Vu (2002, Experiment 1) reported an 18-ms up-right/down-left advantage for keypresses performed with a normal hand placement.

Weeks and Proctor's (1990) third experiment used verbal stimuli and spoken responses. The results showed an 87-ms advantage for the mapping of the words ABOVE to "right" and BELOW to "left" than for the opposite mapping and, similarly, a 55-ms advantage for the mapping of the words RIGHT to "above" and LEFT to "below." Finally, their last experiment indicated a right-"up"/left-"down" mapping advantage for centered left- or right-pointing arrows assigned to the spoken responses "up" or "down." Proctor et al. (2002, Experiment 1) showed in addition that the up-right/down-left advantage occurs for left–right keypresses to the words *above* and *below* and spoken "left"–"right" responses to left and right stimulus locations. Thus, as implied by the salient features coding account, the up-right/down-left advantage is a general phenomenon that is found not only when aimed movement responses are made to physical stimulus locations, but also when responses are keypresses with the index fingers of each hand or spoken location words, when location words are mapped to spoken or keypress responses, and when arrows are mapped to spoken responses. Moreover, at least for words and arrows paired with vocal responses, the advantage generalizes to situations in which the stimuli refer to horizontal location and the responses to vertical location (see Cho & Proctor, 2004b, Experiment 2, where a right–up/left–down advantage was obtained with joystick movements).

8.2.2 Dual-Strategy Hypothesis

The major challenge to Weeks and Proctor's (1990) salient features coding account of the up–right/down–left advantage is whether asymmetric coding is a general property of spatial representation, as they assumed, or is restricted to linguistic representation. Umiltà (1991) originally made this challenge in a commentary on Weeks and Proctor's article. He accepted the central claim of Weeks and Proctor's salient features coding account, which is that the up-right/down-left advantage is due to asymmetric coding of the members of the stimulus and response sets. However, he restricted this asymmetry to verbal codes, arguing that "the salient-features coding hypothesis … applies only to codes that are verbal in nature" (p. 83) and stating, "I believe that those orthogonal S-R compatibility effects are attributable to the use of verbal labels" (pp. 84–85). According to Umiltà, verbal codes have the salient features, or polar referents, of up and right, but spatial codes are symmetric and do not have polar referents. Consequently, when response selection is mediated by verbal codes, the up-right/down-left advantage should appear, but when it is mediated by spatial codes, the advantage should be absent.

Umiltà's (1991) belief was founded primarily on two considerations: (a) The evidence for asymmetries in above–below and left–right from the word–picture verification literature suggested that they were due to linguistic coding, and (b) the effects reported by Weeks and Proctor (1990) tended to be larger when the stimuli or responses were words than when they were not. With regard to the first consideration, Umiltà noted, "The authors that first described the asymmetries in locational judgments had attributed the phenomenon to the use of verbal labels (Chase & Clark, 1971; Olson & Laxar, 1973; Seymour, 1974)" (p. 84). Umiltà was correct in noting that the findings from the word–picture verification literature are often attributed to linguistic encoding for unmarked versus marked words (those such as UP that refer to a positive end of the dimension and those such as DOWN that refer to a negative end), specifically to faster encoding of the marked member than of the unmarked one. However, a thorough review of the literature shows that (a) the primary effects of polarity, or salience, coding in verification tasks are on comparison or response–selection processes, not stimulus encoding, and (b) coding with respect to a polar referent for the dimension is not restricted to words but is a more general property of spatial representation (Proctor & Cho, in press). This latter conclusion is also in agreement with research from many different areas that indicates that one form of spatial representation, called conceptual representations by Logan (1994) and categorical spatial relations representations by Kosslyn (1994), involves coding the position of one object relative to another, reference object. With regard to Umiltà's second consideration, subsequent studies have not routinely found the up-right/down-left advantage to be larger when either the stimuli or responses are words than when they are not. For example, Proctor et al. (2002) obtained an up-right/down-left advantage of 16 ms for location words mapped to keypresses, compared with the 18-ms effect for physical locations mapped to keypresses (see also Cho & Proctor, 2002).

Adam, Boon, Paas, and Umiltà (1998) elaborated on Umiltà's (1991) dual-strategy hypothesis and provided evidence that they interpreted as support for it. In their Experiment 1, subjects responded to vertically arrayed stimuli by pressing one of two keys arrayed horizontally, one in each hemispace, under one of two conditions of trial initiation. In the participant-paced condition, the subject initiated each trial by pressing a right response key, whereas in the computer-paced condition, the computer initiated each trial following a 750-ms intertrial interval (ITI). The results showed 25-ms shorter mean RT in the computer-paced condition than in the participant-paced condition, and the up-right/down-left advantage was evident only in the participant-paced condition. Adam et al.'s Experiment 2 also included conditions in which the responses were the spoken words "left" and "right," as well as ones in which they were keypresses. Overall, in that experiment, an up-right/down-left advantage was evident in the participant-paced condition but not the computer-paced condition, and this pattern did not interact significantly with response modality.

Adam et al. (1998) explained the finding that the up-right/down-left advantage was present in the participant-paced conditions but not in the computer-paced conditions in terms of Umiltà's (1991) proposal that the advantage is due to a verbal coding strategy. They assumed that, as proposed initially by Umiltà, verbal codes are slower to form than visual codes. The relatively long RT in the participant-paced condition would imply that a strategy of using verbal codes was adopted and, for this reason, the up-right/down-left advantage occurred. In contrast, the relatively short RT in the computer-paced condition would mean that a strategy of using visual codes was adopted and, for this reason, no up-right/down-left advantage was found.

There are several problems with Adam et al.'s (1998) interpretation of their results in terms of the dual-strategy hypothesis. First, Dutta and Proctor (1992, Experiment 2) obtained the up-right/down-left advantage for keypress responses with computer-paced presentation, indicating that the advantage is not limited to participant-paced conditions. In fact, Adam et al.'s Experiment 3, in which presentation was computer-paced and intertrial interval was varied, similarly showed "a small up-right/down-left advantage that was independent of ITI" (p. 1589). Second, the linkage of the participant- versus computer-paced manipulation to the verbal–visual coding distinction is tenuous, particularly given that responses were only slightly faster in the computer-paced condition than in the participant-paced condition. Third, the advantage that Adam et al. obtained in the participant-paced condition may have been influenced by the specific trial-initiation method of making a right response.

Proctor and Cho (2001) performed a thorough evaluation of Adam et al.'s (1998) findings and conclusions concerning effects of trial initiation. Their Experiments 1 and 2 were designed as procedural replications of Adam et al.'s Experiments 1 and 2, one using manual left–right keypresses to the up–down stimulus locations and the other using vocal "left"–"right" responses. In contrast to Adam et al.'s results, both experiments showed an up-right/down-left advantage (11 ms in Experiment 1 and 18 ms in Experiment 2) that was of similar magnitude when the trial presentations were computer-paced as when they were participant-paced. Moreover, Cho and Proctor (2001) found that the up-right/down-left advantage was reduced when the subject's initiating action was a left response compared with when it was a right response. This outcome occurred when the initiating action and response were both keypresses, both spoken location names, or one was keypress and the other a spoken name. We will describe an explanation for this effect of initiating action later in the chapter, but for now, the main point is that these effects generalized across verbal and nonverbal response modes. The results of Cho and Proctor's study are consistent with Weeks and Proctor's (1990) salient features coding account, according to which the up-right/down-left advantage is due to asymmetry in coding the alternatives on each dimension. However, they provide no support for Adam et al.'s claim that this asymmetry is restricted to verbal codes.

In their Experiment 4, Adam et al. (1998) used a visual mapping signal, presented in the middle of the screen 1,250 ms before the imperative stimulus, to designate the mapping in effect for that trial. For half of the subjects, the mapping signal was a verbal description (e.g., the words "above-right"/"down-left"), whereas for half it was a visual graphic (see Figure 8.2a). The crucial finding was that an up-right/down-left advantage of 52 ms was obtained with the verbal mapping signals, but no mapping effect was evident with the graphic ones. Kleinsorge (1999) obtained similar results in his Experiment 1, in which subjects responded to up and down stimuli by moving their right hand from a home key to a left or right key. As in Adam et al.'s study, the mapping signal for a given trial was presented prior to stimulus onset. In separate halves of the experiment, the signal was verbal (e.g., down → left, up → right) or pictorial (two 90° segments of a circle; see Figure 8.2b). An up-right/down-left advantage of 52 ms was obtained with the verbal signals but not with the pictorial signals.

Although the results of Adam et al.'s (1998) Experiment 4 and Kleinsorge's (1999) Experiment 1 are consistent with Umiltà's (1991) hypothesis that asymmetric coding of alternatives is restricted to verbal codes, Kleinsorge noted that another difference between the precue conditions could be responsible for the results. Specifically, whereas the two individual S-R assignments could be coded simultaneously with the pictorial precues, they had to be coded sequentially when reading the verbal

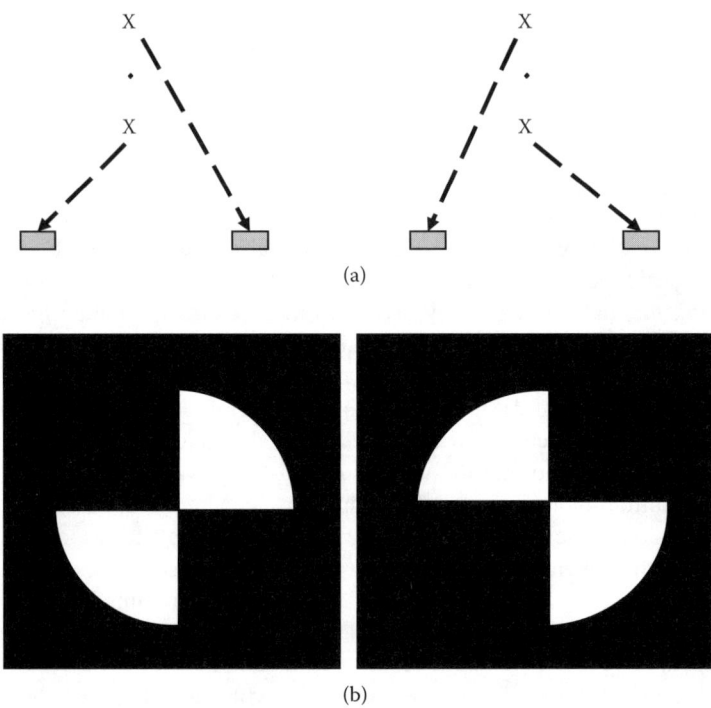

FIGURE 8.2 The graphic and pictorial mapping signals used by (a) Adam et al. (1998) and (b) Kleinsorge (1999) to signal up-right/down-left and up-left/down-right.

instructions. To force sequential coding in both conditions, Kleinsorge presented only half of each of the mapping displays in his Experiment 3 (e.g., down → left for verbal instruction or the lower left quadrant for the pictorial instruction), requiring that the subject generate the other S-R assignment. In this case, both verbal and pictorial signals yielded an up-right/down-left advantage of 18 ms. Based on this and other evidence, Kleinsorge concluded that the orthogonal S-R compatibility effect is not dependent on verbal coding.

Why was the orthogonal S-R compatibility effect absent in the conditions of Adam et al.'s (1998) and Kleinsorge's (1999) experiments in which the mapping was signaled graphically? Cho and Proctor (2003) speculated that the absence can be attributed to subjects basing their responses in those conditions on the cue images. That is, evidence indicates that there is a second form of spatial representation, called perceptual representations by Logan (1994) and coordinate spatial relations representations by Kosslyn (1994), which consists of spatial images with metric properties. Several aspects of Adam et al.'s and Kleinsorge's methods encouraged maintenance of images of the mapping precues, which could have caused the imperative stimulus to be represented in a similar manner. Reliance on coordinate spatial relations may also be responsible for a finding of Proctor and Cho's (2001), that the up-right/down-left advantage was absent under conditions of speed stress.

To summarize, evidence concerning the nature of spatial codes generally and the up-right/down-left advantage specifically indicates that asymmetric coding of the stimulus and response alternatives is not restricted to verbal codes but is a general property of relational coding. Performance is better when the more salient stimulus code is mapped to the more salient response code than when it is mapped to the less salient one. The few exceptions to the up-right/down-left advantage seem to occur under conditions that promote reliance on coordinate spatial representations (i.e., images).

8.3 INFLUENCES OF HAND, HAND POSTURE, AND RESPONSE ECCENTRICITY ON ORTHOGONAL S-R COMPATIBILITY

Bauer and Miller's (1982) study showed orthogonal S-R compatibility for unimanual movement responses to vary as a function of whether the right or left hand was used for responding. Subsequent experiments by Michaels (1989) and colleagues demonstrated effects of response eccentricity (i.e., the location at which the responses were carried out, relative to body midline) and hand posture as well. These effects led Lippa and Adam (2001) to propose that, although the overall up-right/down-left advantage is due to correspondence of asymmetric stimulus and response codes, as Weeks and Proctor (1990) claimed, the orthogonal S-R compatibility effects that vary as a function of hand, hand placement, and hand posture are distinct phenomena that require a different explanation.

8.3.1 MOVEMENT PREFERENCE HYPOTHESIS

Bauer and Miller (1982) proposed a movement preference account for their data because they concluded that a coding account could not explain their finding that orthogonal S-R compatibility effects were influenced by the hand used for responding. Bauer and Miller's specific explanation is as follows: When a stimulus is presented, there is an automatic tendency to respond in its direction, which generates implicit movement commands. Combining the implicit movement commands with the explicit movement commands for the direction of the target response yields a clockwise or counterclockwise movement. For example, if an up stimulus is mapped to the right response and a down stimulus to the left response, onset of the up stimulus generates an implicit movement command to move upward that, when combined with the explicit movement command to respond right, yields a clockwise rotation. The remaining assumption in Bauer and Miller's account is that, due to anatomical asymmetry in the arm and hand, the right hand prefers clockwise rotation and the left hand counterclockwise rotation.

Michaels (1989; Michaels & Schilder, 1991) also advocated the movement preference account and showed that the location of the responding hand affects orthogonal S-R compatibility. In Michaels's Experiment 1, subjects made left–right toggle-switch movements to up and down stimuli with each hand at three different locations: body midline and 30- and 60-cm laterally in the ipsilateral hemispace. Both the left-hand responses and right-hand responses at the body midline were faster with the up-right/down-left mapping than with the up-left/down-right mapping

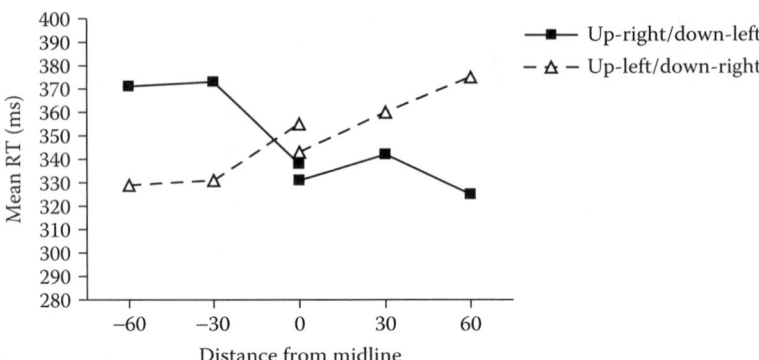

FIGURE 8.3 Reaction time as a function of distance from midline (in cm) and responding hand. Data from Michaels (1989, Experiment 1).

(see Figure 8.3), as in Bauer and Miller's (1982) study. When the right hand was positioned in the right hemispace, the up-right/down-left advantage increased in magnitude. In contrast, when the left hand was positioned in the left hemispace, responses were faster with the up-left/down-right mapping than with the up-right/down-left mapping. Michaels and Schilder (Experiment 1) obtained similar results when the responses were left–right deflections of the index finger.

Michaels's (1989) Experiment 2 was similar to her Experiment 1, except that the stimuli were located along the horizontal dimension and the responses along the vertical dimension, and only the midline and 60-cm positions were examined. The results again showed an interaction with response location. When responding at midline, the left–up/right–down mapping showed an advantage of about 13 ms for the left hand and no advantage for the right hand, whereas at the 60-cm ipsilateral positions, the right hand showed a 25-ms advantage for that mapping and the left hand an 8-ms advantage for the left-down/right-up mapping. Michaels deduced, "It seems clear … that Bauer and Miller are correct in asserting that the characteristics of the motor system figure significantly in the establishment of 'compatibilities'" (p. 271), with her findings showing that the effect of hand "*itself* can depend on yet *another* contingency, hand eccentricity" (p. 270, italics in original). She concluded that S-R compatibility is determined by "states of the action system," which include both the hand used for responding and hand eccentricity.

The movement preference account implies that anatomical factors associated with the responding hand should be the primary determinants of the influence of response eccentricity on orthogonal S-R compatibility. Weeks, Proctor, and Beyak (1995) evaluated this implication by dissociating the effects of response hand and response location through having subjects perform the task with up–down stimuli mapped to left–right toggle-switch responses performed with each hand at all response locations, both ipsilateral and contralateral. They obtained a similar inter-action pattern for the preferred mapping as a function of hemispace, with the up-left/down-right mapping yielding better performance in the left hemispace and the up-right/down-left mapping in the right hemispace. Most important, this pattern did not interact with the responding hand. That is, the position of the toggle switch was

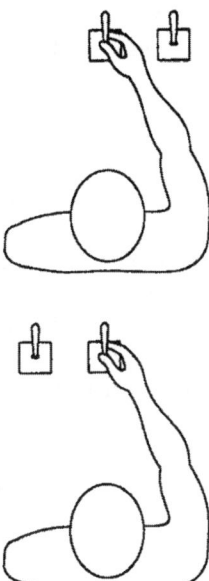

FIGURE 8.4 Illustration of the response arrangement used by Weeks, Proctor, and Beyak (1995). An inactive "dummy" switch was placed to the right (top picture) or left (bottom picture) of the active switch to provide a referent relative to which the active switch could be coded as left or right, respectively. From "Stimulus–response compatibility for vertically oriented stimuli and horizontally oriented responses: Evidence for spatial coding," by D. J. Weeks, R. W. Proctor, and B. Beyak, 1995, *Quarterly Journal of Experimental Psychology, 48A*, p. 377. Copyright 1995, The Experimental Psychology Society. Reprinted with permission.

the primary factor determining whether an up-right/down-left advantage or up-left/down-right advantage was obtained, regardless of whether it was operated by the left or right hand. This result is difficult to explain in terms of different movement preferences for the left and right hands and suggests that the major state of the action system, in Michaels's (1989) words, that determines the eccentricity effect is the location at which the action occurs.

Weeks et al. (1995) provided even stronger evidence in their Experiment 2 that location coding, and not the state of the motor system, is crucial to the response eccentricity effect. In that experiment, responses in all conditions were made with the right hand operating a toggle switch at body midline. An inactive toggle switch was placed to the left or right of the active toggle switch throughout a block of trials (see Figure 8.4), providing a referent object relative to which the active switch would be coded as right or left. Even though responses in all cases were at the same physical location and with the hand in the same posture, an up-right/down-left advantage of 20 ms was evident when the active switch was in the right relative position, and this reversed to a 12-ms up-left/down-right advantage when the active switch was in the left relative position. Proctor and Cho (2003, Experiment 2) replicated this general pattern with unimanual joystick responses instead of toggle switches and, more important, with left–right keypresses made with the index fingers on a response box as a function of the placement of an inactive response box to the left or right. In sum, the evidence indicates that the primary determinant of the response eccentricity effect is the position of the response key relative to other referent objects or frames.

8.3.2 Referential Coding

Lippa (1996) noted that orthogonal S-R compatibility effects could be produced in a manner similar to the spatial compatibility effects obtained with parallel stimulus and response arrays if the representation of the stimulus dimension were converted to match that of the response dimension, or vice versa. She referred to this type of account as the referential coding hypothesis. The basic idea of the referential coding hypothesis is that subjects will code one of the orthogonal sets in a manner that will make the coding parallel to the other set if a frame of reference is available that allows such parallel coding.

Hommel and Lippa (1995) reported evidence for referential coding of the stimulus set with respect to a face background. Specifically, they presented circles in up and down locations on the eyes of a photograph of the late actress Marilyn Monroe, which was rotated 90° clockwise or counterclockwise with respect to the normal, upright orientation (see Figure 4.6). In their Experiment 1, across the two face orientations, responses were significantly faster on average by 7 ms when the mapping was such that the location corresponding to the right eye, as viewed by the subject, was mapped to the right response and the location corresponding to the left eye to the left response than when the mapping was reversed. Proctor and Pick (1999) replicated this face context effect across a variety of faces that controlled for factors such as asymmetries of the photographs and found the effect to be larger and more robust than suggested by Hommel and Lippa's results. The idea that stimuli are coded relative to multiple frames of reference is one that we have emphasized in this book. Proctor and Pick noted that the left–right coding relative to the face context apparently acted in addition to the coding that produces the up-right/down-left advantage because the advantage was also evident in the mean data of most of the experiments.

Ladavas (1987) proposed an account for which the left–right button presses made with the left and right hands were transformed to a vertical code in order to explain the effect of handedness on orthogonal S-R compatibility that she reported. Specifically, Ladavas suggested that the dominant hand is coded as "higher" in position than the nondominant hand, thus providing a basis for correspondence and noncorrespondence on the vertical dimension along which the stimulus locations differed.

Lippa (1996) examined orthogonal S-R compatibility tasks in which the responses were unimanual movements from a home key to a target key, as in Bauer and Miller's (1982) study. She presented evidence to support the proposal that response positions are coded with respect to the fingertip-to-wrist axis of the responding hand. With left and right stimulus locations mapped to up and down response locations, the most natural placement of the responding hand on the home key is diagonally, at a 45° angle for the left hand and a 135° angle for the right hand (see Figure 8.5a). With this placement, for the left hand, the upper response position will be left of the fingertip-to-wrist axis of the hand and the lower position right; for the

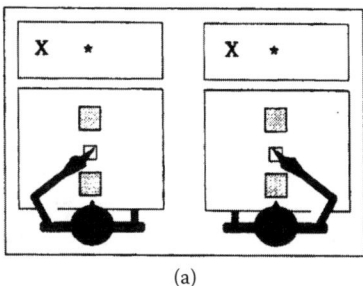 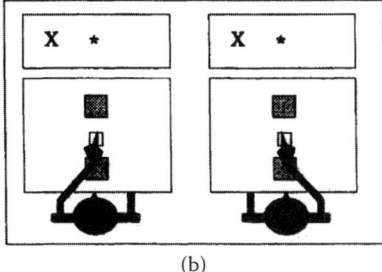

(a)   (b)

FIGURE 8.5 Unimanual aimed movements made with the responding hand positioned naturally (at a 45° angle for the left hand and 135° for the right hand) in the left panel (a) and with the fingertip-to-wrist axis aligned with the vertical response keys in the right panel (b). These placements were used in Lippa's (1996) Experiments 1 and 2, respectively. From "A referential-coding explanation for compatibility effects of physically orthogonal stimulus and response dimensions," by Y. Lippa, 1996, *Quarterly Journal of Experimental Psychology, 49A*, pp. 953 and 957. Copyright 1996, The Experimental Psychology Society. Reprinted with permission.

right hand, the upper position will be right of the fingertip-to-wrist axis and the lower position left. Thus, the predicted result is that a right-up/left-down mapping preference will be evident when responding with the right hand and the opposite preference when responding with the left hand. Lippa in fact obtained this predicted interaction in her Experiment 1. Moreover, she got rid of the interaction of mapping and hand in Experiment 2 by instructing subjects to hold their responding hand in line with the response keys (see Figure 8.5b). With this placement, the possibility of coding the response keys as left and right relative to the fingertip-to-wrist axis of the hand is eliminated because one key is only up and the other only down relative to this axis.

In Lippa's (1996) Experiment 3, the response apparatus was placed 37 cm to one side of the sagittal midline for some trial blocks and 37 cm to the other side for other trial blocks. In all cases, subjects were instructed to hold their responding hand at a 45 to 90° angle such that for both the right and left hand the upper response would be right and the lower one left when responding in the left hemispace and the opposite when responding in the right hemispace. In agreement with the referential coding hypothesis, a right-up/left-down mapping advantage was evident for both hands responding in the left hemispace and a right–down/left–up advantage when responding in the right hemispace.

For up–down stimulus locations mapped to left–right responses, instructing subjects to position their responding hand so that it is aligned with the row of response keys should allow the keys to be coded as up or down with respect to the hand because one is in the direction of the tip of the hand and the other in the direction of the wrist. As predicted, Lippa (1996) demonstrated in her Experiment 4 that performance was better with the up-right/down-left mapping for the left hand and with the up-left/down-right mapping for the right hand when the responding hand was kept in line with the response keys.

Lippa's (1996) Experiment 5 was analogous to her Experiment 3 in using a hand placement that should eliminate coding the response keys along the same dimension as the stimuli. Specifically, the task of responding to up–down stimuli with left and right responses was performed with the responding hand held at a right angle to the row of the response keys, thus preventing the hand from providing a reference frame from which the left or right response keys could be coded as up or down. Although, as predicted, the interaction of hand and mapping was eliminated, an unpredicted up-right/down-left advantage of more than 50 ms was evident regardless of whether the left or right hand was used for responding. Similarly, Michaels and Schilder (1991) obtained orthogonal S-R compatibility effects that varied as a function of response eccentricity for finger deflection responses made with the hand constrained to be at a right angle to the left–right responses. Thus, although referential coding of the type proposed by Lippa may contribute to orthogonal S-R compatibility effects in certain situations, such coding is not necessary for the effects to occur.

8.3.3 END-STATE COMFORT HYPOTHESIS

Lippa and Adam (2001) proposed an end-state comfort hypothesis, which bears similarity to both the movement preference and referential coding accounts, to explain response eccentricity effects of the type reported by Michaels (1989) and Weeks et al. (1995). This hypothesis assumes that "responses are spatially transformed and cognitively mapped onto the stimulus dimension according to relative hand posture, thereby mediating the pattern of facilitation and interference in response selection" (Lippa & Adam, 2001, p. 156). The transformation involves mentally rotating the response keys so that they are brought into alignment with the vertical stimuli. Whether the rotation direction is clockwise or counterclockwise is determined by the physical rotation preference of the hand. For example, when the left hand is positioned in the left hemispace at a right angle to the left–right line of response keys, it is easy to turn the hand inward but not outward. According to the end-state comfort hypothesis, a clockwise mental rotation of the response keys (transforming left to up and right to down) would be preferred in this situation, producing correspondence for the up-left/down-right mapping but not the up-right/down-left mapping.

Lippa and Adam (2001) tested implications of the end-state comfort hypothesis in their Experiment 1 using a vertically oriented stimulus display and varying the position of a response board containing a home key and response keys to the left or right of it. Subjects made unimanual movement responses with the left hand when the response board was in the left hemispace and right hand when it was in the right hemispace, at locations in front of the stimulus panel and behind the stimulus panel in the sagittal plane (see Figure 8.6). Regardless of whether the response board was in front of or behind the stimulus panel, an up-right/down-left advantage of 32 ms was obtained when responding with the right hand in the right hemispace and a reversed up-left/down-right advantage of 31 ms when responding with the left hand in the left hemispace. Lippa and Adam's Experiment 2 was similar to their Experiment 1, except that the orientation of the stimulus panel was left–right and that of the response keys was up–down. In this case, responding in the left hemispace

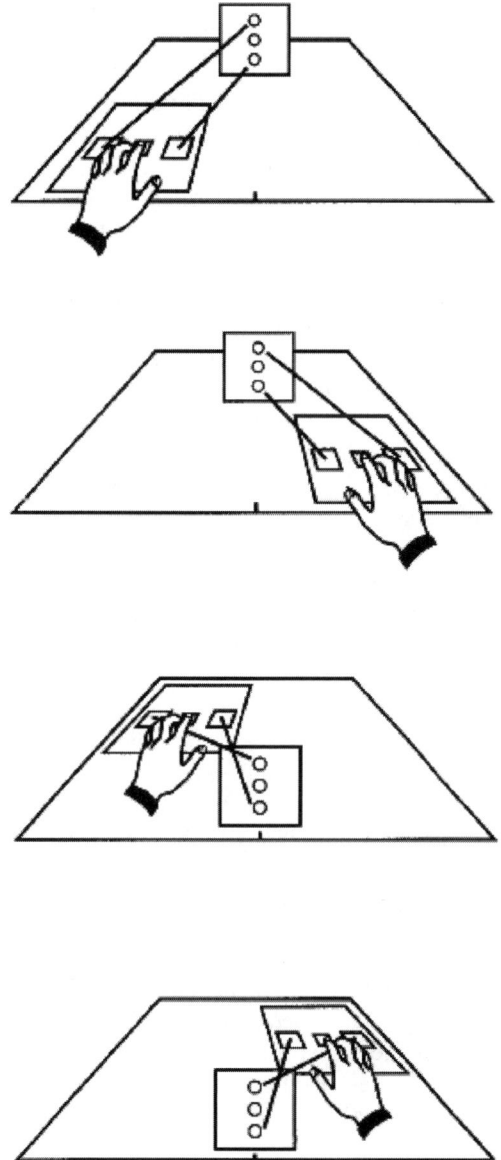

FIGURE 8.6 Unimanual aimed movements made with the response board placed to the left of the display screen and to the right of the display screen, with the display screen positioned behind it (top two panels) or in front of it (bottom two panels). From "Orthogonal stimulus–response compatibility resulting from spatial transformations," by Y. Lippa and J. J. Adam, 2001, *Perception & Psychophysics, 63,* p. 159. Copyright 2001, The Psychonomic Society, Inc. Reprinted with permission.

showed a right-up/left-down advantage and responding in the right hemispace a left–up/right–down advantage, the opposite of that obtained in Experiment 1. Both patterns conform with predictions of the end-state comfort hypothesis.

In Lippa and Adam's (2001) Experiment 4, subjects responded with the index and middle fingers on a computer mouse that was held upright in the left or right hand, with either the back of the hand or the palm oriented toward the person's face. When the back of the hand was up, the right hand showed an up-right/down-left advantage and the left hand an up-left/down-right advantage, and these advantages reversed when the palm was up. Lippa and Adam attributed the reversals of the advantages with the change in hand posture to the rotation preferences for the two hands reversing. However, as Cho and Proctor (2003) pointed out, it seems that clockwise and counterclockwise rotations of the left and right hands, respectively, are most comfortable, regardless of whether the palm or back of the hand is toward the person, in which case no reversal with hand posture would be predicted.

The end-state comfort account is primarily applicable to unimanual responses that logically could be coded relative to the rotation preferences of the single responding hand. Yet, as mentioned previously, the response eccentricity effect occurs as well when the responses are keypresses made with the index fingers of the left and right hands (Proctor & Cho, 2003): The up-right/down-left advantage obtained when the response apparatus is centered at midline increases when the box is placed in the right hemispace and reverses to an up-left/down-right advantage when the apparatus is placed in the left hemispace. Likewise, placement of an inactive response apparatus to the left or right of a centered response apparatus, on which the keypresses are made, influences the mapping preference much as an inactive apparatus does for joystick and switch movements. Given that effects of response eccentricity and relative position of the response apparatus are similar for unimanual movement responses, for which end-state comfort offers an account, and keypresses made with the left and right index fingers, for which it does not, end-state comfort likely contributes little if any to these effects.

8.3.4 Salient Features Coding Account

Weeks and Proctor (1990) originally developed their salient features coding account to explain the overall up-right/down-left advantage that occurs for a variety of stimulus and response sets. As described earlier, the current evidence strongly favors this account of that advantage. However, application of the salient features coding account to effects of the type described in this section is less obvious. The following remarks by Lippa and Adam (2001) reflect these points:

> The applicability of the salient-features hypothesis to orthogonal SRC effects that vary with hand or response position is limited. Of course, this does not negate the fact that the salient-features coding principle accounts for the overall advantage of up-right/down-left mapping across a range of stimulus and response sets. (p. 171)

However, we think that a unitary account of the two types of orthogonal S-R compatibility effects in terms of salient features coding is not only possible but also preferable.

Weeks et al. (1995) extended the salient features coding account to orthogonal S-R compatibility effects that vary with response position (the response eccentricity effect). They proposed that placing the response apparatus in the left or right hemispace increases the salience of the response associated with that hemispace. Thus, the up-right/down-left advantage is increased when responding in the right hemispace and reduced or reversed when responding in the left hemispace, regardless of which hand is used. The previously noted finding that similar results can be obtained when the response apparatus is at body midline by placing an inactive apparatus to the right or left of it to serve as a referent is predicted on the basis of relative location coding. These outcomes are consistent with Weeks et al.'s interpretation of the response eccentricity effects in terms of the salient features coding hypothesis, but they create difficulty for any explanation that relies on movement preferences because the location of the active response switch and the placement of the controlling limb did not vary across placements of the inactive response apparatus.

Cho and Proctor's (2001) finding that the right–left prevalence effect is reduced when preceded by a left initiating action compared with a right initiating action, even when the initiating action and task response are in different modalities, can be explained in a similar manner. Pressing the left or right key, or saying "left" or "right," to initiate a trial increases the salience of the corresponding response relative to the other response. Consequently, the right response is more salient when the initiating action is "right" than when it is "left," and vice versa for the left response, producing a larger up-right/down-left advantage with right initiation than with left initiation.

Cho and Proctor (2003) proposed that the reason why the relative salience of the response alternatives is affected by response eccentricity is that responses are represented in terms of multiple asymmetric codes. The fundamental idea of their explanation is that spatial codes for the responses are formed relative to several reference frames, with the direction and magnitude of the orthogonal S-R compatibility effect depending on the combined contributions of the different codes. When responding to a vertical stimulus display with left–right responses made at body midline, the horizontally arrayed responses are coded with right as positive polarity and left as negative polarity. Because the up stimulus location is positive polarity and the down location is negative polarity, the up-right/down-left mapping yields better performance than the opposite mapping. When the response apparatus is in the right hemispace, the apparatus is right with respect to several frames of reference relative to which it could be coded; consequently, additional positive polarity response codes contribute. In contrast, when the apparatus is in the left hemispace, these additional positive polarity codes are for the left response and oppose the up-right/down-left advantage. One line of evidence consistent with this interpretation is that an overall up-right/down-left advantage is often evident when performance is averaged across the various response locations (e.g., Cho & Proctor, 2002).

Cho and Proctor (2002) provided evidence for multiple asymmetric response codes in a study investigating an effect of hand posture reported by Michaels and Schilder (1991, Experiment 3). The latter authors had subjects make left–right switch movements at midline with a prone or supine hand posture, using the left or right hand. The left hand showed an up-right/down-left advantage of 15 ms in the prone

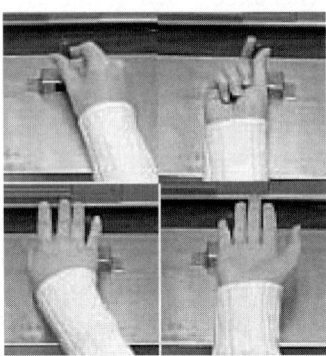

FIGURE 8.7 Grasping the response switch with the right hand, between the thumb and index finger (top row) and the ring finger and pinkie finger (bottom row), in prone (left column) and supine (right column) postures.

posture but an up-left/down-right advantage of 23 ms in the supine posture. The right hand showed a weaker pattern in the opposite direction, that is, an up-left/down-right advantage of 3 ms in the prone posture and an up-right/down-left advantage of 7 ms in the supine posture. Cho and Proctor manipulated hand posture and response eccentricity factorially in their Experiment 1. The response eccentricity effect replicated that of Weeks et al. (1995): The up-right/down-left advantage evident when responding at midline increased in size when the switch was in the right hemispace and reversed when it was in the left hemispace, regardless of which hand was used for responding. Hand posture showed an influence on orthogonal S-R compatibility similar to that reported by Michaels and Schilder. But, importantly, this effect did not interact with the effect of response eccentricity, suggesting separate contributions of each.

Cho and Proctor (2002) noted that the influence of hand posture on orthogonal S-R compatibility could be characterized as the up-right/down-left advantage being larger when the switch was to the right of the body of the hand (prone posture for left hand and supine posture for right hand) than when it was to the left (supine posture for left hand and prone posture for right hand). Consequently, they hypothesized that switch location was coded relative to the body of the hand. To test this hypothesis, in their Experiment 2, all responses were made at midline but with the switch grasped normally, between the index finger and thumb, as in Experiment 1, or between the pinkie and ring fingers (see Figure 8.7). This manipulation shifts the body of the hand relative to the switch. The effects of hand posture on orthogonal S-R compatibility varied as a function of how the switch was grasped. In all cases, the shift in the compatibility effect from the thumb–index finger grasp to the pinkie–ring finger grasp was in the direction expected if switch location were coded relative to the body of the hand.

Cho and Proctor (2005) presented evidence that response position is also coded relative to the display. Their Experiment 1 required left–right responses (for different groups of subjects: vocal location words, keypresses made with index fingers, joystick movements made with the left hand, and joystick movements made with the

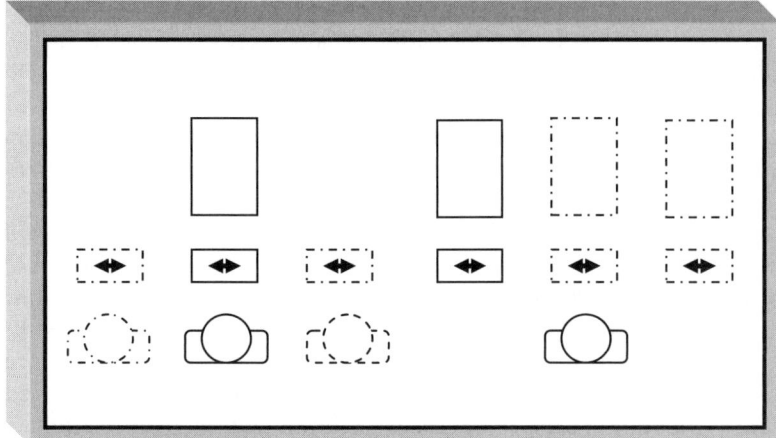

FIGURE 8.8 Position of the display, response apparatus, and subject in the stimulus display referent condition (left side), for which the display position remained constant and the subject's position changed with the lateral position of the response apparatus, and in the body midline referent condition (right side), for which the display position varied with the lateral position of the response apparatus and the subject's position remained constant. From "Representing response position relative to display location: Influence on orthogonal stimulus–response compatibility," by Y. S. Cho and R. W. Proctor, 2005, *Quarterly Journal of Experimental Psychology, 58A*, p. 856. Copyright 2005, The Experimental Psychology Society. Reprinted with permission.

right hand) to a stimulus presented above or below a fixation row of plus signs. This entire display was presented to the left or right side of the display screen, with display position varying randomly from trial to trial. The up-right/down-left advantage was larger when the stimulus set was on the left side of the screen (21 ms), relative to which the response set would be located on the right, than when the stimulus set was on the right side of the screen (8 ms). This difference did not interact with response mode, although it tended to be smaller with the vocal responses. A Simon effect for display side also occurred, with right responses being faster when the display was to the right side and left responses faster when it was to the left side. That these two effects reflect contributions of different codes was supported by the results of Experiment 2, in which the interval between onsets of the row of plus signs (which indicated the display side) and the imperative stimulus was varied. Whereas the Simon effect for display side decreased as this interval increased (which is the typical "decay" function of horizontal Simon effects), the effect of display location on the mapping preference did not.

In their Experiment 3, Cho and Proctor (2005) isolated two possible contributors to the response eccentricity effect: response location relative to the stimulus set (or display) and response location relative to body midline. In each case, responses were made at three positions. In the former case, the response positions were centered with the display screen or to the left or right of it, but the subject's position was changed so that the response apparatus was always at body midline (see Figure 8.8, left side). In the latter case, the position of the subject at the center was held constant, but the position of the display screen was always varied to align with the response apparatus (see Figure 8.8, right side). Different subjects performed with the keypress and unimanual response modes described above.

The results of Cho and Proctor's (2005) study showed that response location relative to the stimulus set was the most important factor. An up-right/down-left advantage of 33 ms relative to the stimulus set was found at the right response location, and this reversed to a 19-ms up-left/down-right advantage at the left response location. Location relative to body midline also tended to show an influence on the orthogonal S-R compatibility effect, primarily for unimanual responses (up-right/down-left advantage of 17 ms in the left hemispace and 40 ms in the right hemispace). Cho and Proctor (2004a, Experiment 2) confirmed both of these findings using a method in which responses were made in the left or right hemispace, with the display positioned relative to the response apparatus (being aligned with it or positioned to the left or right). The results suggest that although the influence of response eccentricity on the orthogonal S-R compatibility effect for keypresses is entirely due to coding location of the response apparatus relative to the display, the influence for unimanual movement responses is also due to coding location of the apparatus relative to body midline.

Finally, although display location along the horizontal dimension influences the up-right/down-left advantage, apparently through providing a reference frame relative to which the location of the response apparatus is coded, display location along the vertical dimension has little if any effect on coding of stimulus location. Cho and Proctor (2004a, 2004b) reported experiments in which the entire stimulus set (fixation row of X characters and imperative stimulus) appeared in the upper or lower half of the display screen on different trials. Although this irrelevant stimulus-set location variable produced an orthogonal Simon effect (response eccentricity and hand posture effects similar to but smaller than those typically obtained for relevant location information; Cho & Proctor, 2004a, Experiment 1), it had no influence on the mapping effect for the relevant location of the imperative stimulus. These results imply that the asymmetric coding of stimulus location is relatively unaffected by the stimulus context.

8.4   CORRESPONDENCE OF ASYMMETRIC CODES AS A GENERAL PRINCIPLE OF BINARY-CHOICE REACTIONS

Asymmetric coding of stimulus and response alternatives extends beyond orthogonal spatial tasks to other binary choice–reactions tasks. Proctor and Cho (in press) reviewed the literature on word–picture verification tasks, in which subjects must indicate whether a spatial word (e.g., ABOVE) inside of a square describes the location of a filled dot relative to the square, and their close relatives, sentence–picture verification tasks. Although results from verification tasks are typically interpreted as supporting polarity coding, such coding is often assumed to be restricted to words (e.g., Umiltà, 1991) and to affect primarily the time to encode or identify the word (less time required to encode a linguistically unmarked term, e.g., ABOVE, than to encode a marked term, e.g., BELOW; Chase & Clark, 1971). Proctor and Cho showed that the extensive database for verification tasks provides unambiguous evidence that polarity coding occurs for both linguistic and nonlinguistic stimuli, and for responses as well. The results conform most closely to a model developed by Seymour (1974) that emphasizes correspondence of polarity codes. In the model, stimulus features and responses are coded as positive or negative polarity, and the combined correspondence of the features determines the speed with which a positive (true) or negative (false) response is made.

Proctor and Cho (in press) summarized evidence that also implicates a role for polarity correspondence in numerical parity (odd–even) judgments. A polarity correspondence account has been accepted explicitly for the linguistic Markedness Association of Response Codes (MARC) effect, which is better performance with the mapping of even to right (unmarked, or positive) and odd to left (marked, or negative) than with the opposite mapping (Nuerk, Iverson, & Willmes, 2004). The Spatial–Numerical Association of Response Codes (SNARC) effect refers to better performance when a right keypress is made to a large number and a left keypress to a small number than vice versa. The SNARC effect has been interpreted as a Simon-type effect that is due to correspondence of position on a horizontal number line with the response position (e.g., Dehaene, Bossini, & Giraux, 1993). Proctor and Cho pointed out that evidence suggests that correspondence of code polarities (plus for large number and right response; minus for small number and left response) is at least a contributing factor to the SNARC effect.

Another phenomenon in which polarity correspondence seems to play a role is the Implicit Association Test (IAT; Greenwald, McGhee, & Schwartz, 1998) effect, which has been popular in the social psychology literature. For the IAT, left and right keypresses are made to stimuli from two attribute categories (e.g., pleasant and unpleasant words) and two target categories (e.g., flower and insect names). Two mappings of the target categories to responses are used in different trial blocks (e.g., pleasant and flower to one response and unpleasant and insect to the other, or pleasant and insect to one response and unpleasant and flower to the other). The IAT effect is that performance is better with one mapping than the other. For example, Greenwald et al. found that responses were approximately 150 ms faster when flower was paired with pleasant and insect with unpleasant than vice versa. The IAT effect has been attributed to correspondence of the target categories with the attribute categories, which in the case of flowers and insects, Greenwald et al. interpreted as "indicating more positive attitudes toward flowers than insects" (p. 1468). However, Rothermund and Wentura (2004) and Kinoshita and Peek-O'Leary (2005) have presented evidence that the data suggest instead that in many situations the IAT effect is due to correspondence of polar codes, or relative saliencies, for the two categories, rather than to implicit associations of the meanings of the categories. Thus, the data from a range of binary-choice tasks suggest that correspondence of stimulus and response codes for features that differ in salience, or polarity, is a significant factor influencing performance.

8.5 TASKS WITH MORE THAN TWO ALTERNATIVES

Although mappings of two stimulus locations to responses for orthogonal spatial dimensions yield compatibility effects that conform to correspondence of asymmetric stimulus and response codes, it is of both theoretical and applied interest to determine whether such effects occur when there are more than two alternatives. As described in Chapter 2, Proctor, Reeve, Weeks, Campbell, and Dornier (1997) examined orthogonal variations of the four-choice precuing task in which pairs of locations are cued prior to a target stimulus to which one of four keypresses (with the index and middle fingers of the hands) is to be made. For the orthogonal arrangements, the leftmost stimulus (or response) location was always paired with the bottom position of the other array, with the remaining locations mapped in order to the remaining positions. As noted, responses were slower when the display and response orientations were orthogonal (806 ms) than when they were parallel (647 ms), but none of the higher order interactions of this factor with precue type were significant. Proctor et al. interpreted these results as suggesting that the task was performed similarly with orthogonal arrays as with parallel arrays, except for an additional transformation to align the orthogonal frames of reference. Proctor et al.'s Experiment 2 provided additional support for this interpretation by showing virtually perfect transfer of the benefits of three sessions of practice with the orthogonal arrangement to performance with a parallel arrangement in a fourth session. In that session, their mean RT and benefits for the individual precue conditions were comparable to those for subjects who performed in all sessions with the parallel arrangement.

In Proctor et al.'s (1997) study, only a single S-R mapping was used for the orthogonal arrays. Lu and Proctor (1998) reported the results of two experiments for which the stimuli were arrayed vertically and the responses horizontally, but with the bottom-to-top/left-to-right mapping used by Proctor et al. compared with that of a bottom-to-top/right-to-left mapping. One experiment used the precuing method of Proctor et al., whereas the other used a basic four-choice method with no precue. Neither experiment showed a significant difference between the two mappings. Biel and Carswell (1993) obtained similar results for a five-choice task in which keypresses were made with the fingers of the right hand placed on a horizontally arranged keypad to stimuli presented in one of five vertical locations. Response time averaged nonsignificantly shorter for the bottom-to-top/left-to-right mapping (775 ms; error rate of 2.75%) than for the bottom-to-top/right-to-left mapping (788 ms; error rate of 2.4%), with both yielding much shorter RTs than a random mapping (1,704 ms; error rate of 5.3%). Thus, at present, there is little evidence that a mapping effect similar to the up-right/down-left advantage for two-choice tasks occurs when there are more than two S-R alternatives.

Andre et al. (1991) examined orthogonal S-R compatibility in three-choice tasks and proposed that it may be influenced by what they called colocation, "the tendency for the display at either end of the display array, which is closest to (colocated with) a control at the end of the control array, to be mapped to that control" (p. 1547). For example, for a row of three stimulus locations mapped to a column of three response locations, the preferred mapping (according to colocation) would be left-to-right to top-to-bottom when the response column is aligned with the leftmost stimulus and right-to-left to top-to-bottom when it is aligned with the rightmost stimulus.

Andre et al. (1991) reported an experiment in which subjects responded to a row or column of three stimuli using the three fingers on a single hand. The response array could be parallel to the stimulus array, orthogonal to the stimulus array, or angled such that the responses had both left-to-right and top-to-bottom components. Not surprisingly, RT was faster when the response array was parallel to the stimulus array (364 ms) than when it was orthogonal to it (415 ms). Responses were just as fast with the angled response array (359 ms) as with the parallel array, indicating that subjects were able to map the response locations to corresponding stimulus locations on the basis of the left-to-right dimension. For the orthogonal arrays, performance depended on both the responding hand and whether the stimulus array or response array was the one that was oriented vertically (or horizontally). For vertically arrayed stimuli and horizontally arrayed responses, the up-right/down-left mapping produced better performance than the alternative mapping for left-hand responses, whereas the opposite held true for right-hand responses. For horizontally arrayed stimuli and vertically arrayed responses, the left–up/right–down mapping produced better performance than the alternative mapping for the left hand, whereas the opposite held true for the right hand. Colocation had only a nonsignificant effect on performance.

The few studies that have examined compatibility effects for orthogonal S-R arrays of more than two alternatives show little evidence for asymmetric coding of the type implicated in two-choice tasks, suggesting that coding asymmetry arises from categorical codes in which one stimulus is coded relative to another. With more than two choices, mental alignment or rotation of the arrays seems to play a larger role than in two-choice tasks. Although Andre et al.'s (1991) results imply that there are preferred directions of rotation that are influenced by several factors, other results show little evidence of preferred directions. More work is needed to specify the conditions under which orthogonal S-R compatibility effects occur when there are more than two S-R alternatives.

8.6 CHAPTER SUMMARY

Current knowledge regarding orthogonal S-R compatibility effects in two-choice tasks can be summarized as follows: With vertical stimuli and horizontal responses, an up-right/down-left mapping advantage is obtained in many situations. The up-right/down-left advantage seems to reflect correspondence of asymmetric stimulus and response codes, as proposed by Weeks and Proctor's (1990) salient features coding account. Such asymmetric coding is not limited to verbal codes (Proctor & Cho, 2001), counter to the suggestion of Umiltà (1991) and Adam et al. (1998). The influences of response hand, response eccentricity, and hand posture on orthogonal S-R compatibility can be accommodated by an asymmetric coding account that assumes that coding of stimuli and responses takes place with respect to multiple frames of reference. Effects based on correspondence of the code polarities for orthogonal dimensions and of the spatial correspondence for parallel dimensions add together to determine the relative RT for various S-R mappings. Asymmetric coding of stimulus and response alternatives seems to be a general property of human information processing that is evident in a variety of binary choice tasks, but there is little evidence at present for its being a significant factor in tasks with more than two alternatives.

9 Population Stereotypes for Direction of Motion and Color, Word, and Picture Associations

9.1 INTRODUCTION

A compatible mapping of displays to controls is in part determined by the natural tendencies of a population. The literature on population stereotypes, which dates as far back as that on S-R compatibility, explores people's tendencies to associate certain control actions with specific display properties. A sample of individuals from the population of interest is typically provided with paper-and-pencil illustrations of various display and control configurations, or with actual display–control devices, and asked to indicate how they would operate the control to produce a specified change in the display. Sometimes they are asked the meaning that a particular display feature signifies. The stereotypical response is the one that is provided most frequently, and the percentage of individuals giving that response is an indication of the strength of the stereotype. The basic idea is that for situations in which there are strong population stereotypes, an interface will benefit by being designed to be consistent with the stereotypes.

As an example, Bergum and Bergum (1981) presented 127 American students with drawings of everyday items such as a light switch, door with the knob on the left, and lever-handle faucet (see Figure 9.1) and asked them to indicate whether (a) the switch was in an on or off position, (b) the door should open inward or outward, and (c) the movement of the faucet lever in the indicated direction would turn the water on or off. For these situations, 88% of the students indicated that the up switch represents on, 72% that a door with the knob on the left should open inward, and 93% that upward movement of the faucet lever should turn on the water. Bergum and Bergum also found 15 other significant stereotypes for motion, order, and location, shown in Table 9.1. Thus, for many everyday items there is much agreement for the associations of certain positions of those items with specific outcomes, although some associations are stronger than others (e.g., the upward lever movement for on stereotype is stronger than inward opening of doors with the knob on the left stereotype).

Certain stereotypes are due to spatial relationships that are inherent in interactions with the physical environment anywhere in the world and thus will generalize broadly. The tendency to make a spatially corresponding response to a stimulus, which is the topic of many studies described in this book, is a good example.

(a) Preferred relation: Light switch is in on position

(b) Preferred relation: Door opens in

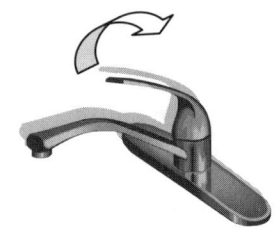

(c) Preferred relation: Turn water on

FIGURE 9.1 Pictures of everyday items that are similar to the ones used in Bergum and Bergum's (1981) study, and their preferred relations.

However, other stereotypes are a function of experience and learning within a particular culture and therefore are culturally specific. For example, through experience, people in the United States associate an up position of a light switch with on, as Bergum and Bergum (1981) found, and a down position with off, whereas people in the U.K. have the opposite association (Broadbent, 1958). As a result of cultural specificity, caution should be exercised when generalizing population stereotypes from one culture to another.

Loveless (1962) noted more generally that "investigations of population stereotypes need to be conducted with considerable care" (p. 361) because there can be differences in preferences both between operators and within a single operator over time. For this reason, Loveless recommended against reliance on the judgments of a few individuals, even if they have considerable expertise or are highly confident about which relations are stereotypical. Instead, a representative sample large enough for statistical testing should be taken from the target population. Humphries and

TABLE 9.1
Population Stereotypes Identified by Bergum and Bergum (1981)

1. Vertical number sequences should progress from top to bottom.
2. Horizontal scales should increase as movement of the control moves from left to right.
3. Vertical scales should increase as movement of the control moves upward.
4. A depressed pushbutton that locks in position is associated with being ON.
5. A vertical lever in the up position represents a high value.
6. A vertical lever door latch that is in the down position is associated with being locked.
7. The right door of double doors is associated with IN.
8. A vertical thumbwheel that is rotated up or away is associated with ON.
9. Clockwise rotation of a shower knob is associated with making the water colder.
10. Clockwise rotation of a dimmer knob is associated with making the light brighter.
11. Clockwise rotation of a rotary switch is associated with ON.
12. Clockwise rotation of a rotary selector with a digital display is associated with increased values.
13. For a double faucet, right is cold.
14. For a vertical traffic light, top is red.
15. For a horizontal traffic light, right is green.

Shephard (1959) showed that age, intelligence, and handedness are among the factors that determine which display–control relations are expected. Moreover, the type of test given to obtain the population stereotype may affect the outcome. For example, responses to paper-and-pencil questionnaires may not match behavioral measures when performing with the actual display–control relation (see, e.g., Vu & Proctor, 2003; Wu, 1997).

Population stereotypes obtained in experimental studies, whether with an actual display and control device or with a paper-and-pencil illustration of it, can also be affected by any prior response or judgment made to another display–control configuration (Loveless, 1962). Yet, when testing only a single display–control configuration with a group of subjects, it cannot be assumed that an observed relation is "reversible." For example, a preference for clockwise movement of a control to move a pointer upward does not necessarily mean that a preference for counterclockwise control movement to move the pointer downward will also hold. In fact, Holding (1957) found that some people prefer clockwise movement of a knob to produce any type of display movement. To establish that a relation is reversible, it must be shown that preferences exist to operate a control in one manner for one control–display effect and in the opposite manner for the other effect (e.g., turning a knob clockwise to move a pointer to the right and counterclockwise to move it to the left).

9.2 LINEAR DISPLAY INDICATORS AND THEIR RELATIONS TO CONTROLS

9.2.1 PREFERENCES FOR DISPLAY–CONTROL RELATIONS

When a display and its control move linearly along the same plane, the preferred direction of motion is that of the control in the same direction as the display

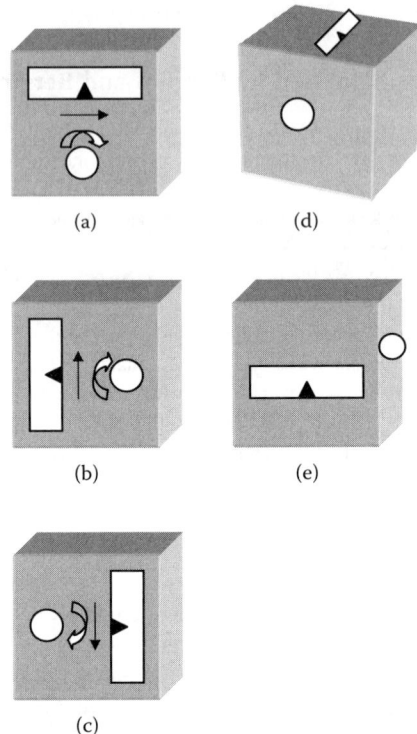

(a)   (d)

(b)   (e)

(c)

FIGURE 9.2 Depictions of the display and control arrangements used in Warrick's (1947, as described by Loveless, 1962) study. The arrows indicate the direction-of-motion stereotypes for arrangements a through c.

movement that it operates (Loveless, 1962). This preference is not surprising given that correspondence of locations and/or directions of motion is the most basic spatial compatibility effect. Stereotypic relationships are less obvious when the control is rotary and the display movement linear. Warrick (1947, as described by Loveless) had operators perform with five linear display–rotary control relations, depicted in Figure 9.2. When the control was in the same plane as the display and aligned with its center (conditions a, b, and c), there was an expectancy for the indicator to move in the same direction as the part of the control that was nearest to the display. This finding has come to be known as Warrick's principle. However, when the control was arrayed on a different plane than the display (conditions d and e), expectancies for direction of motion relations did not emerge, and operators tended to produce clockwise movements to produce any display movement, consistent with Holding's (1957) clockwise-for-anything stereotype. Loveless noted that the stereotypes observed in conditions a and b were replicated by Gibbs (1949), who used a large handwheel instead of a small rotary knob. The fact that Warrick's principle held for a large handwheel as well as a small knob provides evidence that the direction of the control movement is indeed what gives rise to the preference rather than the direction of the hand movement. This outcome is in agreement with the results from

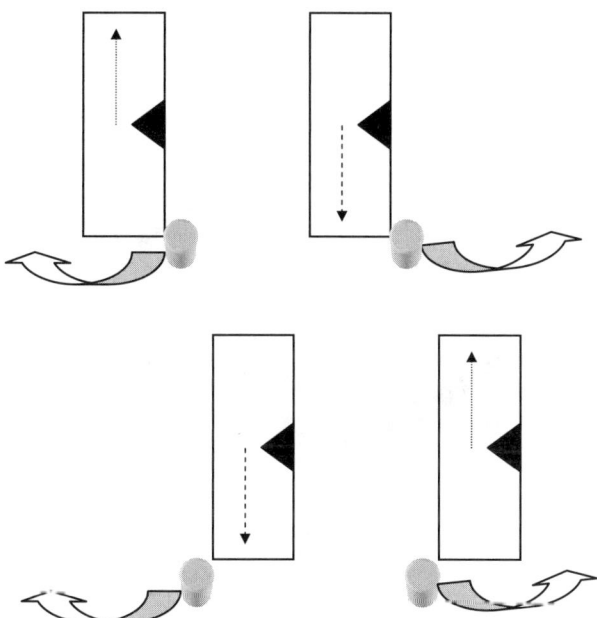

FIGURE 9.3 Illustration of the display and control arrangement used by Mitchell (1948). The lever controls extended outward, perpendicular to the vertical display indicator. The block arrows represent the clockwise or counterclockwise direction of motion when moving the lever to the left or right.

compatibility experiments described earlier in the book indicating that the action goal is more important for response selection than the physical way in which the goal is achieved.

Mitchell and Vince (1951) reported a study conducted by Mitchell (1948) that showed a relation similar to Warrick's principle for lever controls. Mitchell examined the preferred direction of lever movement to control a vertical display pointer placed above the control to form a right angle with it (see Figure 9.3). To move the pointer up, the majority of subjects favored moving the lever to the left (or clockwise) when it was located to the right of the vertical display and operated by the right hand, and to the right (or counterclockwise) when it was to the left of the display and operated by the left hand.

Ross, Shepp, and Andrews (1955) investigated display–control relationships for a dot moving in the horizontal (left/right) or vertical (top/bottom) dimension of a circular display, controlled by a rotary knob that could be rotated clockwise or counterclockwise, a push–pull knob, or a lever that could be moved in the up, right, down, and left directions (see Figure 9.4). A paper-and-pencil test was used, and the possible directions in which the control could be moved were printed on the diagram. Subjects were asked to circle the arrow corresponding to the direction in which they would move the knob to produce the indicated movement of the dot. The three knobs were drawn in three different planes (horizontal, frontal, and

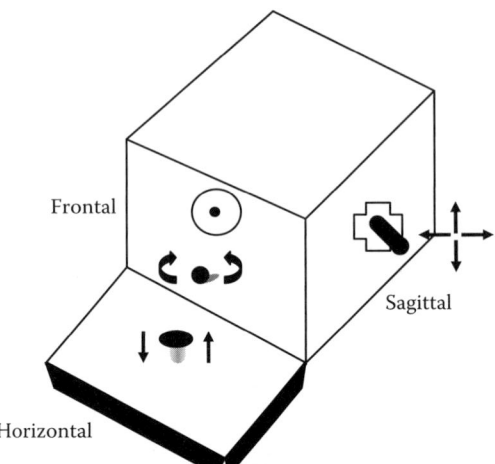

FIGURE 9.4 Illustration of the display and control apparatus depicted in Ross, Shepp, and Andrews's (1955) paper-and-pencil survey. The dot can move in the horizontal or vertical dimension of a circular display, and the control is a rotary knob, a push–pull knob, or a lever arrayed in the horizontal, frontal, or sagittal plane. The arrows indicate the movement directions for each control.

sagittal), yielding nine possible combinations. Each subject was presented with only three test diagrams, one with each control type.

For the rotary control, Ross et al. (1955) found strong stereotypes of clockwise-to-right and its reverse relation, counterclockwise-to-left, regardless of the plane in which the control knob was placed. Similarly, the counterclockwise-to-down stereotype was evident in all three planes, although the clockwise-to-up relation was only significant when the knob was in the sagittal plane. For the push–pull knob, stereotypes were less apparent. The in-to-right stereotype was significant for the horizontal plane, but an in-to-left stereotype emerged in the sagittal plane. The in-to-up stereotype was significant in the horizontal and frontal planes, but the out-to-down stereotype was only significant in the horizontal plane and was accompanied by a significant in-to-down stereotype when the knob was in the sagittal plane. For the lever control, the right-to-right stereotype was evident in all three planes. Although the reverse, left-to-left, stereotype was obtained for the horizontal plane, a right-to-left stereotype was found for the frontal plane. The only other significant stereotype was the up-to-up relation obtained in the frontal and sagittal planes. For the three controls that Ross et al. examined, a strong, reversible clockwise-to-right and counterclockwise-to-left stereotype was evident only for the rotary control.

As mentioned earlier, Holding (1957) showed that not all stereotypes are reversible. He examined the preferred direction of motion for seven arrangements of a linear display placed orthogonally to a rotary control. The display pointer moved at a right angle relative to the plane of rotation of the control knob. Five of the display–control arrangements were for right-handed persons and two for left-handed persons. Within each arrangement, half of the displays moved away from the control and half moved toward it. Subjects were asked to rotate the control with their dominant hand in a direction that would produce motion of the pointer in the instructed dimension. Each subject performed with only one combination of the display–control configuration and direction of pointer movement. Holding found that 64% of participants responded with a clockwise turn of the knob. For the arrangements designed for right-handers, 69% of responses were clockwise, but for the arrangements designed for left-handers, a reverse pattern was not evident. However, left-handers did produce more counterclockwise responses than right-handers. Moreover, there were significantly more counterclockwise responses for configurations in which the display moved toward the control. Thus, participants tended to prefer clockwise movement to move the pointer in a direction away from the control and showed a more general preference of making a clockwise movement for any display–control relation.

In summary, population stereotypes for direction of motion can be established by asking people their preferences for various display control relations. Although principles developed from these results can be relatively robust, such as Warrick's principle, other relations may change significantly as a function of the type of control, the position control relative to the display, and the hand used for responding.

9.2.2 Using Error Rate as a Measure of Preferred Relations

Performance with various combinations of display–control relations can also be used to determine preferred direction of movement for a population. Mitchell and Vince (1951) reported a study conducted by Vince (1945) in which subjects had to make a series of discrete responses to separate ink marks that were above or below a horizontal line on a scrolling roll of paper by moving a control knob in the same or opposite direction as the ink mark. Subjects made many more errors when they were to move the knob in the opposite direction than when they were to move it in the same direction, particularly when the interval between two successive stimuli was less than 2 s. The preference for the same over opposite direction held across tasks in which the stimuli were spaced irregularly (Vince & Mitchell, 1946) and in which a secondary task was performed concurrently with the other hand (Mitchell, 1947). These latter two studies also included linear controls that moved left to right or forward and backward, for which the relation to the upward–downward pointer movement is less clear-cut. Forward and backward movements yielded intermediate numbers of errors compared to the same, upward and opposite, downward control mappings, with the ordering of performance from best to worst being

- Control movement up to move display pointer up, and down for down
- Control movement forward to move display pointer up, and backward for down
- Control movement right to move display pointer up, and left for down
- Control movement backward to move display pointer up, and forward for down
- Control movement left to move display pointer up, and right for down
- Control movement down to move display pointer up, and up for down

Mitchell and Vince (1951) also examined movement preferences for continuous pursuit tracking tasks in which subjects were to keep a moving pointer on a horizontal line by moving a control handle up–down, forward–backward, or left–right, as in the discrete response task. With the continuous task, there was no significant difference in performance for the six types of control movement–display movement relations when the control was operated by the subject's dominant hand (in this case, the right hand). This good performance for all direction relations with the dominant hand is likely due to the visual feedback available for subjects to make corrective adjustments. For the nondominant hand, the difference between the best of the six control–display relations (the up–up/down–down relation) and the two worst (the left–up/right–down and down–up/up–down relations) was significant. Moreover, when the continuous tracking task was performed simultaneously with each hand, the differences in performance for the various types of control–display direction relations with the dominant right hand became more apparent. Thus, Mitchell and Vince conclude, "When the situation is made more complex, performance again appears to be affected by the directional relationship between control and display" (p. 29).

Loveless (1962) noted that the conclusion of Mitchell and Vince (1951), that the directional relationship between the control and display had no significant influence for simple continuous tasks controlled with the dominant right hand, may be misleading. This is because in that study the controls were levers, "whose rotary motion may well have affected the operator's expectations" (p. 358). Moreover, Loveless noted that although different movement relations may not affect the overall performance of simple, continuous tasks, natural tendencies may contribute to the initial response made by the person. Even though the initial response can be corrected easily, it may have consequences for sensitive or dangerous equipment that is controlled.

Vince (1950, as described by Mitchell & Vince, 1951), showed that there were also marked individual differences and differences in learning rates for nonpreferred display–control movement relations, but not much difference for preferred ones. Although accuracy of responses with some nonpreferred relations examined by Mitchell and Vince, but not all, can become as high as that with the preferred relations after 11 sessions of practice with the display–control relation, under stressful conditions people tend to revert back to the preferred relation. Support for this latter point was obtained by Garvey (1957, as described by Loveless, 1962), who trained two groups to equivalent levels of performance with an acceleration–control system and an acceleration-aided–control system, the latter of which was initially easier to use. When the operators were asked to perform with the systems in stressful environments, performance decreased for both systems, but the decrease was larger for the one that was initially more difficult to use.

Overall, Mitchell and Vince (1951) showed that errors can be used as a performance measure to specify population stereotypes for direction of movement because more natural display–control relations will yield fewer errors. Although practice with nonpreferred mappings can lower the errors to a minimal level, response time may still be slower than with a preferred mapping (see Fitts & Seeger, 1953), and when the person is stressed or fatigued, natural response tendencies may reappear.

FIGURE 9.5 Picture of the drill-loading machine device used in Simpson and Chan's (1988) study. From "The derivation of population stereotypes for mining machines and some reservations on the general applicability of published stereotypes," by G. C. Simpson and W. L. Chan, 1988, *Ergonomics, 31*, p. 330. Copyright 1988, Taylor & Francis Ltd. Reprinted with permission.

9.2.3 COMPLEX DISPLAY–CONTROL RELATIONS

Simpson and Chan (1988) examined 24 complex control–response relations that are commonly found on mining machines to determine whether population stereotypes exist for them. They used groups of subjects from three occupations (engineers, fitters and operators, and administrative and clerical staff) of the British Coal Company to determine whether the stereotypes differ for the populations with different technical backgrounds. The engineers design the machines and display–control layouts and relations, the operators and fitters are the people who actually use the mining machines in their work, and the clerical and administrative staff should have little or no experience with the machines themselves or their design. The apparatus was a model of a drill-loading machine device that used a ball-headed lever as the control, which could be moved in the horizontal and vertical planes (see Figure 9.5). The 12 control–response arrangements are listed in Table 9.2, and each subject performed with only one directional condition (i.e., the subjects were told to use their dominant hand to move the control lever to move the machine in the instructed direction, but not in the reversed direction later on).

Simpson and Chan (1988) compared performance of experienced users (i.e., the designers and operators) with that of inexperienced users (i.e., the clerical staff) and found no difference in the 24 control–response relationships. There was also no significant difference in responses between the three groups themselves, so the data were analyzed for stereotypes without occupational group as a factor. Seventeen of the 24 control–response relations showed significant stereotypes. Moreover, Simpson and Chan found that handedness influenced only one of the stereotypes: Right-handers showed a stronger stereotype for moving the control fore/aft to slew (turn) the drill carriage to the right/left than did left-handers. Thus, preferred control–response relations exist for complex machinery as well as for simple display–control configurations. The

TABLE 9.2
The Twelve Control–Response Arrangements
Investigated by Simpson and Chan (1988) with
the Drill-Loading Machine Shown in Figure 9.5

Machine Responses	Control-Motion
Bucket rolled backward/forward	Up/down
Bucket side-discharged CCW/CW	Fore/aft
Main boom lowered/raised	Up/down
Main boom lowered/raised	Fore/aft
Main boom slewed to right/left	Fore/aft
Drill-tip raised/lowered	Up/down
Drill-tip raised/lowered	Fore/aft
Drill carriage slewed to left/right	Fore/aft
Drill carriage slewed to left/right	Left/right
Bucket rolled forward/backward	Fore/aft
Bucket side-discharged CCW/CW	Left/right
Main boom rotated CW/CCW	Fore/aft

CW = clockwise; CCW = counterclockwise

absence of occupational differences in this case implies that the stereotypical relationships identified by Simpson and Chan reflect general response tendencies rather than ones specific to experience with mining. However, in other cases, the "mental model" learned through experience with a system may be a more important factor (e.g., Hoffmann, 1997).

9.2.4 DIRECTION OF MOTION PRINCIPLES AND THEIR COMBINED EFFECTS

Brebner and Sandow (1976) examined direction-of-turn preferences for rotary controls, located in different positions, mapped to linear displays that consisted of a neutral or directional pointer (see Figure 9.6). Each subject was shown diagrams of all 64 combinations of the display–control layout and control–display movement relation and was instructed to indicate whether the control knob should be turned clockwise or counterclockwise in order to move the pointer to the target, which was the value of 15 on the printed scale. For positions a and b, in which the control was located above or below the display, the results clearly indicated that subjects expected to rotate the control knob so that the side of the control on the same side as the scale moved in the same direction as the pointer. This resulted in subjects expecting clockwise rotation for position a and counterclockwise rotation for position b. The use of a directional indicator that pointed to the values on the scale enhanced the stereotypes compared with a neutral indicator. The stronger effect obtained with the directional pointer is likely due to its providing a cue as to which side of the control knob is to be associated with the pointer and expected to move in the same direction.

When the control was located to the left or right of the display (positions c through h), subjects tended to adhere to Warrick's principle, with this stereotype

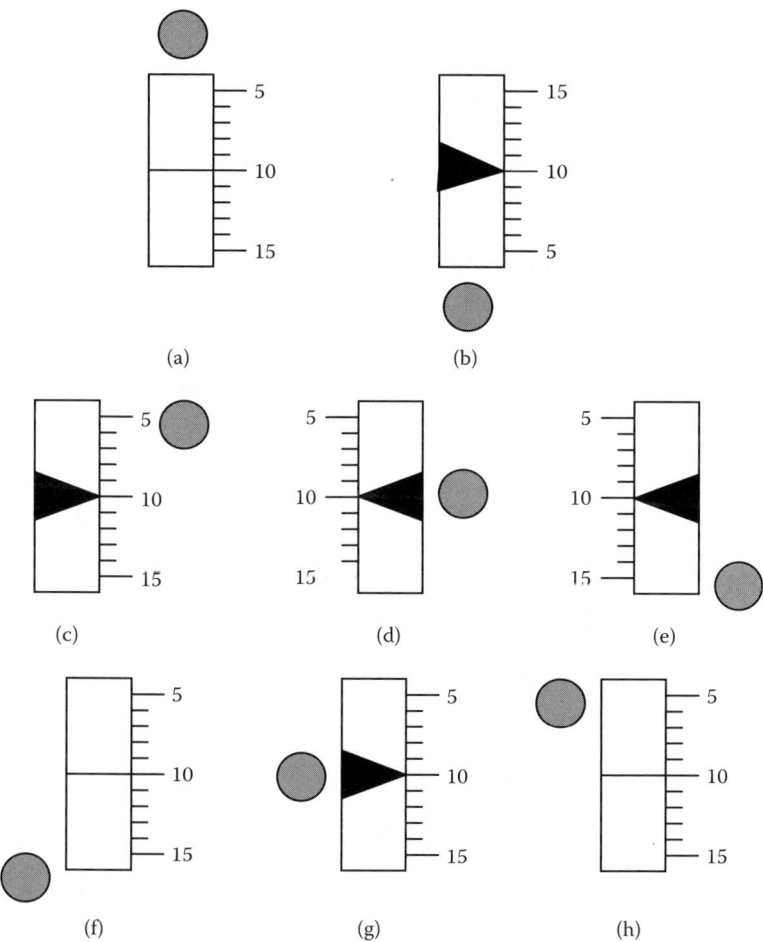

FIGURE 9.6 Examples of the display–control configurations used in Brebner and Sandow's (1976) study. The circles represent the rotary controls located in different positions. Displays a, f, and h consist of neutral pointers and displays b, c, d, e, and g consist of directional pointers.

being stronger when the scale was on the side of the display that was opposite to the control knob. Moreover, males tended to exhibit the stereotypes more strongly than did females. Brebner and Sandow (1976) concluded that two principles contribute to the direction-of-motion stereotypes:

- Warrick's principle (proximity): The scale side of the display and the side of the control nearest to it should move in the same direction.
- Scale-side principle (spatial similarity): The spatially similar right–left half of the control moves in the same direction as the side of the display with the scale markings.

They noted, though, that when the scale is on the same side as the control knob, the two principles conflict. In this case, Warrick's principle is dominant, but its strength is reduced.

Petropoulos and Brebner (1981) conducted a similar study using actual knobs located on the screen where the display indicator was presented, instead of a paper-and-pencil test. They also examined whether the distance (100 mm or 500 mm) between the display and control elements influenced performance and whether the scale printed on the indicator had an effect. Four combinations of scale and pointer direction were examined with each distance, yielding 64 combinations. Each display indicator was presented on the screen for 1.5 s, and subjects were instructed to turn the control knob so that the pointer would move to the top of the display as quickly as possible. They were also told to make a response for each display–control combination presented.

Petropoulos and Brebner (1981) examined the preferred direction of motion as a function of the three following principles:

- Clockwise-to-increase/anything principle
- Warrick's principle (because the displays are oriented vertically, this principle applies only when the control is placed to the left or right of the display)
- Scale-side principle

The distance between the display and control and the presence of the scale had no effect on the preferred control–display movement relation. That is, the same directional relations were obtained regardless of whether the scale was printed on the indicator, suggesting that it was not the scale that was important in Brebner and Sandow's (1976) study, but the direction of the pointer. For the conditions in which the pointers were placed above or below the display, Petropoulos and Brebner (1981) found that people tended to follow the scale-side principle, which replicates Brebner and Sandow's findings. When the scale-side and clockwise-to-increase/anything principles conflicted, the former was slightly dominant, but its preferred movement direction was weakened compared with when both principles coincided. When the control knob was located to the left or right of the display, Warrick's principle could be applied in addition to the scale-side principle. When the two principles were in opposition, Warrick's principle again dominated. The stereotype was strengthened, though, when the scale-side principle was in agreement with Warrick's principle.

Overall, Petropoulos and Brebner (1981) found that subjects turned the knob clockwise 70% of the time, supporting the principle that subjects expect a clockwise movement to result in an increase in the display indicator. An analysis of response times also showed that subjects were faster at making clockwise responses than counterclockwise responses, but this result may be due to the fact that clockwise movement tends to be executed more easily than counterclockwise movement with the right hand, which was used in this study. Based on these findings, Petropoulos and Brebner emphasized the importance of taking into account the different principles rather than adhering to only one. When all principles indicate the same direction of movement relation, a strong response tendency in the predicted direction is exerted, but when the principles conflict, the stereotype is weakened, even if one principle dominates.

Hoffman (1997) also noted that certain principles are only applicable with certain display–control layouts and that the different principles can predict the same or opposing outcomes. He wanted to determine the strength of the various principles and use these weights in a model to predict the stereotype for situations in which different principles are in agreement or are in opposition to each other. In developing his model, Hoffman made two assumptions:

1. The linear sum of the strength of the principles will give the probability that the control will be turned clockwise.
2. If all of the applicable principles are included, the sum of their strengths will equal .50, because by chance alone, the probability of making a clockwise turn will be .50.

Hoffman (1997) discussed four different principles: clockwise-to-increase (CI), Warrick's principle (W), scale-side principle (SS), and clockwise-to-right (CR). He depicted how the different principles would come into play when the linear display is oriented horizontally, the controls are located above or below the center of the display indicator (see Figures 9.7a and Figure 9.7b), and the goal was to move the indicator to the number 10. The formulas for determining the weights are

$$CR + CI + W + SS + .50 = p_1, \text{ for the Figure 9.7a configuration.}$$
$$CI + W + SS - CR + .50 = p_1, \text{ for the Figure 9.7b configuration, because the}$$
$$CR \text{ principle predicts that the pointer should move right, when it needs to be moved to the left to move the indicator to 10.}$$

In the formula, p_1 represents the proportion of times the knob should be turned clockwise, and .50 represents the chance probability of a clockwise turn if no stereotype exists.

In Hoffman's (1997) Experiment 1, 64 slides were made to represent all combinations of control–display relations depicted in Figure 9.7a and Figure 9.7b. The specific display–control combinations were presented randomly, with each being shown for 5 s. Subjects (14 psychology and 23 engineering students) were asked to indicate whether the control should be turned clockwise or counterclockwise to increase the pointer to position 10. Overall, the engineering students tended to indicate fewer clockwise rotations than did the psychology students. Engineering students were more likely to use Warrick's principle, which implies a mechanical linkage, and the psychology students were more likely to use the clockwise-to-right principle.

The likelihood of making a clockwise response for the different display–control configurations was then used to estimate the magnitude of each principle. Hoffmann (1997) formulated mathematical expressions relating the extent to which the relative contributions of the different stereotypes influence the probability that the knob

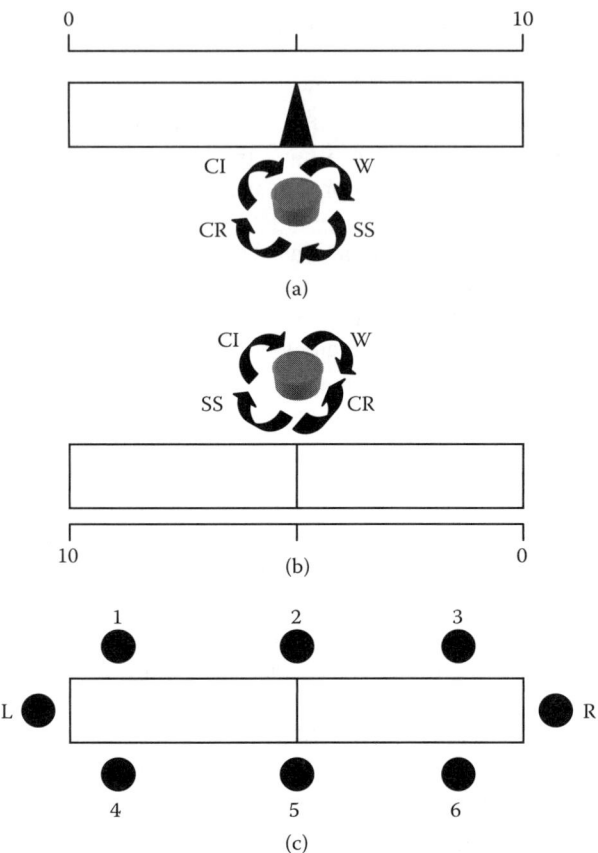

FIGURE 9.7   Illustration of the display and control arrangement used by Hoffman (1997). A directional pointer is used in panel a, and neutral pointers are used in panels b and c. The arrows surrounding the circular control knob depict the expected direction of motion relation for the four stereotypes of clockwise-to-increase (CI), Warrick's principle (W), scale-side principle (SS), and clockwise-to-right (CR). Panel c illustrates the different positions of the control knob relative to the display.

would be turned clockwise for the different configurations. If a stereotype predicted clockwise movement, then it would be given a positive value; if not, it would be given a negative value. As aforementioned, the sum of the relative strength of each stereotype should be .50. For seven of eight display–control combinations examined, the sums of the magnitudes were between .464 and .512; with the other one, the sum was .348. Thus, Hoffman concluded that the model provided a good fit to the data. Overall, the strength of the clockwise-to-right principle was larger for psychology students (.43) than for engineering students (.23). However, the strength of Warrick's principle was smaller for psychology students (.07) than for engineering students (.27). Interestingly, the clockwise-to-increase principle was found to be weak, but reversible (i.e., clockwise = increase; counterclockwise = decrease). Hoffmann also showed that the model was able to provide a good fit of unpublished data, provided by Brebner, for a more general subject population.

In Hoffman's (1997) Experiment 2, one group of subjects viewed three-dimensional drawings of 24 display–control relationships. Four of these display–control drawings were shown to subjects from four different viewpoints, 12 from two different viewpoints, and 8 from a single viewpoint. Another group of subjects was tested with an actual device for which they had to turn a knob to move a pointer to the target location. In addition to the four principles examined in Hoffman's Experiment 1, a new principle was added: the unfolded scale-side principle. This principle is similar to the scale-side principle except that it arises from an "'unfolding' of the device about the line joining the planes containing the control and display" (p. 209). For the paper-and-pencil test, Hoffman found that the viewpoint had an effect on the expected movement of the control. This influence was strongest for the clockwise-to-increase/anything principle, for which the proportion of clockwise responses changes systematically from .08 to .72 as the angle of view changed from 70° left to 70° right. The viewpoint had less of an effect when Warrick's principle was applicable, changing only from .33 to .35 as the angle of view changed from 30° left to 30° right. The influence of viewpoint for paper-and-pencil tests indicates that caution needs to be exerted when such tests are designed to capture the direction-of-motion stereotypes for three-dimensional display–control devices. Hoffman's model was able to provide a good fit for the data, accounting for 70% of the stereotype when Warrick's principle was applicable and between 32% and 66% in the other cases. When Warrick's principle was not applicable, no other dominant stereotype emerged.

For performance with the actual device, Hoffman (1997) found that for arrangements in which Warrick's principle was not applicable, the clockwise-to-increase/anything preference was evident (a) when the display was on top of the rectangular device, (b) with an increasing scale, and (c) when the control was in front of it. When Warrick's principle was applicable, it was dominant for most cases, accounting for 86% of the total strength, and showed reversibility when the decreasing scale was used. The other principles yielded negligible effects. However, there were two exceptions to this finding: When the display was increasing from left to right and the display or control was located in either the top or front location, the clockwise-to-right principle accounted for 23% of the stereotype and Warrick's principle for 68%. Based on these findings, Hoffman concluded that three-dimensional display–control relations should be designed in accordance with Warrick's principle and care should be taken to ensure that the clockwise-to-right principle does not oppose Warrick's principle because it decreases the strength of the stereotype.

Hoffman's (1997) Experiment 3 used response time to measure the natural relation between 64 two-dimensional display–control relations that were projected onto a screen, illustrated in Figure 9.8. All subjects were engineering students, and they were instructed to decide as quickly as possible in which direction the control should be turned to move the pointer to 10, the target position, and then to rotate the control knob in that direction. Four principles were examined: clockwise-to-right/up, Warrick's, scale-side, and clockwise-to-increase/anything. Overall, Warrick's principle, when applicable, was the most important for determining the overall

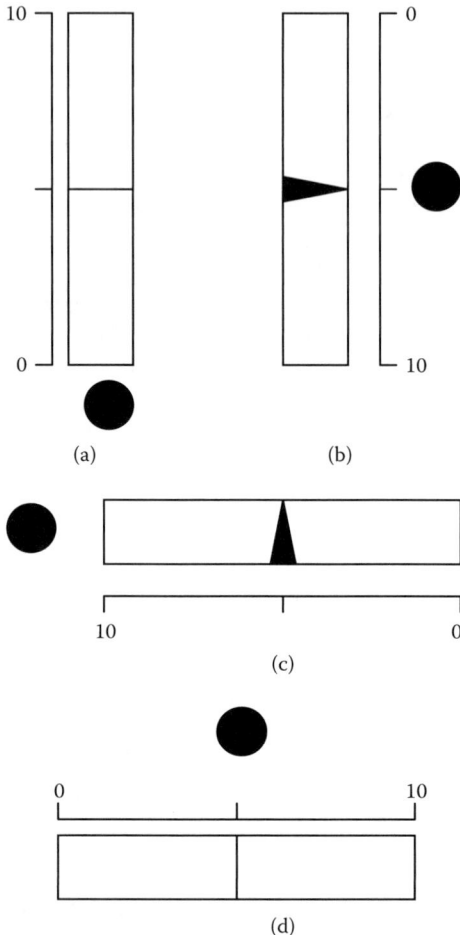

FIGURE 9.8 Horizontal and vertical display–control configurations used in Hoffman's (1997) Experiment 3. The circle represents the control knob, the pointers in configurations a and d are neutral, and b and c pointers are directional.

strength of the stereotype. Moreover, Hoffman's model provided a better fit for the horizontal displays than for the vertical displays. The influence of the scale-side principle was small for horizontal displays, but was larger for vertical displays, especially when Warrick's principle was not applicable. The clockwise-to-right principle was important for horizontal displays, but the clockwise-to-up principle was not as strong for vertical displays. The strength of the stereotype, as estimated by Hoffman's model, showed a linear relationship with response time only for horizontal displays when the controls were located at the top or bottom. A subsequent analysis showed that response time correlated well ($r = .96$) with the principle that had the greatest strength for the display–control configuration, indicating that the other stereotypes were ignored. Thus, speeded choices are primarily a function of the dominant stereotype, whereas unspeeded choices are also influenced by the non-dominant stereotypes.

9.2.5 CROSS-CULTURAL COMPARISONS

It is important to determine which stereotypes hold across cultures and which do not. A product or system that is designed to conform to the stereotypes of one culture may not conform to those of another. Because the increasingly global nature of the economy implies that people from different cultures may use products originally designed for one specific culture, differences in expectancies may lead to poor usability for populations other than the one for which the product was originally designed.

Wong and Lyman (1988) compared preferred direction-of-motion relations of American and Japanese right-handed subjects for a vertical display pointer and a rotary control knob by using a paper-and-pencil test similar to Brebner and Sandow's (1976) study. The instructions on the test were in the native language of the subjects, and they were asked to indicate which direction they would turn the control knob to move the display pointer in the instructed direction. Wong and Lyman examined the strength of the stereotype when the clockwise-to-increase/anything principle, Warrick's principle, and scale-side principle were in agreement versus when they were not. Of the 24 control–display arrangements tested, only one yielded a stereotype that was the same for both the American and Japanese subjects. For this arrangement, for which all three of the aforementioned principles were applicable and in agreement, subjects indicate a reversible stereotype of clockwise movement to increase the pointer value and counterclockwise movement to decrease it. When the scale-side and Warrick principles conflicted, both groups of subjects also showed a stereotype that was consistent with Warrick's principle, although the dominance of Warrick's principle was stronger for American than Japanese subjects.

Courtney (1988) examined direction-of-motion stereotypes with a Chinese population. He used a paper-and-pencil test similar to that described earlier by Ross et al. (1955), with three-dimensional drawings of display–control devices. When the control was rotary and arrayed in the horizontal and frontal planes, strong reversible stereotypes were obtained for clockwise-right/counterclockwise-left, replicating Ross et al. A clockwise-for-anything stereotype also emerged for responses requiring an up/down movement of the dot, except in the sagittal plane, where subjects preferred the counterclockwise-to-down relation. Courtney also confirmed that Warrick's principle was supported for configurations to which it was applicable. When Warrick's principle did not apply, there was a strong clockwise-for-anything response. Courtney compared his findings with those reported by Ross et al. and Loveless (1962) and found that Chinese subjects preferred the clockwise-for-anything stereotype more than did their American counterparts.

Courtney (1994b) conducted a similar study with Hong Kong Chinese subjects. When responses were made with a rotary control, a clockwise-to-right stereotype was obtained in all three planes, but a counterclockwise-to-left stereotype was evident only when the knob was in the horizontal and sagittal planes. The clockwise-to-up stereotype was obtained for control knobs in all three planes, but a similar clockwise-to-down stereotype was evident when the knob was in the horizontal and frontal planes. Thus, similar to the Chinese subjects in Courtney's (1988) study, the Hong Kong Chinese preferred the clockwise-for-anything stereotype. Where Warrick's principle was applicable, the results were consistent with it. These studies also examined preferred direction of movement for push–pull knobs and lever controls. However, the findings obtained with these controls were less clear-cut and were highly dependent upon the plane in which the control was arrayed.

Courtney (1994a) conducted a study with horizontal and vertical displays and rotary controls to examine the preferred direction of motion of Hong Kong Chinese students when the Warrick's, scale-side, and clockwise-to-increase/anything principles coincide and when they do not (see also Courtney, 1992). Subjects performed the task on an acrylic box apparatus in which the display was projected and the control mounted. Eight slides were used in which the pointer type (neutral or directional), scale side (left or right of the display), and direction of increase on the scale (ascending or descending) were used for all three knob locations (horizontal, frontal, or sagittal plane), yielding 24 combinations. Subjects were instructed to turn the knob to increase the value on the display (e.g., from 10 to 15). For both horizontal and vertical displays, strong stereotypes emerged in the expected direction when Warrick's principle and the scale-side principle were in agreement. However, when the principles conflicted and neither principle dominated, there was a tendency to make the clockwise movement for anything. This finding, though, is in contrast to that of an earlier study by Courtney (1992) in which, with vertical displays, Warrick's principle was weakened, but still dominated when the two principles conflicted.

Chan, Shum, Law, and Hui (2003) used a control panel with eight knob positions, similar to that used by Hoffmann (1997; see Figure 9.7c, with L corresponding to position 8 and R to position 7 in the present study), and a horizontal display with a directional or neutral pointer. The scale was marked at the top or bottom of the display. Chinese subjects were asked to rotate the knob to move the pointer in the center of the display to the left or right. Direction of motion and response time were recorded. Overall, there was a strong tendency to respond clockwise to move the pointer right, and this stereotype was somewhat stronger than its opposite, counter-clockwise to move the pointer left. When the clockwise-to-right and Warrick's principles were in agreement, strong stereotypes were obtained and responses were made quickly. However, when the two principles conflicted, or when Warrick's principle was not applicable, the response times for clockwise-to-right stereotypes were shorter due to the fact that the clockwise-to-right principle was in effect.

In summary, certain population stereotypes, such as Warrick's principle, hold across different cultures. However, the dominance of Warrick's principle when other principles conflict with it tends to be stronger for American subjects than for Chinese subjects, with the latter preferring the clockwise-for-anything principle. Moreover, for both Chinese and Japanese subjects, there was a stronger preference for the clockwise-for-anything principle when Warrick's principle was not applicable. For horizontal linear displays and rotary controls, American (Ross et al., 1955), Chinese (Courtney, 1988), and Hong Kong Chinese subjects (Courtney, 1994a) all showed reversible stereotypes of clockwise-to-right and counterclockwise-to-left. However, whereas the American subjects showed a counterclockwise-to-down stereotype, the Chinese subjects preferred the clockwise-for-anything stereotype. Thus, designers need to be aware that when they use a specific population of users to determine direction-of-motion guidelines, some of the results may not generalize to different populations.

9.3 ROTARY DISPLAYS AND THEIR CONTROLS

9.3.1 FUNDAMENTAL PRINCIPLES

Warrick (1948, as described by Loveless, 1962) used a paper-and-pencil test to examine whether clockwise movement of a control knob is expected to yield clockwise movement of a circular display. The test consisted of 32 diagrams, with a 90° arc of circular display, which represented one fourth of a circle, and a circular control knob. Each of the four quadrants of the circular display was examined, with the control knob being position to the left, right, above, or below the display. Within each arrangement type, one diagram required the pointer to be moved in the clockwise direction to reach the target reference mark, and the other counterclockwise. For all arrangements, there was a clockwise-to-clockwise stereotype. Subsequent behavioral studies, though, found that certain quadrants of the circular dial yielded larger clockwise-to-clockwise stereotypes than others (Fitts & Simon, 1952; Graham, 1952).

Warrick (1949, as reported by Loveless, 1956) had subjects switch systematically between controlling two knobs that varied with respect to their positions relative to the display indicator (see Figure 9.9) in three different experiments. There were four display–control relationships for the pair of displays and controls: direct (clockwise movement of the control produces clockwise movement in the display)–direct; reversed (clockwise movement of the control produces counterclockwise movement in the display)–reversed, direct–reversed, and reversed–direct. Operators performed faster and more accurately with consistent display–control relations between pairs than with inconsistent relations. Also, within the consistent relation, better performance was evident for the direct–direct pair than for the reversed–reversed pair. The advantage for maintaining consistent relations is similar to the benefit on dual-task performance of maintaining consistent mappings for the two tasks, as described in Chapter 10.

In Experiment 4, the display arrangements were different from each other, and this resulted in a different pattern of results (see Figure 9.9): The reversed-reversed relation yielded the worst performance, and there was no difference between the other three combinations. Warrick (1949) interpreted this finding as support for his principle: A display pointer is expected to move in the same direction as the part of the control nearest to the display. However, Murrell (1952, cited in Loveless, 1956) pointed out that in Warrick's Experiment 4, the best performance was obtained with the reversed–direct relation, which does not necessarily conform to Warrick's principle. Because this condition did not differ significantly from the direct–direct condition, Loveless noted that it does not discredit Warrick's conclusions.

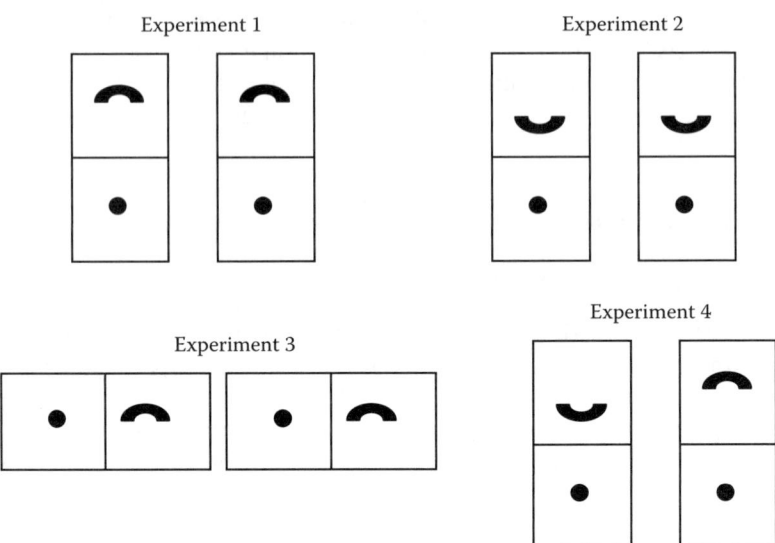

FIGURE 9.9 Depiction of the pairs of displays and controls used in Warrick's (1949, as reported by Loveless, 1956) Experiments 1 through 4. Displays are depicted by half circles and control knobs by solid dots.

Simon (1954, described in Loveless, 1962) examined performance of the upper and lower quadrants of a circular display when the display was controlled by an upward-pointing lever that could be rotated clockwise or counterclockwise. Results showed that, for the upper quadrant, performance was better when the pointer moved in the same direction as the lever, but for the lower quadrant, there was no difference between the control–display relations. This latter finding suggests that there is ambiguity regarding the directional relationship between circular pointer displays and lever controls in that lever controls can be regarded as translatory or rotary.

Fitts and Simon (1952, as described in Loveless, 1956) suggested that when the display–control relation is ambiguous, responses can be made with respect to a *rectangular-coordinate hypothesis* (i.e., clockwise rotation at 9 o'clock is up, whereas clockwise rotation at 3 o'clock is down) or a *polar-coordinate hypothesis* (i.e., angular rotation). When the target is at the top of the scale, the pointer movement along both coordinates leads to the same response, but when it is at the bottom of the scale, then conflicting responses may arise depending upon rotary or translatory (up/down or left/right) coding tendencies.

Loveless (1956) conducted an experiment to compare performance using different display–control configurations to examine whether people have rotary or translatory coding tendencies. Tracking error scores were obtained for four conditions (see Figure 9.10 for illustrations of the conditions), in which the targets were placed at the top or bottom of a circular scale, and the control knob was placed directly below it. Periodically, the pointer shifted position in a step-like manner, and the subject was to counteract the movement by rotating the control to keep the pointer on the target. Four relations were tested:

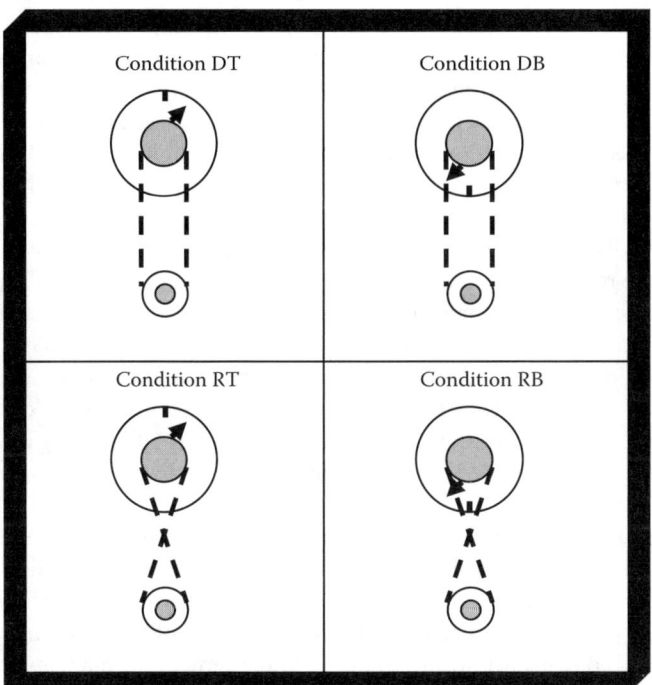

FIGURE 9.10 Illustration of the display and control configuration used by Loveless (1956). The control knob (depicted by double circles) was designed to move the pointer from the target, and the subject was to counteract the movement to keep the pointer on the target. There were four directions of motion conditions: DT (direct top), DB (direct bottom), RT (reverse top), and RB (reverse bottom).

- DT (direct top), in which the target is at the top and the pointer moves to the right/clockwise; clockwise or right movement of the knob increases the error. In this case, the rotary and translatory coding tendencies operate in the same direction.
- DB (direct bottom), in which the target is at the bottom and the pointer moves to the left/clockwise; clockwise movement of the knob increases the error and left movement decreases the error. In this case, the rotary and translatory coding tendencies conflict.
- RT (reverse top), in which the target is at the top and the pointer moves to the right/clockwise; clockwise or right movement of the knob decreases the error. In this case, the rotary and translatory coding tendencies operate in the same direction.
- RB (reverse bottom), in which the target is at the bottom and the pointer moves to the left/clockwise; clockwise movement of the knob decreases the error and left movement increases it. In this case, the rotary and translatory coding tendencies conflict.

Loveless (1956) found that performance was better for the DT condition than for the other three conditions. Estimates of component strength indicated that subjects responded primarily to the rotary motion of the pointer, with a lesser propensity to respond with respect to left and right. There also was a tendency for better performance when the pointer was at the top of the scale than when it was at the bottom.

In Loveless's (1956) second experiment, the targets were placed on the left or right of the scale, instead of in top or bottom positions as in Experiment 1. In this situation, the DL (direct-left) condition yielded the best performance, DR (direct-right) intermediate performance, and the two reversed relations the worst performances. These results again indicate that there is a strong tendency to expect a direct clockwise–clockwise movement relation. There was also a smaller preference for clockwise control movement to move the pointer upward. Based on both of his experiments, Loveless concluded that (a) performance improves as the target moves from the bottom of the scale to the top, (b) clockwise movement of a control is expected to yield clockwise movement in a pointer, and (c) depending on the location of the target, clockwise movement of a control is expected to yield upward or right movement in a pointer.

Loveless (1956) conducted a third experiment to determine whether the preferred direction-of-motion relations described above generalize to linear displays as well as rotary ones. The control knob was directly below the display. There were four control–display relations in which clockwise movement of a control moved a display pointer upward, downward, to the right, or to the left. Performance was best when the pointer moved right, intermediate when it moved up, and worst when it moved left or down. Loveless also showed that clockwise rotation of a control knob is better associated with upward movement or movement to the right than downward movement or movement to the left. He compared the results obtained for horizontal and vertical scales when the control knob was located directly under the display indicator and found that the difference between the preferred (i.e., clockwise = right or up) and nonpreferred (i.e., clockwise = left or down) control–display relation was larger for the horizontally arrayed than vertically arrayed displays.

Loveless (1959) conducted another experiment to determine whether the preferred association of clockwise with right and up varies as a function of the position of the display pointer and control knob. The control knob was placed to the right of the pointer display, and subjects performed a tracking task with one of the four different display–control relationships (i.e., clockwise control movement resulted in right, left, upward, or downward pointer movement). The results were similar to those obtained when the control knob was located directly under the display pointer, with clockwise knob movement being better for upward pointer movement or movement of the pointer to the right. Loveless (1959) reported that Warrick (1947) obtained similar results for the vertical scale. The major difference in findings obtained with the control knob placed to the right of the display instead of below it was that there was no significant difference in the strength of preferred and nonpreferred movements for horizontally and vertically arrayed displays.

One reason there was a greater difference in the strength of the preferred display–control relation when the control knob was placed directly below the display is that with the horizontal display, the line of movement passed to the side of the knob, whereas with the vertical display, it passed through the center of the knob. Loveless (1959) suggested that this result pattern could be explained by Warrick's principle, in which an indicator of a display is expected to move in the same direction as the nearest point of the control.

9.3.2 THUMBWHEEL STEREOTYPES WITH A CHINESE POPULATION

Chan, Courtney, and So (2000) examined direction-of-motion stereotypes for circular displays mapped to horizontal or vertical thumbwheel responses arrayed in four different planes (see Figure 9.11). Each display–control layout was presented indi-

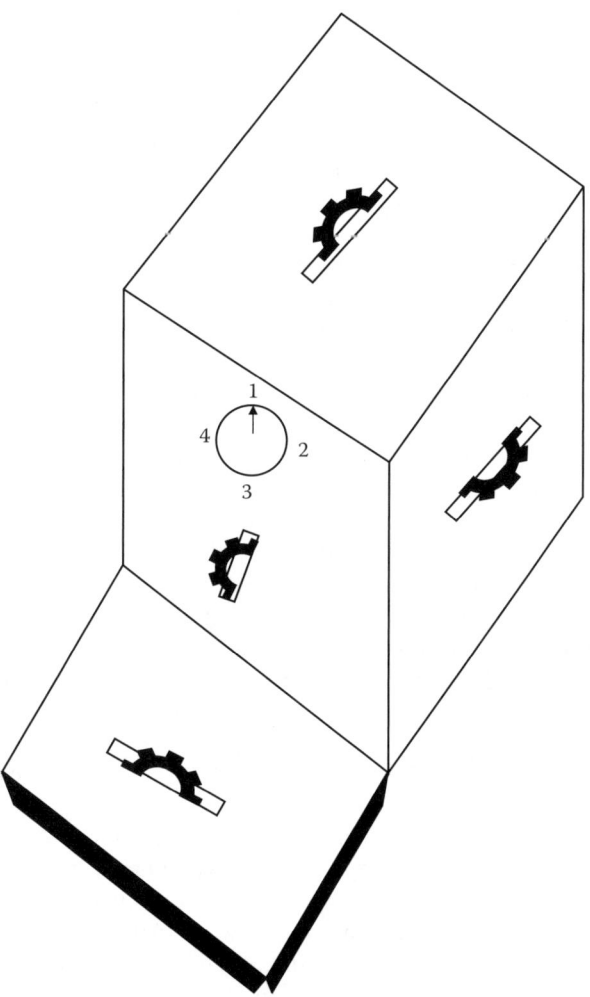

FIGURE 9.11 Illustration of the display–control layout used by Chan, Courtney, and So (2000). The black gears represent the thumbwheels that could be arrayed in one of four planes.

vidually on a computer screen and subjects were to indicate which way to rotate the thumbwheel by pressing the left or right arrow key for horizontally oriented thumbwheels or the up or down arrow keys for vertically arrayed thumbwheels. The pointer was placed at one of four cardinal positions (12, 3, 6, or 9 o'clock) and was to be moved clockwise or counterclockwise. Forty Hong Kong Chinese students participated, and half were given the vertical thumbwheel first and half the horizontal one first. They were tested with all 32 combinations of planes (4), cardinal positions (4), and direction-of-movement instructions (2). Response times averaged 279 ms for horizontal thumbwheels and 240 ms for vertical thumbwheels. Responses were faster when the subjects were instructed to turn the display clockwise than counterclockwise.

For the horizontal thumbwheel responses, Chan et al. (2000) found no difference in the total number of left and right control movements. There was a left-for-counterclockwise and right-for-clockwise stereotype when the pointer was positioned at 12 o'clock. The reverse pattern (left-for-clockwise and right-for-counterclockwise) was obtained when the pointer was positioned at 6 o'clock, except for when the thumbwheel was positioned at the top of the box. When the pointer was at the 3 o'clock position, a left-to-clockwise stereotype emerged, but it was reversible only when the thumbwheel was positioned at the top. Similarly, when the pointer was at the 9 o'clock position, there was a strong stereotype for left-to-clockwise that was reversible only when the thumbwheel was arrayed on the base of the apparatus in the horizontal plane or on the side in the sagittal plane.

For the vertical thumbwheel responses, when the pointer was positioned at 3 o'clock, up-to-clockwise and down-to-counterclockwise stereotypes were found in all planes. The opposite was true when the pointer was positioned at 9 o'clock: Reversible stereotypes of down-to-clockwise and up-to-counterclockwise were found in all planes. When the pointer was at the 12 o'clock position, there was a tendency to move the thumbwheel up for anything. Similarly, when the pointer was at the 6 o'clock position, there was a tendency to move the thumbwheel down for anything. Thus, for both horizontal and vertical thumbwheels, strong and reversible stereotypes were obtained when the pointer was perpendicular to the thumbwheel's axis of motion. Designers should use a horizontal thumbwheel when the pointer position is resting at 6 or 12 o'clock, and a vertical thumbwheel at 3 or 9 o'clock, at least when the pointer needs to move 90° clockwise or counterclockwise.

9.4 OTHER POPULATION STEREOTYPES

9.4.1 Color Associations

Color coding can be used to convey meaning (e.g., red = danger) about the state of a system or display. Bergum and Bergum (1981) examined color coding using a paper-and-pencil test to assess the meaning that 127 Americans assigned to colors. They gave subjects 12 concepts (caution, cold, danger, far, go, hot, near, off, on, radiation, safe, and stop) and asked them to indicate which of six colors (yellow, blue, red, green, orange, and blue) best represented the concept. Results showed that the color red was highly associated with stop (100%), hot (94.5%), danger (89.8%), radiation (59.1%), and on (50.4%), and the color green with go (99.2%) and safe (61.4%). The color blue was associated with cold (96.1%) and off (31.5%), yellow with caution and near (81.1% and 38.6%, respectively), and purple and blue with far (34.6% and 30.7%, respectively). It is important to note that the association of red with stop and green with go was 100%, or close to it. Moreover, the three colors of red, green, and blue were the colors that were associated with the greatest number of concepts. Bergum and Bergum suggested that this bias toward using these colors might be a result of the sensitivity of the cones to the colors red, green, and blue.

Courtney (1986) conducted a study to examine color associations for a Chinese population. Subjects were given eight colors (red, orange, yellow, green, blue, purple, black, and white) and nine concepts (safe, cold, caution, go, on, hot, danger, off, and stop) and were asked to indicate the association between the concepts and colors. Results showed that the colors red and green yielded the highest associations, with red being most strongly associated with danger (64.7%), stop (48.54%), and hot (31.1%), and green with safe (62.2%), go (44.7%), and on (22.3%). Moreover, for each other concept, there was also at least one color that was strongly and significantly associated with it: cold–white (71.5%), caution–yellow (44.8%), and off–black (53.5%).

In comparison with the American population stereotypes for color associations obtained by Bergum and Bergum (1981), the Chinese population stereotypes were similar for safe–green, caution–yellow, go–green, hot–red, danger–red, and stop–red. It is interesting to note, though, that even though both populations shared the aforementioned color associations, they were stronger for the Americans than the Chinese in most cases. The most notable examples are for stop and go, which the Americans unambiguously associated with red (100%) and green (99%), respectively; these associations were weaker in strength for the Chinese population (being around 50%).

Chan and Courtney (2001) conducted a study of color associations with Hong Kong Chinese subjects using an elaborated version of Courtney's (1986) test, with 10 colors (addition of pink and gray) and 16 concepts (addition of potential hazard, radiation hazard, soft, hard, weak, strong, and normal). Subjects were presented with a paper-and-pencil test. Consistent with prior studies, the colors red and green yielded the most associations. Red was significantly associated with stop (66%), danger (63%), caution (40%), potential hazard (26%), radiation hazard (27%), and strong (27%). Green was significantly associated with go (63%), safe (38%), and on (24%). The colors yellow and purple were not significantly associated with any of the concepts, but white was associated with two concepts: normal (25%) and off (23%). The remaining color–concepts associations were hot–orange (28%), cold–blue (23%), soft–pink (21%), weak–gray (26%), and hard–black (46%).

The colors of red, green, and blue were associated with over half of the concepts tested in Chan and Courtney's (2001) study. More importantly, with the Hong Kong Chinese, red was used as a high associate for danger and caution or potential danger. This finding suggests that the Hong Kong Chinese do not differentiate between these two concepts, whereas the Yunnan Chinese (Courtney, 1986) and Americans (Bergum & Bergum, 1981) do tend to differentiate these concepts by associating danger with red and caution with yellow. Across all three populations surveyed, red is associated with danger and stop, and green is associated with go and safe. Green is also highly associated with on for the two Chinese populations, but not with the American population, who associated on with red. It should be noted, though, that for both Chinese populations, red was also associated with on, but to a lesser degree than green.

9.4.2 WORD AND PICTURE ASSOCIATIONS

The chapter began with examples of display–response associations for pictures of everyday objects (Bergum & Bergum, 1981). Smith (1981) also evaluated population stereotypes for words and pictures, but included three different groups of subjects: 92 male engineers, 80 women (wives, daughters, friends of the engineers, and secretaries), and 55 human factors specialists (mostly male; attendees of the Human Factors Society's 1977 annual meeting). He found that some stereotypes were robust across the three groups. These stereotypes included the counterclockwise-to-left stereotype and the clockwise-to-increase stereotype for rotary knobs, and the preference for numbers ascending from left to right when numbered keys were to be pressed with all 10 fingers of the left and right hands.

However, Smith (1981) also found marked differences between picture and word associations for the three groups. For example, when asked what the ordering should be for the four quadrants of a circle, engineers preferred a clockwise ordering from the upper right (33%) or counterclockwise ordering from the upper right (34%). About half of the human factors specialists also preferred the ordering of clockwise from the upper right (45%), but the other half preferred a reading order (left-to-right, top-to-bottom; 43%). Moreover, the majority of women (54%) preferred the reading order for labeling of the quadrants. Another example involved the statement "left wing down." A majority of the women (64%) indicated that the words represented a command to lower the left wing, as did approximately half of the engineers and human factors specialists (48% and 49%, respectively). However, the other half of the latter two groups interpreted the statement as an error of the left wing being too low, for which the proper response of the pilot should be to raise the left wing, whereas only 36% of the women did.

As a final example, when the subjects were given a depiction of highway lanes (see Figure 9.12) and asked to indicate whether A or B was the outside lane, Smith (1981) found that 80% of human factors specialists indicated that B was outside, whereas half of the engineers and women subjects picked A and half picked A and B. These findings are in agreement with Hoffman's (1997) findings that different groups of professionals may show different preferences due to their understanding of different perception–action relations and their experience with different display–control configurations. Taken together, these findings indicate that population stereotypes may be specific not only to a culture, but to groups of individuals within the culture as well.

9.5 CHAPTER SUMMARY

Population stereotypes that emerge for specific display–control configurations can be used to help designers ensure that the display–control relations for a particular

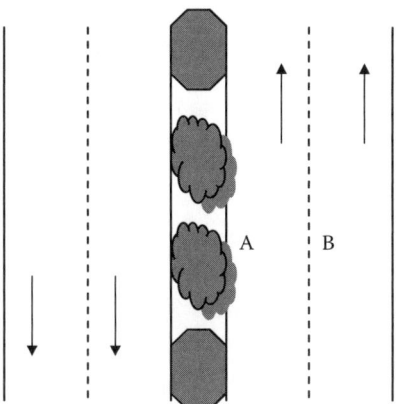

FIGURE 9.12 Picture of highway lanes similar to the one used by Smith (1981). Subjects were to indicate whether Lane A or B is the outside lane.

design are in agreement with the expectancies and preferences of the target population. Although much of the literature established the stereotypes by examining the stated preferences of samples from target populations, some studies have used performance measures, such as error rates. It should be noted, though, that self-reports of strong preferences for certain display–control relations may not always guarantee better performance. As discussed in Chapter 12, people sometimes indicate a preference for display–control layouts that do not result in the best performance. Moreover, stereotypes obtained with a specific population may not generalize to another population. Designers should take into account well-established population stereotypes, but also measure performance with the possible alternative display–control configurations to ensure maximal usability.

10 Stimulus–Response Compatibility Effects in Dual-Task Performance

10.1 INTRODUCTION

Many jobs involve performing two or more tasks concurrently. For example, air traffic controllers must coordinate the movements of many airplanes as they move through the controllers' designated sectors so that the planes are at safe separations from one another and are following the correct flight paths. At the same time, the controllers have to ensure that the pilots are receiving proper instructions about detours to avoid bad weather and turbulence and that the pilots are able to take off and land in an orderly fashion (Freudenrich, 2005). Multiple tasks not only increase the information processing demands placed on the operator, but they also provide an opportunity for various interactions between tasks to occur. Moreover, performance of multiple tasks requires coordination of processes for the tasks that is not necessary when each task is performed in isolation. Because response selection and other central-processing requirements are more complex in situations that require multiple-task performance, issues of S-R compatibility become more multifaceted. Maintaining compatibility of the stimuli and responses within each task remains important, but, in addition, the relation between the stimuli and responses across the tasks is critical and can affect performance in several ways.

The literature on multiple-task performance is immense, and we will not report a comprehensive review of it here. Our focus is primarily on compatibility effects, which have been studied mainly in the context of dual-task performance. Issues of concern include the extent to which highly compatible tasks can be performed together with little or no cost, whether emergent features that arise from the combination of task components can be used to simplify response selection, and the occurrence of Simon-like cross-task correspondence effects between tasks. Much of our discussion centers on what is called the psychological refractory period (PRP) effect, which has a long history of intensive research within experimental psychology (see Pashler, 1994, and Pashler & Johnston, 1998, for reviews).

10.2 THE PRP EFFECT AND THE CENTRAL BOTTLENECK MODEL

The PRP effect was given its name by its discoverer, C. W. Telford (1931), in an article titled, "The Refractory Phase of Voluntary and Associative Responses." Telford noted that considerable evidence indicated that neural tissue exhibits a period of decreased excitability to stimulation, called the refractory period, and that a refractory phase had been found for several reflexes. However, he pointed out, "So

223

far, no standard technique has been developed for investigating the refractory period of voluntary processes. It is not known whether such a phenomenon exists and plays a part in complex behavior" (p. 6).

Telford's (1931) study had the goal of demonstrating the existence of a refractory period effect for voluntary processes. In his first experiment, Telford measured simple RT to tones presented at onset intervals of 500 ms, 1 s, 2 s, and 4 s. His primary finding was that RT was longer at the 500-ms interval (335 ms) than at the other three intervals (241, 245, and 276 ms, respectively). Telford concluded,

> These results indicate that immediately after responding to a simple auditory stimulus by pressing a telegraph key there is a period of intrinsic unreadiness for response as shown by lengthened reaction times to stimuli applied during this interval. This period is comparable to the refractory phase of more elementary systems. (p. 12)

In another experiment, Telford (1931) demonstrated a higher error rate for shorter–longer judgments of relative line length when successive pairs of lines were presented at 500-ms intervals than when the intervals were longer. Although Telford was studying sequential effects under conditions in which the shortest interval allowed the response to one stimulus to be completed before the onset of the next stimulus, rather than dual-task performance, and his interpretation of his results as analogous to neural refractoriness has not stood the test of time, his work spawned a plethora of studies on the PRP effect.

A variety of accounts of the PRP effect have been proposed, but the most resilient over the years has been the central bottleneck model, sometimes called the response–selection bottleneck model. Vince (1947) proposed such an account to explain results from experiments that used a method that is a step closer to the contemporary PRP procedure, although her task still examined sequential effects. Subjects were to track, with a pointer, a continuous horizontal line moving through a window, of which only a small portion was visible at any moment. Periodically, a pair of "stimuli" occurred, consisting of a discrete step of the line up or down, followed during an interval of 50 to 1,600 ms by a horizontal line, and then a step in the opposite direction. RT to the second stimulus in the pair was shortest when the interval was 1,600 ms and became progressively longer as it decreased to 50 ms, with RT being nearly twice as long at that interval (510 ms) as at the 1,600-ms interval (260 ms).

To explain her results, Vince (1947) said,

> A possible hypothesis is that the delay caused by the reaction time in sensori-motor action is largely occupied by some central 'computing' system, during the course of which the appropriate response is selected and organized. If, during this time, further sensory impulses were received, the 'computing' process would be disturbed, and the response disorganized; this appears to be prevented from happening by the fact that the reception of one stimulus renders the sensori-motor system temporarily refractory to stimuli of a similar kind. (p. 156)

The central bottleneck model is associated most strongly with Welford (1952), who reviewed the existing evidence on the PRP effect and concluded that it was most consistent with a central bottleneck:

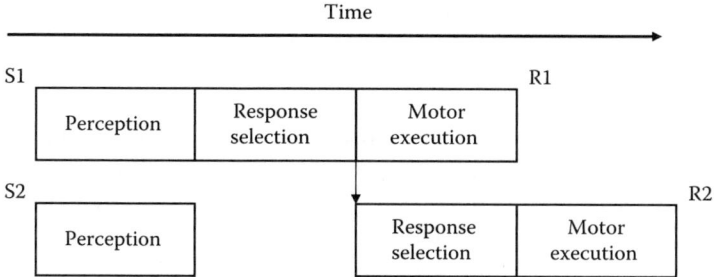

FIGURE 10.1 The central response–selection bottleneck model. The downward-pointing arrow indicates that response selection for Task 2 cannot proceed until response selection for Task 1 is completed.

> The evidence would seem clearly to indicate that the refractoriness is in the central mechanisms themselves, and the theory has been put forward that it is due to the central processes concerned with two separate stimuli not being able to co-exist, so that the data from a stimulus which arrives while the central mechanisms are dealing with data from a previous stimulus have to be 'held in store' until the mechanisms have been cleared. (p. 3)

Although the results considered by Welford (1952) in his classic article advocating the central bottleneck model were based on single-task performance, the model has continued to receive emphasis in the now-standard dual-task paradigm for investigating the PRP effect. Elithorn and Lawrence (1955) introduced this paradigm, but because their tasks involved simple rather than choice reactions, we will not describe their study here. In the PRP paradigm, two different tasks, Task 1 and Task 2, are performed on each trial. The interval between the onset of the stimulus for Task 1 (S1) and that for Task 2 (S2), called the stimulus onset asynchrony (SOA), is varied (either randomly or between blocks of trials). Reaction time for the first task (RT1) often does not vary much with SOA, but that for the second task (RT2) typically increases at shorter SOAs, that is, as S1 and S2 occur closer together in time. This increase in RT at short SOAs is defined as the PRP effect.

The central response–selection bottleneck model has been advocated most vocally in recent years by Pashler (1984, 1994) and colleagues. According to this model (see Figure 10.1), stimulus identification and response execution for Tasks 1 and 2 can occur in parallel, but response selection for the two tasks must operate serially. The response–selection bottleneck model implies the following: Because Task 1 has priority, processing of S1 begins immediately upon its presentation. Identification of S2 also begins upon its presentation, regardless of how soon it occurs after the onset of S1. At long SOAs, response selection for Task 1 is completed by the time that identification of S2 is finished, so response selection for Task 2 can begin immediately. However, at short SOAs, response selection for Task 1 may not yet be completed, in which case initiation of response selection for Task 2 must be delayed. Consequently, at short SOAs, there is "slack" in the processing sequence for Task 2 between the completion of stimulus identification and initiation of response selection (Schweickert, 1983). This slack can absorb, at least in part, an increase in time to identify S2.

The central bottleneck model makes several predictions that have been confirmed in at least some studies. Because response selection for Task 1 must be completed before that for Task 2 can begin, any lengthening of the time for Task 1 stimulus identification or response selection will lead to a larger PRP effect; conversely, any reduction of the total duration for those Task 1 processes will reduce the PRP effect. For example, providing subjects with practice at Task 1 reduces RT for that task (RT1), at least in part by improving response–selection efficiency, resulting in a smaller PRP effect when subsequently paired with a Task 2 that has not been practiced (Ruthruff, Van Selst, Johnston, & Remington, in press).

Similar results occur when S-R compatibility is manipulated for Task 1. Schweizer, Jolicœur, Vogel-Sprott, and Dixon (2004) paired tone identification for Task 1 (press one of four keys with the left-hand fingers to tones of 200, 500, 1250, and 3125 Hz) with visual letter-identification for Task 2 (press one of three keys with right-hand fingers to the letter H, O, or S). For Task 1, the mapping of low- to high-pitch tones to responses was ordered from left to right in one condition and random in another. RT1 was approximately 90 ms longer with the random mapping than with the ordered mapping. This S-R compatibility effect for Task 1 carried over to Task 2, slowing RT2 by about the same amount. As predicted by the central bottleneck model, the effect of Task 1 compatibility on RT2 interacted with SOA, being evident primarily at the short SOAs (see also Ruthruff et al., in press, Experiment 2).

The central bottleneck model also predicts that when the difficulty of identifying S2 is increased, the effect of this difficulty on RT2 should be less at short than long SOAs, due to the fact that at least some of the additional time to identify S2 can be absorbed in the slack available at short SOAs. In contrast, for variables that affect Task 2 response selection or response execution, which have their influence after the bottleneck, the extra time cannot be absorbed by the slack. Therefore, their effects should be additive with those of SOA (Schweickert, 1983).

The predicted patterns of results for Task 2 manipulations have been found for several variables of the respective types. For example, Pashler and Johnston (1989) had subjects respond to a low- or high-pitch tone for S1 by pressing one of two keys with the index and middle fingers of the left hand and to a visual letter A, B, or C for S2 by pressing the index, middle, or ring finger of the right hand. In each dual-task trial block, half of the letters were presented at low intensity (gray on dark background) and half at high intensity (white), with each of three SOAs (50, 100, and 400 ms) used equally as often. The variable of S2 intensity interacted with SOA on RT2. At the longest SOA, RT2 was approximately 30 ms shorter for the high-intensity stimuli than for the low-intensity ones, but this intensity effect decreased as SOA decreased. It was not evident at all at the 50-ms SOA. In contrast, repetition of S2 (and R2) from the previous trial, a variable whose influence is typically attributed to response selection (e.g., Soetens, 1998), had an additive effect with SOA. RT2 was 25 ms shorter when S2 repeated than when it did not, and this effect was additive with those of SOA and S2 intensity. Thus, the intensity manipulation, which should affect the duration of S2 identification, showed the predicted underadditive interaction with SOA on RT2, whereas the repetition manipulation, which should affect the duration of S2 response selection, showed the predicted additive effect with that of SOA.

McCann and Johnston (1992) found that manipulations of S-R compatibility for Task 2 also affected the PRP effect in the manner predicted by the response–selection bottleneck model. For Task 1, the stimuli were two tones, separated by a 300-ms intertone interval; the first tone was a standard relative to which the second was to be classified as higher or lower in pitch by saying "high" or "low," respectively. After an SOA of 50, 150, 300, or 800 ms from the onset of the comparison tone, a rectangle or triangle of one of three sizes was presented in the center of the display screen. The form designated the left or right hand, and the size indicated whether the left, middle, or right response key (on which the index, middle, and ring fingers were placed) should be pressed. The Task 2 mapping of size to finger was consistent for one shape (a left-to-right ordering of small to large) and arbitrary for the other. Although RT2 was 60 ms longer with the less compatible arbitrary mapping than with the ordered mapping, this S-R compatibility effect did not interact significantly with that of SOA. McCann and Johnston obtained similar results in their Experiment 2, for which Task 2 required a left or right keypress with the index finger of the left or right hand to either an arrow pointing to the left or right or the letter M or T. Responses were 59 ms slower to the less compatible letter stimuli than to the arrows, but this compatibility effect did not interact with SOA. Thus, both of the manipulations of S-R compatibility for Task 2 affected RT2 in a manner consistent with the hypothesis that their effects were postbottleneck.

10.3 SIMON EFFECTS FOR IRRELEVANT STIMULUS LOCATION

In McCann and Johnston's (1992) Experiment 2, the arrow and letter stimuli for Task 2 were not presented in a centered location but in left or right locations, with location varying randomly from trial to trial. Thus, stimulus location, though irrelevant, could either correspond or not correspond with the response for the trial. A Simon effect of 16 ms overall was obtained for RT2, indicating faster responding when S2 location corresponded with R2 than when it did not. In contrast to the effects of S-R compatibility for the relevant stimulus dimension, this effect of irrelevant location correspondence did interact with SOA (see Figure 10.2). The Simon effect was 35 ms at the 800-ms SOA, and it gradually disappeared as SOA decreased. McCann and Johnston suggested that this underadditive interaction could be given one of two interpretations. It could be taken as evidence that the Simon effect has its locus in stimulus-identification processes, as Hasbroucq and Guiard (1991) and Stoffels, Van der Molen, and Keuss (1989) have suggested. Alternatively, the interaction could be a consequence of rapid dissipation of the activation of the corresponding response produced automatically at onset of S2. Their logic was that because response selection for Task 2 is delayed at short SOAs, this automatic response activation would have dissipated by the time that S2 response selection was performed.

The decay account suggested by McCann and Johnston (1992) bears some similarity to the two-process account proposed by De Jong, Liang, and Lauber (1994) to explain the temporal courses of the Simon effect and its reversal with an incompatible relevant S-R mapping in the Hedge and Marsh variant (see Chapter 6). De Jong et al.'s model includes an unconditional component in which the corresponding

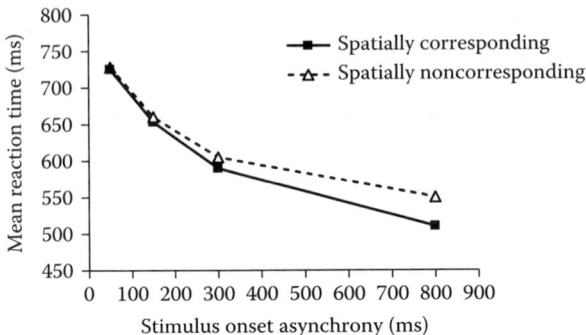

FIGURE 10.2 RT2 as a function of SOA and spatial compatibility (correspondence) in McCann and Johnston's (1992) Experiment 2.

response is automatically activated and then decays, and a conditional component in which the instructed response–selection rule is applied. Because the conditional component is independent of the amount of time that intervenes between stimulus presentation and response selection, its contribution is constant. According to the model, in the case of a typical Simon task, the Simon effect decreases with delay in responding because the automatic activation component decays, leaving only the contribution of the conditional component. For the Hedge and Marsh reversal, De Jong et al.'s model predicts that the magnitude will become larger with time after stimulus onset because the automatically activated tendency to make the corresponding response will decay, leaving only the reversed effect caused by misapplying the "respond opposite" rule. De Jong et al. provided evidence that when a single task is performed, although the Simon effect decreases across the RT distribution, the Hedge and Marsh reversal in fact increases.

Lien and Proctor (2000) reasoned that if the underadditive interaction of the Task 2 Simon effect with SOA observed by McCann and Johnston (1992) in the PRP paradigm is due to decay of activation, then the Hedge and Marsh reversal should show an overadditive interaction. That is, whereas the Simon effect decreases at short SOAs when responding compatibly to arrow direction, the Hedge and Marsh reversal obtained when responding incompatibly to arrow direction should increase. Lien and Proctor reported three experiments modeled after McCann and Johnston's study, with their Experiment 1 using similar tasks to those of McCann and Johnston's Experiment 2. In Lien and Proctor's Experiments 2 and 3, the vocal responses for Task 1 were replaced with index- and middle-finger keypresses, and only arrows were used as Task 2 stimuli. Experiment 3 differed from Experiment 2 in that the Task 1 stimuli were visual letters rather than auditory tones. All three experiments showed a compatibility effect for the relevant arrow-direction dimension, with RT2 being longer for the incompatible direction-response mapping than for the compatible mapping. In Experiments 2 and 3, mapping had an overadditive interaction with SOA, indicating that the PRP effect was larger for the incompatible mapping than for the compatible mapping.

In all three experiments, the Simon effect results for the compatible arrow mapping replicated the underadditive interaction found by McCann and Johnston (1992): The Simon effect for RT2 decreased as SOA was reduced. The incompatible arrow mapping showed the Hedge and Marsh reversal (shorter RT2 when stimulus location and response location did not correspond than when they did). The pattern of results as a function of SOA was less clear for the Hedge and Marsh reversal than for the normal Simon effect, but it also tended to be underadditive. There was no sign of an overadditive interaction of the type implied by De Jong et al.'s (1994) two-process model. Although the results are inconsistent with that specific model, they do not completely rule out that the decrease in Simon effect for Task 2 at short SOAs could be due to dissipation of automatic activation.

10.4 CONSISTENCY OF MAPPINGS

Duncan (1979) noted, "A divided attention situation is more than the sum of its component single tasks. *Emergent* aspects of the whole situation must be considered" (p. 216). He demonstrated this point using the PRP paradigm in his Experiment 2, in which subjects performed a pair of three-choice tasks (see Figure 10.3). The stimuli for Task 1 were vertical lines at three different locations on the left side of the screen, and the responses were keypresses made with the ring, middle, and index fingers of the left hand. Task 2 was identical to Task 1, except that the stimuli were presented on the right side of the screen and responses were made with the right hand. S1 preceded S2 by an SOA of 50, 250, or 450 ms, varying randomly from trial to trial, with both stimuli remaining visible until both R1 and R2 were made. For each task, the mapping could be spatially compatible (each stimulus was assigned to its corresponding response) or incompatible (each stimulus was assigned to the response at the mirror opposite location). All combinations of the mappings for each task were used in different trial blocks, with the Task 1 and Task 2 mappings being consistent (compatible–compatible or incompatible–incompatible) or inconsistent (compatible–incompatible or incompatible–compatible). Duncan analyzed only the outer S-R positions for each task because the response to the middle stimulus was the middle finger for both mappings.

RT1 and RT2 were both shorter when the S-R mappings for the two tasks were consistent than when they were inconsistent (see Figure 10.4), with RT1 being 530, 650, 695, and 675 ms for the compatible–compatible, incompatible–incompatible, compatible–incompatible, and incompatible–compatible conditions, respectively, and RT2 being 635, 775, 845, and 910 ms. Note that the benefit of consistency extended not only to the compatible–compatible pairing but also to the incompatible–incompatible pairing. In other words, responses were faster when both tasks had the same spatially incompatible mapping than when they did not, regardless of whether the mapping for a particular task in the mixed condition was compatible or incompatible. For the consistent combinations, both Task 1 and Task 2 showed large S-R compatibility effects of over 100 ms that were evident at all SOAs. However, the inconsistent combinations showed no S-R compatibility effects, as is often the case when mappings are mixed (see Chapter 7): There was little difference between the

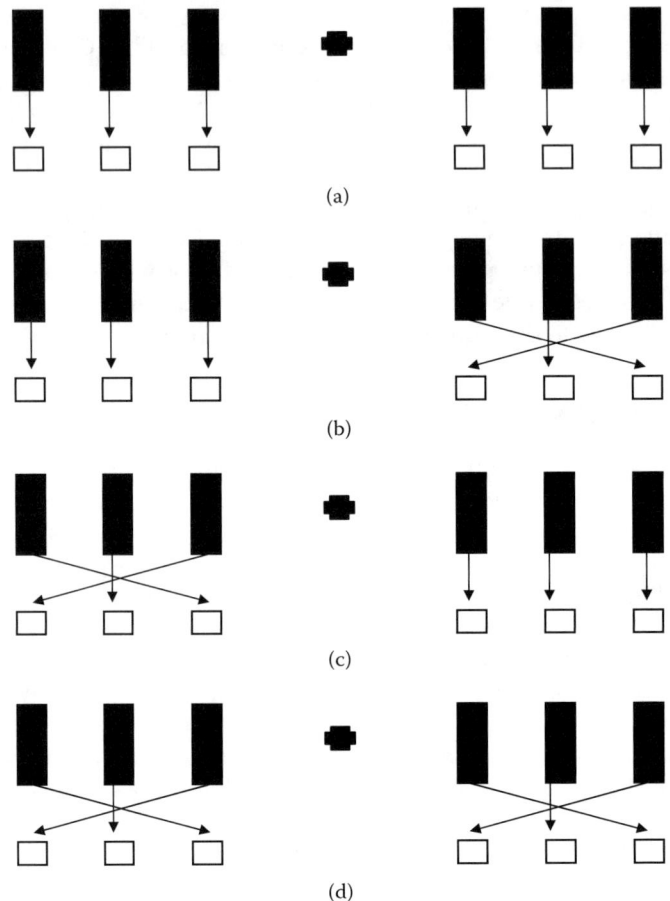

(a)

(b)

(c)

(d)

FIGURE 10.3 The mappings used for Duncan's (1979) three-choice tasks. Panels a and d are consistent mapping conditions (compatible and incompatible, respectively) and b and c are inconsistent mapping conditions for which one task uses a compatible mapping and the other an incompatible mapping.

compatible and incompatible mappings for RT1, and RT2 was slower for the compatible mapping than for the incompatible mapping.

    If subjects make errors when performing with inconsistent mapping combinations by misapplying the mapping to one task that was appropriate for the other one, the majority of errors for each task should be responses that would have been correct with the other mapping. Indeed, Duncan (1979) found that the percentage of errors of pressing the opposite key when the alternative mapping was incompatible and the corresponding key when the alternative mapping was compatible was much larger than that of pressing the adjacent, center response key. Duncan said the following with respect to both the RT and error data from his experiment: "These data provide rather strong support for the idea that inconsistent situations involve some emergent choice between abstract S-R mapping rules" (p. 225). If the more

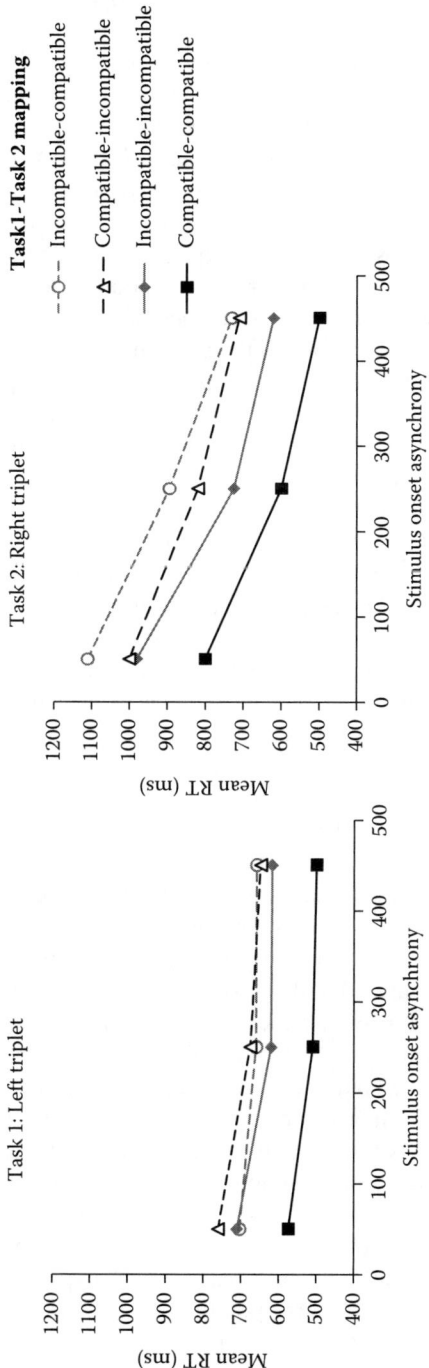

FIGURE 10.4 Mean RT for Task 1 and Task 2 as a function of SOA and mapping consistency in Duncan's (1979) study.

standard case in which the two tasks do not share mappings is taken as the baseline, the emergent feature could be characterized as a benefit of being able to apply the same mapping rule to each task.

A question arises as to whether the emergent choice identified by Duncan (1979) is also a factor when a pair of two-choice tasks is performed. Duncan briefly described a control experiment in which similar results were obtained for two-choice tasks in which Task 1 was like the three-choice task described (but only involved two choices), but Task 2 used a vertically oriented stimulus array and up–down movement of a toggle switch. Even though the two tasks involved different dimensions, performance of inconsistent combinations was difficult, in agreement with Duncan's emergent choice interpretation. Ivry, Franz, Kingstone, and Johnston (1998, Experiment 2) also examined performance of a pair of two-choice tasks, but in their case both S1 and S2 varied along the vertical dimension. Subjects responded with the index and middle fingers of each hand, with the hands placed at a 45° angle so that one finger was above the other in the response plane (the middle finger was above the index finger for each hand). The SOAs in their study were 50, 150, 400, and 1,000 ms, with the longest SOA being much longer than that in Duncan's study. A compatible mapping for Task 1 was paired with a compatible or incompatible mapping for Task 2 in different trial blocks. Ivry et al.'s results showed a consistency effect for RT1 at all SOAs, but the effect was significantly smaller at the longer SOAs than at the shorter ones. Because mapping was only varied for Task 2, Ivry et al.'s study allowed examination of just one of the four possible comparisons for consistency benefits, RT1 for compatible Task 1 mapping.

Ivry et al.'s (1998) primary intent, as described in Chapter 1, was to examine performance of a patient with a severed corpus callosum. This patient showed elevated RT for the inconsistent combinations only on Task 2. Also, whereas the effect of consistency on RT2 was approximately additive for control subjects, it was underadditive for the callosotomy patient. These and other results led Ivry et al. to conclude that the bottleneck for the callosotomy patient was subsequent to response selection.

Duncan (1979) emphasized that an emergent mapping choice was only one of many possible processes that could emerge in dual-task situations. Vu and Proctor (in press) obtained evidence for a role of perceptual emergent features in two-choice tasks. As illustrated in Figure 10.5, they noted that, with a pair of two-choice tasks for which the stimulus and response locations varied along the horizontal axis, the stimulus locations grouped into filled and unfilled pairs when the stimuli for each task were presented together. Vu and Proctor argued that this emergent perceptual feature should benefit performance under consistent mapping conditions, particularly when the mapping rule was incompatible. Their reasoning was as follows: Segregating the subgroup of filled squares should be of some benefit with the compatible–compatible condition because subjects can apply the "respond corresponding" rule to the two members as a whole. For the incompatible mapping, the possible benefit is much greater, though, because the tasks can be converted from "respond opposite to filled locations" to "respond corresponding to unfilled locations," allowing a simpler mapping rule to be used. Because this emergent perceptual grouping feature is available only when the two stimuli are both present, its effect should decrease as SOA increases.

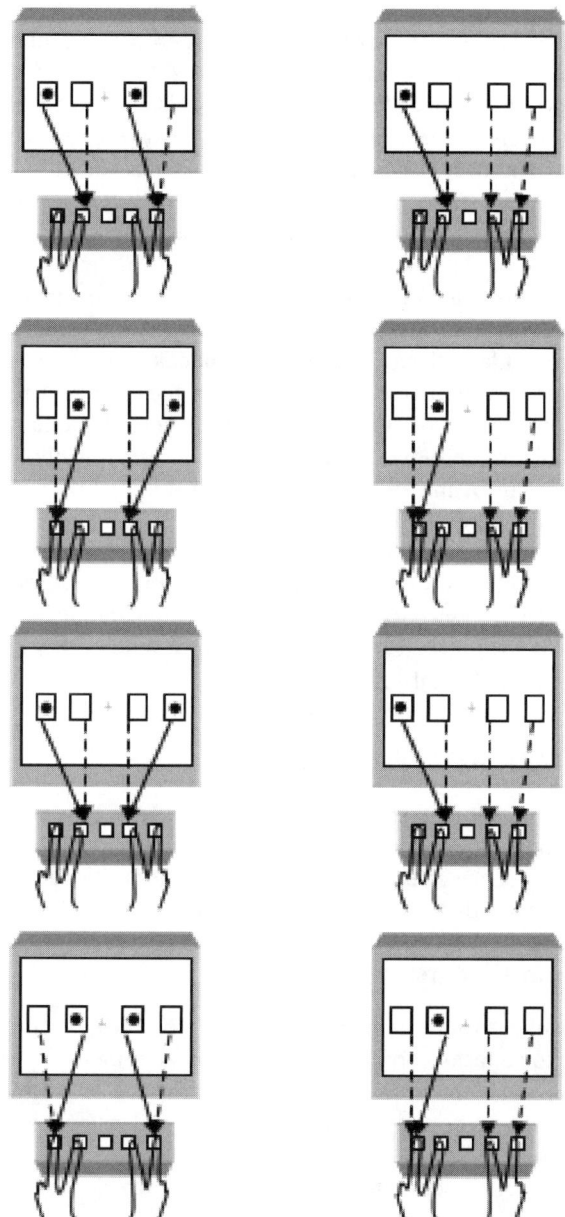

FIGURE 10.5   Perceptual emergent blank feature in two-choice tasks when both stimuli are present (left column) that is absent when only one is (right column). A solid arrow represents the response to be made to the stimulus based on the assigned incompatible mapping. A dashed arrow represents the compatible response to the blank emergent feature. From "Emergent perceptual features in the benefit of consistent stimulus–response mappings on dual-task performance," by K.-P. L. Vu and R. W. Proctor, in press, *Psychological Research*. Copyright 2006, Springer Verlag. Reprinted with permission.

Results of several experiments reported by Vu and Proctor (in press) were in agreement with the hypothesis that this emergent perceptual grouping feature benefited performance for the consistent mapping conditions. As in Duncan's (1979) three-choice tasks, performance benefited from consistent compatible mappings and even more so for consistent incompatible mappings. However, the benefit was largest at the 50-ms SOA and decreased sharply as SOA increased. Although it is possible to account for the reduction in consistency benefit for RT2 as SOA increases in terms of emergent mapping choice, it is not possible to account for the reduction in consistency benefit for RT1 with increasing SOA in this manner. Vu and Proctor reported similar results when the responses for one task were with the inner response keys and those for the other task with the outer response keys, and when responses for one task were made with the left and right hands, and those for the other task made with the left and right feet. However, when the emergent perceptual feature was removed by using left or right tones as the stimuli for one task and left and right visual stimuli for the other, the consistency benefit was eliminated.

The primary point of the studies described in this section is that performance may benefit from consistent S-R mappings for two tasks in at least two ways. One is through elimination of uncertainty about which mapping is appropriate for a particular task. The other is through attending to emergent perceptual features when they allow a simpler response–selection strategy that encompasses both tasks to be adopted. It remains for future research to establish whether other emergent features influence performance in dual-task contexts.

10.5   CROSSTALK BETWEEN TASKS

When the stimulus and/or response sets for Task 1 and Task 2 overlap, the opportunity exists for inadvertent activation across tasks to occur. Such activation can cause cross-task correspondence effects, much as the activation produced by an irrelevant stimulus dimension yields the Simon effect when a single task is performed.

10.5.1   Way and Gottsdanker's Study

Way and Gottsdanker (1968) demonstrated such cross-task correspondence effects for spatial two-choice tasks carried out concurrently. In their Experiment 1, Task 1 required the subject to move a hand-held lever toward or away from the body, according to whether the bottom or top half of the circular upper surface of the lever was illuminated (see Figure 10.6). In different blocks of trials, this task was paired with a similar task performed with the right hand for which the axis was parallel to the first (i.e., toward or away) or perpendicular to it (i.e., left or right movement in response to the left or right half of the upper surface being illuminated). The SOA between S1 and S2 was 100 ms in some trial blocks and 900 ms in others. The mean PRP effect (RT2 at 100-ms SOA minus RT2 at 900-ms SOA) tended to be slightly larger with the perpendicular orientation (75 ms) than with the parallel orientation (68 ms). More interesting, within the parallel orientation, the PRP effect was smaller for trials on which the stimuli (and responses) were in the same direction (54 ms) than in opposite directions (81 ms).

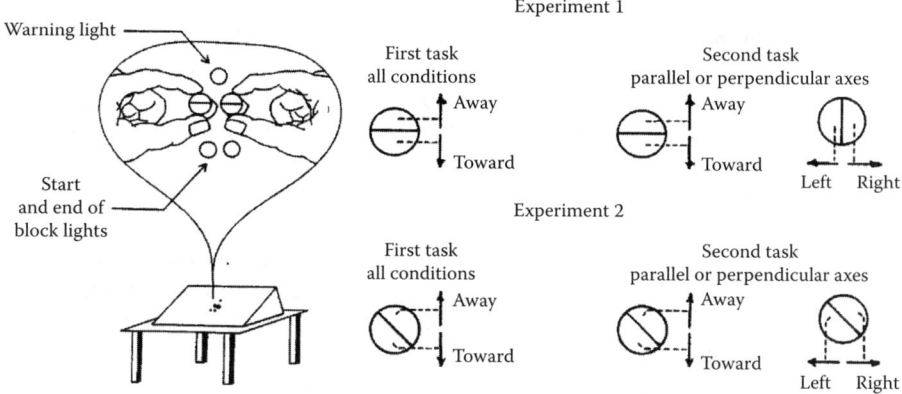

FIGURE 10.6 Apparatus in Way and Gottsdanker's (1968) study (left side), and stimuli for Experiments 1 and 2 (right side). From "Psychological refractoriness with varying differences between tasks," by T. C. Way and R. Gottsdanker, 1968, *Journal of Experimental Psychology, 78*, pp. 40, 42. Copyright 1968, American Psychological Association. Reprinted with permission.

In Experiment 2, the method was modified so that the same stimuli could be used for the two tasks (see Figure 10.6). This was accomplished by illuminating halves of the upper surface of the lever that were below the negative diagonal (toward– left) or above it (away–right). This manipulation allowed separate estimates of the amount of the crosstalk effect that was due to overlap among the stimuli for the two tasks or overlap among the responses. With the perpendicular response assignment (toward–away for Task 1 and left–right for Task 2), the PRP effect was 16 ms smaller when the stimulus was the same for the two tasks than when it was different. With the parallel response assignment, the PRP effect was 35 ms smaller when the stimulus and response were the same than when they were the opposite, suggesting a 19-ms contribution of the response overlap.

Another important finding in Way and Gottsdanker's (1968) study was that RT1 also showed cross-task correspondence effects, though smaller than those for RT2, in both experiments. That is, RT1 was longer at the 100-ms SOA than at the 900-ms SOA, with this increase at the short SOA being larger when the stimulus and response were opposite those of Task 2 than when they were the same (12 ms in Experiment 1 and 16 ms in Experiment 2). This greater elevation at the short SOA for RT1 was not evident with the perpendicular arrangement in Experiment 2, for which only the stimuli were opposite. Way and Gottsdanker concluded, "Since the elevation did not occur when only the signals were opposite, this can be considered a within-trial response antagonism effect" (p. 44). Thus, Way and Gottsdanker not only established that cross-task correspondence effects occur in the PRP paradigm for tasks that have dimensional overlap, but they also provided evidence that the effects have two components, one stimulus related and the other not, and that the effects on RT1 are likely due to the second component.

10.5.2 More Recent Reaction-Time Studies

Hommel (1998a) brought cross-task correspondence effects to researchers' attention more recently. He had subjects make a left or right keypress to the color of a red or green rectangle for Task 1 and to say "red" or "green" to the letter S or H, presented after an SOA of 50, 150, or 650 ms, for Task 2. Both RT1 and RT2 showed correspondence effects at short SOAs. At 50-ms SOA, the keypress response for Task 1 was 28 ms shorter, and the color-naming response for Task 2 was 63 ms shorter, when the color of S1 corresponded with the color of the vocal response for Task 2 than when it did not. This result establishes that cross-task correspondence effects can occur on the basis of S1–R2 overlap, as well as on the basis of S1–S2 overlap and R1–R2 overlap, as demonstrated by Way and Gottsdanker (1968). That R2 had an effect on RT1 again implies that R2 received some activation prior to completion of Task 1.

Lien and Proctor (2000) reported cross-task correspondence effects for their Experiments 2 and 3, described earlier, in which Task 1 required left–right keypresses with one hand to low- or high-pitch tones (Experiment 2) or centered visual letters (M or T; Experiment 3) and Task 2 required left–right keypresses with the other hand to compatibly or incompatibly mapped left–right arrow directions (for which location to the left or right of fixation was irrelevant). Strong R1–R2 correspondence effects were found on RT1 at short SOAs, particularly when the Task 2 mapping of arrow direction to responses was compatible. Moreover, a correspondence effect for S2 irrelevant location with R1 was also evident for RT1, which was independent of the R1–R2 correspondence effect. Thus, a cross-task Simon effect for location of S2 on RT1 was obtained.

Lien, Schweickert, and Proctor (2003) showed that R1–R2 correspondence effects vary as a function of the foreknowledge that subjects have about the task transitions. In three experiments, subjects classified digits as even or odd and letters as consonant or vowel, using keypresses of the left middle and index fingers to a stimulus left of fixation for Task 1 and of the right index and middle fingers to a stimulus right of fixation for Task 2. SOA between S1 and S2 was varied between blocks at values of 50, 150, 300, and 800 ms. In Experiment 1 subjects had no foreknowledge as to whether Task 1 or Task 2 would require odd–even or consonant–vowel judgments, whereas in Experiment 2 they knew in advance whether the required judgments would be the same or different for the two tasks, but not which specific judgment(s) would be required. Cross-task correspondence effects of similar magnitude were obtained in the two experiments, where both RT1 and RT2 were shorter at short SOAs when the response locations signaled by the two stimuli corresponded than when they did not. These correspondence effects were equally evident for task-repeat and task-switch trials. However, in Experiment 3, in which subjects had full foreknowledge in each block of trials as to which task would be performed for Task 1 and which for Task 2, cross-task correspondence effects were evident only when the two tasks were the same. When subjects knew that Task 1 would require one type of judgment, say, odd–even, and Task 2 the other type, consonant–vowel, they were able to isolate the processing for each task such that the stimulus for one produced little if any activation of the corresponding response for the other.

Logan and Schulkind (2000) demonstrated cross-task correspondence effects for the categories of Task 1 and Task 2 stimuli in a variety of tasks. In their Experiment 1, subjects classified a stimulus (S1) presented in an upper position as a letter or digit with a keypress made by the index or middle finger of the right hand and a stimulus (S2) presented in a lower position as a letter or digit with a keypress made with the index or middle finger of the left hand. All combinations of assignments of Task 1 categories to right-hand responses and Task 2 categories to left-hand responses were used. Task 2 showed a standard PRP effect, with RT2 lengthening at short SOAs. More important, both tasks showed correspondence effects at SOAs of 300 ms or less, with RT1 and RT2 being 100 to 200 ms shorter when S1 and S2 were from the same category (e.g., both letters) than when they were from different categories. Logan and Schulkind obtained similar results in their Experiments 3 and 4, in which lexical decisions (word–nonword) were made for strings of five letters. Moreover, they showed in Experiment 4 that more specific associative priming between tasks occurs, with both RT1 and RT2 being shorter when the two stimuli in the lexical decision tasks were semantically associated than when they were not.

In Logan and Schulkind's (2000) Experiment 2, the stimuli for both Task 1 and Task 2 were digits, and the digits had to be classified as odd–even or high–low (greater than 5 or less than 5). Subjects performed blocks for all four pairings of the two types of Task 1 with the two types of Task 2, and in this experiment, high and even responses were made with the index fingers and low and odd responses with the right index fingers. The primary finding of interest was that large category-match effects for RT1 and RT2 were evident when Task 1 and Task 2 were the same, with RT1 showing no such effect and RT2 only a small effect when Tasks 1 and 2 were different, as in Lien et al.'s (2003) study. Thus, even when the stimuli for the two tasks are from the same set (digits), little cross-task crosstalk occurs if the subjects know that different judgments will be required for each task.

Logan and Gordon (2001, Experiment 1) had subjects perform high–low judgments for only the top digit in the display (single task) or for both the top and bottom digits (dual task). As in Logan and Schulkind's (2000) study, Task 1 responses were made with the right hand and Task 2 responses with the left hand, using the index fingers to designate high and the middle fingers to designate low. Cross-task correspondence effects were evident for Task 1 as well as Task 2 in the dual-task condition, but no effects due to stimulus congruence alone were evident for Task 1 in the single-task condition. Logan and Gordon partitioned the corresponding trials into those for which both the digits and responses were the same for the two tasks (identical trials) and those for which the digits were different but the responses were the same (repetition trials). Both RT1 and RT2 showed a considerable benefit for repetition trials compared with noncorresponding trials (746 versus 760 ms for RT1 and 759 versus 835 ms for RT2), with identical trials showing an even larger benefit (719 ms for RT1 and 706 ms for RT2). Thus, response correspondence was sufficient to produce a benefit, although the size of the benefit was larger when stimulus correspondence was also present.

Logan and Gordon's (2001) Experiment 2 tested whether similarity of stimulus mode (i.e., perceptual overlap) was necessary for cross-task correspondence effects to occur. This was accomplished by centrally presenting a color word (RED,

GREEN, or BLUE), with colored bars (red, green, or blue) above and below it. For half the subjects the color word was S1 and the colored bar was S2, whereas for the other half this relation was reversed. The Task 1 responses were made with three fingers of the right hand and the Task 2 responses with three fingers of the left hand. As in their Experiment 1, single-task conditions were examined as well in which both stimuli were presented but only the Task 1 response was made. In this case, the single-task performance showed a small cross-task correspondence effect. However, the effect was an order of magnitude larger when both tasks were performed. The authors note that the cross-task correspondence effects on RT1 are strong evidence that Task 2 response activation began before Task 1 response selection was completed.

In Logan and Gordon's (2001) Experiment 3, the stimuli were pictures that depicted animals or nonanimals and words that named animals or nonanimals. Two different tasks were performed on these: form judgments, in which subjects had to decide whether the stimulus was a picture or a word, and animacy judgments. In different sessions, all combinations of the tasks were performed for Task 1 and Task 2. Large cross-task correspondence effects were again obtained, but only when the task set was the same for the two tasks. When the task set was different, the correspondence effects were absent.

10.5.3 Tasks with Unspeeded Responses

Cross-task correspondence effects occur not only when both tasks require choice reactions but also when one task does not require a speeded response. Koch and Prinz's (2002) Experiment 1 presented a red or blue X at the center of the screen as S1, to indicate whether the required response would be left or right. The response for this task was executed by moving the index finger of the dominant hand from a home key to a left or right key. However, the response was not to be made until a tone, as a go signal, occurred 1,300 ms later. At delays of 100, 900, 1,200, or 1,300 ms after the colored X, a small circle appeared briefly at the center of the display and then moved to the left or right and was masked. The subject had to indicate the direction of dot movement, 500 ms after response execution for Task 1, by pressing the left or right key. The perceptual task showed no effect on response accuracy of correspondence with the reaction-task response, but the reaction task showed shorter RT when the visual target direction corresponded with the response than when it did not. This cross-task correspondence effect occurred only when the visual target was presented simultaneously with or 100 ms before the auditory go signal.

Koch and Prinz's (2002) Experiment 2 showed that physically making the same unimanual response was not the critical factor in the cross-task correspondence effect by showing a similar effect on RT when a vocal response "left" or "right" was made for the perceptual task. In their third experiment, instructions given at the start of one fourth of the trials indicated that no response would be required for the perceptual task. Although the overall dual-task interference was reduced greatly on the no-response trials, the cross-task correspondence effect on RT was not.

Azuma, Prinz, and Koch (2004) modified the procedure so that it was closer to the more typical PRP task. The visual stimulus was presented first, followed after an SOA of 100 or 1,200 ms by a high- or low-pitch tone, to which a speeded left–right response was to be made. The percentage of no-report trials for Task 1 was increased to 50%. In this case, not only did the reaction task show a cross-task correspondence effect at 100-ms SOA, but so did the perceptual judgment task on the report trials. In Experiment 2, the report and no-report conditions were blocked, to try to ensure that subjects were not attending to the visual stimulus on the no-report trials. Again, cross-task correspondence effects were observed for both types of trials that did not interact with whether a perceptual judgment was required or not.

Koch, Metin, and Schuch (2003) examined cross-task correspondence effects for an unspeeded Task 1 (report at the end of the trial whether a briefly displayed dot moved to the left or right) and a speeded Task 2 in which a high- or low-pitch tone was classified by moving the index finger of the dominant hand from a home key to a left or right key. Task 1 was performed on only half of the trials, with a message at the beginning of each trial indicating whether dot-movement direction would have to be reported. An SOA of 100 or 1,200 ms separated the visual and auditory stimuli; in Experiment 1, SOA was varied within subjects, either randomly within trial blocks or between trial blocks, and in Experiment 2, between subjects. Cross-task correspondence effects were evident in higher accuracy for Task 1, and both higher accuracy and shorter RT for Task 2, when the movement direction and R2 corresponded than when they did not. The correspondence effect for RT2 interacted with SOA, being evident primarily at the 100-ms SOA. When SOA was varied within subjects (Experiment 1), the correspondence effect for RT2 was of similar magnitude when movement direction for S1 was to be reported and when it was not, whereas when SOA was varied between subjects (Experiment 2), the effect was larger for the report condition than for the no-report condition. This result suggests that performance using only a single SOA may allow subjects to develop strategies to minimize the impact of Task 1 on performance of the speeded Task 2.

10.5.4 Summary

Cross-task correspondence effects are now well documented. They occur in a variety of situations and can be quite large. The fact that the correspondence effects are evident in RT1, as well as RT2, implies that the S2 is translated into response activation prior to the completion of response selection for Task 1. Hommel (1998a) has proposed that such translation of stimulus information into response activation is automatic, with the bottleneck being only in the final decision about which response to make for each task.

10.6 CAN THE BOTTLENECK BE BYPASSED?

Given the widespread occurrence of the PRP effect and its attribution to a central-processing bottleneck, a natural question to ask is whether the bottleneck can be bypassed in certain situations. Two factors that have been proposed to allow bottleneck bypass and "perfect timesharing" are ideomotor compatibility and extended practice.

10.6.1 DOES IDEOMOTOR COMPATIBILITY WITHIN TASKS ALLOW PERFECT TIMESHARING?

The concept of ideomotor compatibility, introduced by Greenwald (1970a, 1970b), was described in Chapter 3. Ideomotor compatibility, according to Greenwald (1972), refers to "the dimension denoting the extent to which a stimulus corresponds to sensory feedback from its required response" (p. 52). He further proposed, "The stimulus of highly ideomotor-compatible combinations should effectively select the response without burdening limited-capacity decision processes of the central nervous system" (p. 52). This reasoning leads directly to the prediction that if at least one of two tasks in the PRP paradigm is ideomotor, then no PRP effect should occur.

Greenwald and Shulman (1973) tested this proposition in two experiments. In both experiments, Task 1 was visual–manual and Task 2 was auditory–vocal. For each task, there were two versions, one ideomotor compatible and the other S-R compatible but not ideomotor compatible. The ideomotor compatible version of Task 1 involved moving a switch left in response to a left-pointing arrow in a left position or right in response to a right-pointing arrow in a right position, and the S-R compatible version involved making the switch movements in response to the visual words LEFT and RIGHT. The ideomotor version of Task 2 involved saying "A" and "B," respectively, to the auditory letters A and B, whereas the S-R compatible version involved saying "one" or "two," respectively, to the same stimuli. Each of the four combinations of Task 1 and Task 2 was performed by a different group of subjects. SOAs between Task 1 and Task 2 of 0, 100, 200, 300, 500, and 1,000 ms were used and varied between trial blocks in which SOA was held constant.

The results are shown in the top half of Figure 10.7. Greenwald and Shulman (1973) noted, "The magnitude of the PRP effect, as indexed by the ISI [SOA, in our terminology] factor, was, contrary to predictions, not significantly affected by Task 2 compatibility" (p. 72). However, Greenwald and Shulman interpreted the results more favorably with respect to perfect timesharing of ideomotor compatible tasks than this outcome might imply because RT1 increased as SOA increased for the three conditions in which at least one of the tasks was ideomotor compatible. Consequently, Greenwald and Shulman averaged RT1 and RT2 and noted, "These combined results more closely resembled the predicted effects, with the PRP effect being nearly absent when both tasks were IM compatible, and most strongly present when neither task was IM compatible" (p. 73). However, they noted that even for the combined RT data, the condition for which both tasks were ideomotor compatible showed 18-ms longer RT at 0-ms SOA than at 1,000-ms SOA that "seemed large in light of the no-difference prediction" (p. 73).

Greenwald and Shulman (1973) reasoned that because the instructions for Experiment 1 stressed that Task 2 always followed Task 1, subjects may have imposed a constant order on the tasks, rather than trying to perform them simultaneously. Consequently, in their Experiment 2, subjects were told that most often the two signals would be simultaneous, and SOAs were concentrated at short values (0, 100, 200, and 1,000 ms). In this case, RT2 showed substantial PRP effects for all but the condition for which both tasks were ideomotor compatible, which showed no effect

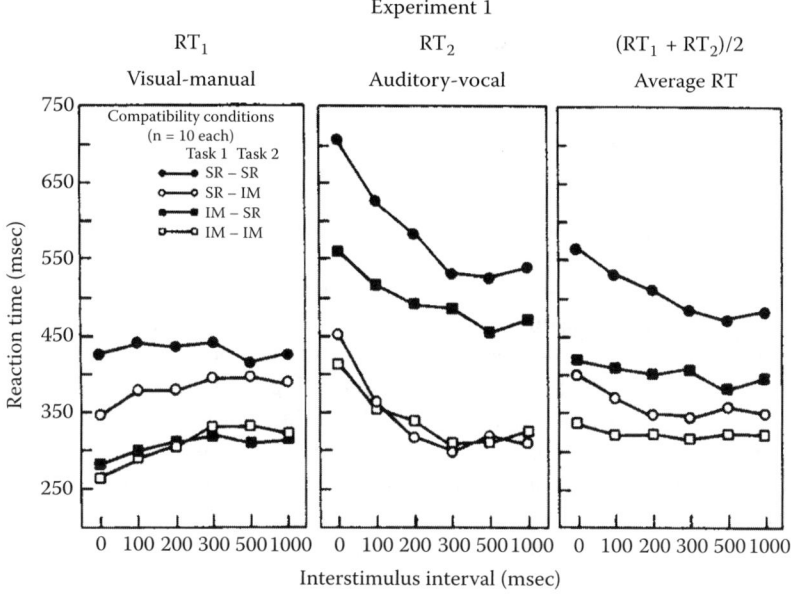

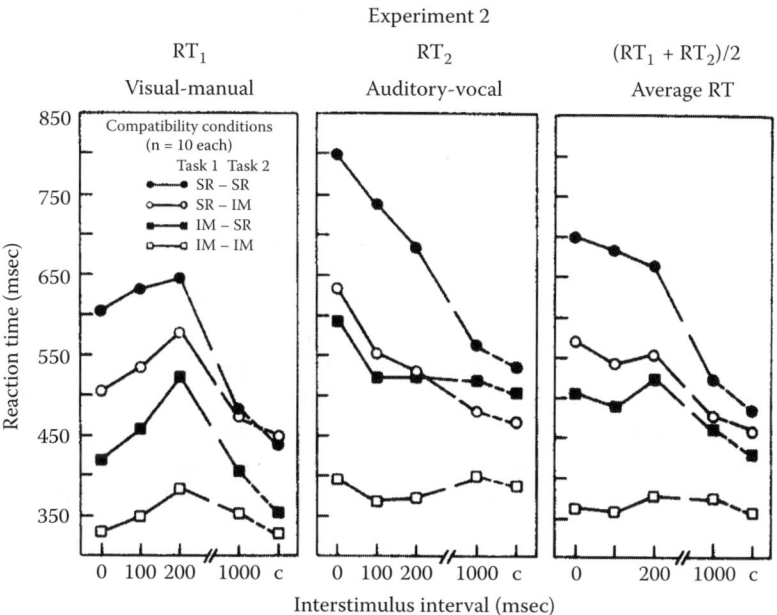

FIGURE 10.7   Greenwald and Shulman's (1973) results: RT for Task 1, Task 2, and the combined tasks as a function of interstimulus interval (called SOA in the chapter) and compatibility (SR = S-R compatible only; IM = ideomotor compatible). From "On doing two things at once: II. Elimination of the psychological refractory period effect," by A. G. Greenwald and H. G. Shulman, 1973, *Journal of Experimental Psychology, 101,* p. 74. Copyright 1973, American Psychological Association. Reprinted with permission.

(see bottom half of Figure 10.7). Greenwald and Shulman summarized their findings for the two experiments as follows:

> The psychological refractory period (PRP) effect of interference between 2 choice–reaction tasks at short intertask intervals was eliminated when both of the tasks were ideomotor compatible .... It was concluded that a major source of the PRP effect is a limited capacity mechanism that (a) translates between an encoded stimulus and a response code, and (b) is not needed when a task is ideomotor compatible. (p. 70)

One restriction to Greenwald and Shulman's (1973) description of their results and one serious problem with their interpretation of them are readily apparent from considering the results of their experiments. First, when both tasks were ideomotor compatible, the PRP effect was apparent in their Experiment 1 but not their Experiment 2. This suggests at a minimum that elimination of the effect requires instructions that the signals are most often simultaneous and/or use of the specific SOAs used in their Experiment 2 as opposed to those in their Experiment 1. Regarding their explanation, Greenwald and Shulman's finding of PRP effects when only one of the two tasks was ideomotor compatible does not seem to support the view that the limited-capacity mechanism is bypassed when a task is ideomotor compatible. If this were the case, response selection for the other task should have been able to proceed without any interference from the ideomotor compatible task, or vice versa, which clearly was not the case.

Lien, Proctor, and Allen (2002) noted the discrepancies between Greenwald and Shulman's (1973) data and their conclusions and attempted to replicate their findings in a series of four experiments. Their experiments used the same tasks as Greenwald and Shulman, with the exception that in Experiments 1–3 the arrow stimuli were displayed in the center of the screen, rather than in left and right positions. The SOAs were those used in Greenwald and Shulman's Experiment 1 (0, 100, 200, 300, 500, and 1,000 ms) in Experiment 1 and those of their Experiment 2 (0, 100, 200, and 1,000 ms) in Experiments 2–4. As in Greenwald and Shulman's Experiment 2, the instructions stated that most often the two stimuli would be presented simultaneously. In all experiments, the responses were left–right deflections of a joystick. Substantial PRP effects were found for the condition in which both tasks were ideomotor compatible, as well as for the other three task combinations. Thus, even with the instructions used in Greenwald and Shulman's Experiment 2, the results were like those of their Experiment 1 in showing a PRP effect when both tasks were ideomotor compatible.

Greenwald (2003) questioned several aspects of Lien et al.'s (2002) methods, focusing on their specific instructions as possibly being the reason Lien et al. were not able to replicate the perfect timesharing results of Greenwald and Shulman's (1973) Experiment 2. Specifically, Lien et al. used instructions of the type most often used in RT studies, telling subjects to respond as quickly and accurately as possible, and also instructed their subjects not to wait for the Task 2 stimulus before responding to the first task. Greenwald indicated that the instructions used in Greenwald and Shulman's Experiment 2 had stressed speed of responding over accuracy, although no mention of those instructions is included in their article. He stated, "To the best of the present author's recollection, however, G&S's [Greenwald and Shulman's] instructions for Experiment 2 not only stressed the simultaneous occurrence of the stimuli but also encouraged subjects to respond both rapidly and simultaneously to the simultaneous stimuli" (p. 860).

Greenwald's (2003) Experiment 1 therefore included a manipulation of instructions in an experiment similar to Lien et al.'s (2002) Experiment 4 but with keypresses instead of joystick movements used as responses for the visual–manual task. Some subjects received instructions comparable with those used by Lien et al., which stressed speed and accuracy of responding, whereas other subjects received instructions that stressed response speed and simultaneous responding (which we call speed–stress instructions). The instruction manipulation was effective, as responses on both tasks were faster and slightly less accurate with the speed–stress instructions than with those mentioning speed and accuracy. Whereas a 43-ms PRP effect was evident with speed-and-accuracy instructions, none was evident in the RT2 data with the speed instructions. However, in that instruction condition, the error rate for Task 2 did show a PRP effect, with more errors being made at 0-ms SOA than at 1,000-ms SOA. Thus, the speed–stress instructions do not seem to have eliminated the PRP effect but rather to have shifted it from the RT data to the error-rate data (Lien, Proctor, & Ruthruff, 2003).

Greenwald's (2003) study also suffered from several limitations of his apparatus. Most important, the software used to control the experiment could only record one response at a time, which meant that at the 0 SOA, one response had to be randomly selected. Also, voice-recognition software used to identify the vocal responses miscategorized them on a significant portion of trials. The end result was that error feedback, provided for both tasks on every trial, was highly inaccurate. Shin, Cho, Lien, and Proctor (2005) recently conducted procedural replications of Greenwald's instructional experiment. The overall RT and error rate data showed that the speed instructions were approximately as effective as in Greenwald's experiments at inducing faster responding than the speed-and-accuracy instructions. However, in three different experiments using both joystick movements and keypresses for Task 1 responses, a substantial PRP effect was evident for RT2, even when response speed was emphasized. On the whole, at this point, the data indicate that the pairings of visual–manual and auditory–vocal ideomotor compatible tasks yield PRP effects and that Greenwald and Shulman's (1973) Experiment 2, suggesting that they do not, is an anomaly.

Greenwald (1972) also reported evidence for perfect timesharing of ideomotor compatible tasks in a dual-task procedure that used only simultaneous presentation of the two stimuli. Greenwald varied the stimulus and response modalities in this study, as in Greenwald and Shulman's (1973), but the stimuli and responses all referred to left–right spatial locations. Stimuli were left- and right-pointing arrows positioned to the left and right, respectively, and the auditory words "left" and "right." Responses were left–right movements of a switch and the spoken words "left" and "right." For the high ideomotor compatibility condition, the manual responses were paired with the visual stimuli and the vocal responses with the auditory stimuli, and for the low ideomotor compatibility condition, these S-R pairings were reversed. In addition to performing two-decision trial blocks in which choices had to be made for both tasks, subjects performed zero-decision blocks in which the stimuli (and responses) for both tasks remained the same throughout a trial block (in other words, they were simple RT tasks) and one-decision blocks in which the stimulus for one task was constant and that for the other varied (analogous to single-task trials).

Responses were considerably slower for the low ideomotor compatibility pairing than for the high one for both the one-decision and two-decision blocks. Moreover, RT for both tasks was considerably longer in the two-decision block than the one-decision block when the tasks were low ideomotor compatible (187 and 275 ms longer for the manual and vocal tasks, respectively) but only slightly longer when the tasks were high ideomotor compatible (63 ms and 10 ms, respectively). Greenwald (1972) interpreted the relatively small dual-task cost when ideomotor compatibility was high as consistent with the hypothesis of perfect timesharing. One interesting detail of Greenwald's results was that although cross-task correspondence effects were large when the two tasks were of low ideomotor compatibility (118 and 139 ms for the manual and vocal response tasks, respectively), they were very small when the tasks were of high ideomotor compatibility (11 and 10 ms). Thus, in the latter case, response activation was restricted primarily to the correct task, and there was little crosstalk between tasks.

Lien, McCann, Ruthruff, and Proctor (2005b) used a PRP paradigm, in which SOA between S1 and S2 was varied, to examine all four combinations of low and high ideomotor compatible versions of Tasks 1 and 2. In four experiments, Task 1 used visual stimuli mapped to left–right joystick movements, and Task 2 used auditory stimuli mapped to "left"–"right" vocal responses. Across experiments, a nonideomotor compatible Task 1, for which stimuli were the visual letters A and H, and an ideomotor compatible Task 1, for which the stimuli were left- and right-pointing arrows (located to the left or right of center, respectively) were paired with a nonideomotor compatible Task 2, for which the stimuli were auditory words "one" and "two," and an ideomotor compatible Task 2, for which the stimuli were the words "left" and "right." In agreement with Lien and colleagues' other studies, all conditions showed PRP effects, although the effect size was largest when both tasks were nonideomotor compatible (228 ms), intermediate when one task was ideomotor compatible and the other not (120 ms when Task 1 was ideomotor compatible and 131 ms when Task 2 was), and smallest when both tasks were ideomotor compatible (52 ms). Replicating Greenwald (1972), RT2 showed a significant 30-ms correspondence effect at the three shortest SOAs (0, 50, and 150 ms) when both tasks were not ideomotor compatible and a nonsignificant 8-ms correspondence effect when both tasks were ideomotor compatible. Although the PRP effects were of similar magnitude for the conditions in which one task was ideomotor compatible and the other not, when Task 1 was ideomotor compatible, RT2 showed a cross-task correspondence effect of about 30 ms, whereas when Task 1 was not ideomotor compatible, RT2 showed only a nonsignificant 8-ms effect. Thus, Task 2 was evidently protected from crosstalk from Task 1 when it was of high ideomotor compatibility.

Lien et al. (2005b) evaluated fits of several possible models to numerous aspects of their results and concluded that the results were most consistent with an "engage bottleneck later" model. According to this model, when Task 2 is ideomotor compatible, processing occurs further into response selection before a bottleneck is encountered. This conclusion implies a partial bottleneck bypass, although there ultimately is still a bottleneck in response selection.

In sum, dual-task performance is best when both tasks are ideomotor compatible. Ideomotor compatibility for Task 1 likely reduces the PRP effect primarily through reducing RT1 and the time for Task 1 response selection. Ideomotor compatibility for Task 2 seems to protect the task from crosstalk from Task 1 and allow processing to occur further into the system before a bottleneck is reached. The evidence for complete bottleneck bypass with ideomotor compatible tasks is not compelling, but whether there are conditions under which perfect timesharing of such tasks can occur is not clear (see, e.g., Greenwald, 2005; Lien, McCann, Ruthruff, & Proctor, 2005a).

10.6.2 Does Extended Practice Allow Perfect Timesharing?

Although perfect timesharing does not seem to occur for unpracticed subjects even when both tasks are ideomotor compatible, it is possible that it could occur when subjects have more practice at the tasks. Initial studies provided little evidence that the PRP effect disappears with extensive practice. Gottsdanker and Stelmach (1971) had a subject practice the tasks used by Way and Gottsdanker (1968), described earlier, which required toward–away lever movements with the left and right hands, for 87 days using a 100-ms SOA. Both RT1 and RT2 decreased with practice, as did the PRP effect (as estimated by comparing RT2 to a single-task control). However, even at the end of practice, a PRP effect of 20 to 25 ms was evident. When the procedure was subsequently changed so that SOA was a variable 50, 100, 200, 400, or 800 ms, there was little transfer of this benefit to the other SOAs: RT was longer at both the 50- and 200-ms SOAs than at the practiced 100-ms SOA. Gottsdanker and Stelmach concluded that this indicated that the subject "had learned a special skill rather than a generally 'less refractory' mode of response" (p. 301). Unfortunately, they did not report cross-task correspondence effect data for this study, as Way and Gottsdanker did.

Recent years have seen a spate of practice studies. Much of the interest in practice has come from a challenge to the response–selection bottleneck model mounted by Meyer and Kieras (1997). They argued that evidence supporting the bottleneck model reflects a strategy adopted by subjects when the instructions state or imply that R1 should be made before R2. They propose that there is no capacity limitation in processing other than a bottleneck for response execution when the tasks require responses from the same output system (e.g., keypresses for Task 1 and Task 2). Thus, according to their strategic response deferment model, it should be possible for people to perfectly timeshare two tasks as long as they do not require the same response modality.

Van Selst, Ruthruff, and Johnston (1999) had subjects perform tone identification for Task 1, with subjects responding "low" to tones of 80 or 200 Hz and "high" to tones of 1250 or 3125 Hz. Task 2 required responding by pressing one of four keys, with the fingers of the right hand, to a visual stimulus from the set {1, 2, 3, 4, A, B, C, D}. For half the subjects, the digits were mapped in a left-to-right order to the four response keys (compatible mapping) and the letters arbitrarily (incompatible mapping), whereas for the other half, the letters were mapped in a left-to-right order and the digits arbitrarily. S1 preceded S2 by a variable SOA of 17, 67, 150, 250, 450, or 850 ms. Six subjects performed this dual task for 36 sessions of 400 trials each. The PRP effect decreased from 352 ms in the first session to approximately 40 ms in the later sessions, with five subjects showing an effect that averaged 50 ms and one subject showing no PRP effect. Most of the reduction in RT and in the PRP effect occurred over the first 18 sessions, and the size of the PRP effect showed an essentially perfect correlation with RT1 ($r^2 = .96$; slope = 1.02). These and other findings led Van Selst et al. to conclude, "Overall, it would appear most reasonable to conclude that the great bulk of the reduction in the PRP is produced by stage shortening" (p. 1281), that is, a shortening of response–selection duration for Task 1.

The S-R compatibility effect on RT2 for ordered versus arbitrary Task 2 mappings was 232 ms in the first three sessions but only 25 ms after practice. Van Selst et al. (1999) noted, "This dramatic reduction was consistent with previous evidence that practice serves primarily to decrease the duration of the response–selection stage" (p. 1278). The interaction of Task 2 compatibility with SOA was nonsignificant both early and late in practice, consistent with the prediction of the central response–selection bottleneck model.

Ruthruff, Johnston, and Van Selst (2001) conducted additional experiments on the five subjects from Van Selst et al.'s (1999) study who continued to show a PRP effect after practice. Experiment 1 was similar to that of Van Selst et al., except that subjects responded "same" or "differ" to a pair of tones for Task 1. The PRP effect in this experiment was 194 ms initially, compared with the 50-ms effect shown by these subjects at the end of the previous study, indicating that the new, unpracticed Task 1 created difficulty in performing the practiced Task 2. RT2 decreased little across four sessions, because Task 2 was already highly practiced, but the PRP effect decreased to 121 ms as RT1 decreased. Unlike in Van Selst et al.'s study, Task 2 compatibility showed an underadditive interaction with SOA on RT2, with the compatibility effect being 27 ms at the longest SOA and 10 ms at the shortest, suggesting that some of Task 2 response selection had become automatized.

In Experiment 2, Ruthruff et al. (2001) had their subjects perform the same Task 1 used by Van Selst et al. (1999) and a different Task 2 (pressing a left key to the visual letter X and a right key to Y with their right hand). The initial PRP effect was only 98 ms, which Ruthruff et al. attributed in part to the prior practice with Task 1, and decreased only to 78 ms over sessions, coinciding with a similar decrease in RT1. In Experiment 3, Ruthruff et al. returned to the original tasks used by Van Selst et al. for both Task 1 and Task 2, but used left-hand keypress responses for the high–low judgments of Task 1 instead of vocal responses. An initial PRP effect of 359 ms was obtained similar to that found at the beginning of Van Selst et al.'s study, and although the PRP effect decreased across 12 sessions, the decrease was not as great as when the responses for the two tasks were in different modalities.

Although Van Selst et al. (1999) and Ruthruff et al. (2001) found evidence for a PRP effect after extensive practice that was consistent with a response–selection bottleneck, Schumacher et al. (2001) argued that this outcome was due in part to their instructions emphasizing speed of Task 1 responses. Schumacher et al. reported evidence for "virtually perfect timesharing" in a similar experiment after modest amounts of practice. Their Experiment 1 used one task in which the stimuli were

220 Hz, 880 Hz, and 3520 Hz tones mapped to the vocal responses "low," "medium," and "high" and another task in which a visual O in one of three positions in a row was mapped to a corresponding keypress made with the index, middle, and ring fingers of the right hand. Performance for dual-task trials, on which the two stimuli were always presented simultaneously, was compared with performance on single-task trials. Over five sessions, each subject received a total of 2,064 dual- and single-task trials. They were instructed to perform each task quickly and accurately and not to constrain the order of the responses, and fast correct responses on both tasks were rewarded. Although RTs were slower initially in the dual-task context than for the single tasks, by the last session the dual- and single-task RTs did not differ significantly: For the auditory–vocal task, dual-task RT was 11 ms slower than single-task RT, and for the visual–manual task, it was 4 ms slower.

For Experiment 2, the same subjects participated in another session with a more standard PRP procedure in which the auditory–vocal task was Task 1 and the visual–manual task was Task 2. The stimuli for the two tasks were presented at varying SOAs of 50, 150, 250, 500, and 1,000 ms. In this experiment, unlike Experiment 1, instructions designated the auditory–vocal task as primary, and the reward structure supported compliance with the instructions. A PRP effect of slightly less than 200 ms was obtained, which Schumacher et al. (2001) interpreted as showing that dual-task interference that is not due to a structural bottleneck can be produced by task instructions.

Schumacher et al.'s (2001) Experiment 3 was similar to their Experiment 1, except that a new group of subjects performed a four-choice visual–manual task for Task 2, for which the mapping of stimulus locations to keypresses (executed by the four fingers of the right hand) was random. In the sixth (and last) session, consid-erable dual-task interference was evident in the mean RT data. However, although some subjects showed a lot of dual-task interference (> 150 ms), others showed little (20 ms or less), leading Schumacher et al. to conclude that even with an incompatible S-R mapping "some participants learn to time-share almost perfectly" (p. 107).

Hazeltine, Teague, and Ivry (2002) replicated Schumacher et al.'s (2001) Exper-iment 1 findings using a slightly different procedure in which spatially compatible responses were made with fingers of the right hand to circles that could occur in three locations and saying "one," "two," or "three" to low-, middle-, and high-pitch tones. They presented only six of the nine possible specific S1-S2 pairings during practice sessions and then used both those and the three "new" pairings in a transfer session. Seven of nine subjects met a criterion of little or no dual-task interference within eight sessions of practice. In the transfer session, RT for both types of dual-task trials did not differ from those for single-task trials, suggesting that the lack of dual-task costs was not due to forming compound S-R associations.

Because RT was faster for the visual–manual task than for the auditory–vocal task, Hazeltine et al. (2002) slowed the onset of response selection for the former task in Experiment 2 for the same subjects by increasing the difficulty of the target discriminability. This was accomplished by presenting distractor circles of different brightness in the locations not occupied by the target. The results still suggested little dual-task cost. In Experiment 3, the tasks from Experiment 1 were used, but after subjects reached a criterion of little dual-task cost, the visual mapping was changed. The leftmost stimulus was still responded to with the leftmost response, but the responses for the other two stimuli were switched. Although RT slowed with the mapping change, dual-task costs still were not evident.

Ruthruff, Johnston, Van Selst, Whitsell, and Remington (2003) argued that little or no dual-task interference after practice, as in Schumacher et al.'s (2001) and Hazeltine et al.'s (2002) studies, does not necessarily imply the absence of a response–selection bottleneck. Rather, they noted that the bottleneck may just be latent. The concept of a latent bottleneck can be understood by realizing that if Task 1 can be performed quickly, such that selection of R1 is completed before completion of stimulus identification for Task 2 even at short SOAs, then no PRP effect will be evident. Anderson, Taatgen, and Byrne (2005) confirmed the possibility of a latent bottleneck for Hazeltine et al.'s study, showing that their results could be simulated in the Adaptive Control of Thought – Rational (ACT-R) architecture, which possesses a central bottleneck.

Ruthruff et al. (2003) provided evidence that the bottleneck was latent for Van Selst et al.'s (1999) subject who had shown no PRP effect after practice. They first established that this subject still did not show a PRP effect when tested under the same conditions. They then changed the SOAs to include negative values, so that S2 occurred earlier in time relative to S1, and a PRP effect became evident. In a final phase, they changed Task 1 to one in which the subject had to indicate whether the third tone in a stream of three was higher or lower in pitch than the first by saying "gate" or "pike." This change to a new Task 1 reintroduced a PRP effect of 167 ms, which was explainable entirely in terms of the increase in RT1.

Task 2 compatibility had a similar effect on this subject's performance as on that for the other subjects from Van Selst et al.'s (1999) study. In Phase 3 of Ruthruff et al.'s (2003) investigation, Task 2 compatibility showed an underadditive interaction with SOA for RT2, being 19 ms at the three positive SOAs and 8 ms at the most negative SOA. In Phase 4, the interaction tended toward overadditivity, but nonsignificantly. Ruthruff et al. tentatively suggested that the bottleneck was relatively late when Task 1 was highly practiced but was pushed earlier in the sequence when it was changed to one that had not been practiced. If correct, this conclusion suggests that practice may cause nonideomotor compatible tasks to be performed like ones that are ideomotor compatible.

Ruthruff et al. (in press) recently reported evidence of bottleneck bypass in one situation. They had subjects perform the two tasks used by Van Selst et al. (1999): "low"–"high" vocal responses to tones of low or high pitch and pressing one of four keys to a visual stimulus from the set {1, 2, 3, 4, A, B, C, D}, with an ordered, compatible mapping for one subset and a random, incompatible mapping for the other. Before the dual-task test sessions, subjects practiced for more than 4,000 trials over eight sessions with either Task 1, Task 2, or both. When Task 1 was pitch identification and Task 2 visual symbol identification, all subjects showed evidence of a response–selection bottleneck regardless of whether practice had been with one or two tasks. However, when a different group of subjects performed visual identification as Task 1 and pitch identification as Task 2, 4 subjects out of 18 showed several indications that the prior practice allowed them to bypass the bottleneck. Thus, a few subjects showed evidence of bottleneck bypass under the favorable conditions of no input and output conflicts, preferred modality pairings, and extensive practice, but only for one order of the two tasks. Dual-task interference is clearly the rule, rather than the exception.

10.7 CHAPTER SUMMARY

Compatibility manipulations have been used often in studies of the PRP effect because evidence has consistently suggested that central response–selection processes are the source of much of the effect. Compatibility has several different consequences for dual-task performance. Increases in S-R compatibility for Task 1 typically reduce the PRP effect because Task 1 response selection can be completed sooner. Compatibility manipulations for Task 2 often have additive or overadditive effects, meaning that the cost of incompatibility on RT2 is at least as high at short as at long SOAs. Dual-task performance can benefit when mappings for the two tasks are consistent, even when the mappings themselves are spatially incompatible, and performance benefits when the stimuli and responses for the tasks are the same compared with when they are not (cross-task correspondence effects). Dual-task performance can be quite good when both tasks are ideomotor compatible or have been practiced in such a way as to reduce processing demands. For both situations, however, there typically is a residual dual-task cost. Whether this cost is due entirely to the strategies adopted by subjects to comply with the task instructions or represents at least in part a structural limitation is an issue of ongoing debate.

11 Models of Stimulus–Response Compatibility Effects

11.1 INTRODUCTION

Most research on S-R compatibility is conducted with the goal of clarifying the mechanisms responsible for such effects. More generally, the concern is to specify the nature of the relation between perception and action. We introduced several models of S-R compatibility in Chapter 1 and have periodically discussed those and other models with respect to specific issues throughout the book. In the present chapter, we provide more detailed consideration of several of the most influential models and approaches in contemporary research. Our intent is to provide a thorough treatment of the alternative views, including their strengths and weaknesses. We also discuss points of agreement and disagreement among the theories and issues in need of resolution.

11.2 THE ALGORITHMIC MODEL OF S-R COMPATIBILITY

Rosenbloom (1986; Rosenbloom & Newell, 1987) provided the first formal model of S-R compatibility and incorporated it into a model of practice. As described in Chapter 1, this S-R compatibility model is in the form of algorithms that characterize the response–selection process for different tasks and mappings. Variations of the model have been developed by Newell (1990) in Soar, his candidate model for a unified theory of cognition, and applied to issues in human–computer interaction by John and Newell (1987, 1990) and John, Rosenbloom, and Newell (1985).

11.2.1 ALGORITHMS FOR S-R COMPATIBILITY AND GOAL HIERARCHIES

Rosenbloom's (1986) objective was to produce "generalized, task-independent models of performance and learning" (p. 136). He took as his focus the class of S-R compatibility tasks, noting, "Though these tasks are important in their own right — especially in the areas of human factors engineering and man-machine interaction — no adequate model for the phenomenon currently exists" (pp. 136–137). Rosenbloom was most concerned with developing a chunking theory of practice, but he based this around his algorithmic model of S-R compatibility. He noted that from the algorithmic model, "We can derive both a metric model of stimulus–response compatibility, and the type of task performance-models that are required for the application of the chunking theory" (p. 137).

Rosenbloom's (1986) algorithmic model of S-R compatibility "is based on the supposition that people perform reaction-time tasks by executing *algorithms* (or programs) developed by them for the task" (p. 156). This algorithmic framework is similar to that proposed by Card, Moran, and Newell (1983), with which most human factors specialists are familiar, in which tasks are analyzed according to their goals, operators, methods, and selection rules (GOMS analysis). In fact, John et al. (1985) referred to the algorithmic theory as the GOMS theory. As noted in Chapter 1, within the algorithmic framework, two steps are required to provide a metric of S-R compatibility. First, the task situations must be analyzed to determine the algorithms employed. This can be accomplished by performing an abstract task analysis or by examining the behavior of subjects performing the task. Second, a complexity analysis of the algorithms must be performed, predicting mean RTs by assigning a time to each primitive step in the algorithm and ascertaining how many of each type of step would be executed when performing a trial of the task.

One of the basic ideas in the development of the algorithmic model was to specify in more detail how response selection could be based on a general response–selection rule, as Duncan (1977b) suggested. Rosenbloom (1986) noted,

> General rule-like behavior, as described by Duncan (1977b), occurs in this model when the same algorithm is used for the different trials in a condition. The more branches that exist to subalgorithms, the less rule-like the behavior is, and the more the model looks like a bunch of individual stimulus–response associations. (p. 158)

Rosenbloom generated the algorithms shown in Figure 11.1 to explain Duncan's (1977b) results for RT in a four-choice spatial task in which the stimulus on each trial was a vertically oriented line in one of four horizontal locations, and the response was a keypress at one of four horizontal response locations (see Chapter 7). The task was performed with a corresponding mapping (in which the response corresponding to the stimulus was to be made), an opposite mapping (in which the response at the mirror opposite location was to be made), or a mixed mapping (the mapping for the two inner locations was opposite and that for the two outer locations corresponding, as in the mixed algorithm shown in Figure 11.1, or vice versa).

Note that the opposite algorithm has one more step than the compatible algorithm, which accounts for why RT is longer with the opposite mapping. Also, the mixed mapping is more complex than either of the pure mappings, accounting for longer RT when the mappings are mixed than when they are not. Rosenbloom emphasized that the algorithmic model was not limited to spatial compatibility and provided a similar account of the results obtained by Morin and Forrin (1962) with digits and symbols mapped to vocal number-naming responses (see Chapter 7).

Rosenbloom (1986) integrated the algorithmic model of S-R compatibility with the chunking theory of learning, based on the concept of a *goal hierarchy* as a control structure. According to Rosenbloom, "The goal hierarchy can be related to the compatibility algorithms ... by imagining a fine-grained procedural hierarchy superimposed on top of the algorithms" (p. 177; see Figure 11.2). The top node in the hierarchy represents the whole task, the intermediate nodes represent segments of the task, and the terminal nodes are the operators that can be executed.

Algorithm *Duncan-Corresponding*:

```
BEGIN
Stimulus ← Get Stimulus ("Line")
Horizontal-Location ← Get Value (Stimulus, "Horizontal Location")
Make Response ("Press-Button", Horizontal-Location)
END
```

Algorithm *Duncan-Opposite*:

```
BEGIN
Stimulus ← Get Stimulus ("Line")
Horizontal-Location ← Get Value (Stimulus, "Horizontal-Location")
Horizontal-Location ← Negate (Horizontal-Location)
Make Response ("Press-Button", Horizontal-Location)
END
```

Algorithm *Duncan-Mixed*:

```
BEGIN
Stimulus ← Get Stimulus ("Line")
Horizontal-Location ← Get Value (Stimulus, "Horizontal-Location")
IF-SUCCEEDED In-Middle? (Horizontal-Location) THEN
    BEGIN
    Horizontal-Location ← Negate (Horizontal-Location)
    Make Response ("Press-Button", Horizontal-Location)
    ELSE
        IF-SUCCEEDED Outside-of-Middle? (Horizontal-Location) THEN
            Make Response ("Press-Button", Horizontal-Location)
END
```

FIGURE 11.1 Algorithms for the corresponding, opposite, and mixed mappings in Duncan's (1977b) task, developed by Rosenbloom (1986).

Initially, performance of the task requires decomposition of the goal into the subgoals and application of the operations for the terminal goals. During the course of performing the task, learning occurs through chunking of the subgoals. Each chunk replaces the processing required initially with a direct link between the parameter values for a specific instance and results. Creation of a chunk improves performance by reducing the computational requirements: "A chunk reduces the effective number of goals in the hierarchy without increasing the amount of computation performed by the remaining goals" (Rosenbloom, 1986, p. 191). In the model, a chunk can be created for a goal when that goal has just been successfully achieved and all of its subgoals have been processed by chunks. This restriction causes lower-level chunks in the goal hierarchy to be created prior to higher-level chunks. With enough practice, the whole goal hierarchy can become chunked and performance will attain an asymptotic level. A benefit of the integrated model is that it allows changes with practice to be modeled based on chunking within the goal

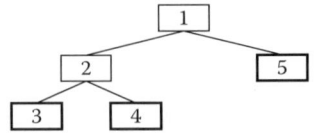

1. Press-Button-Under-Stimulus-Line
2. Get-Horizontal-Location-Of-Stimulus-Line
   3. Get-Stimulus-Line
   4. Get-Horizontal-Location-Of-Stimulus
   5. Press-Button-At-Horizontal-Location

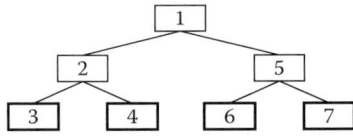

1. Press-Button-Opposite-Stimulus-Line
2. Get-Horizontal-Location-Of-Stimulus-Line
   3. Get-Stimulus-Line
   4. Get-Horizontal-Location-Of-Stimulus
5. Press-Button-Opposite-Horizontal-Location
   6. Compute-Opposite-Horizontal-Location
   7. Press-Button-At-Opposite-Horizontal-Location

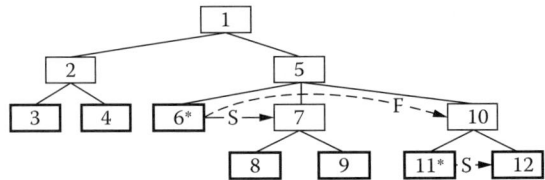

1. Press-Button-At-Or-Opposite-Stimulus-Line
2. Get-Horizontal-Location-Of-Stimulus-Line
   3. Get-Stimulus-Line
   4. Get-Horizontal-Location-Of-Stimulus
5. Press-Button-At-Or-Opposite-Horizontal-Location
   6. Is-Horizontal-Location-In-The-Middle?
   IF-SUCCEEDED *Is-Horizontal-Location-In-The-Middle?* Then
      7. Press-Button-Opposite-Horizontal-Location
         8. Compute-Opposite-Horizontal-Location
         9. Press-Button-At-Opposite-Horizontal-Location
   IF-FAILED *Is-Horizontal-Location-In-The-Middle?* Then
   10. Possibly-Press-Button-At-Horizontal-Location
       11. Is-Horizontal-Location-Outside-Of-Middle?
          IF-SUCCEEDED *Is-Horizontal-Location-Outside-Of-Middle?* Then
          12. Press-Button-At-Horizontal-Location

FIGURE 11.2 Goal hierarchies for the corresponding, opposite, and mixed mappings in Duncan's (1977) task. From "The chunking of goal hierarchies: A model of stimulus–response compatibility and practice," by P. S. Rosenbloom, 1986. In J. Laird, P. Rosenbloom, and A. Newell, *Universal subgoaling and chunking: The automatic generation and learning of goal hierarchies*, pp. 178–180. Copyright 1986 by Kluwer Academic Publishers. Reprinted with permission.

hierarchy for the task. Rosenbloom showed that the model provides good fits for practice data from Fitts and Seeger's (1953) and Duncan's (1977b) studies.

11.2.2 APPLICATIONS TO HUMAN–COMPUTER INTERACTION

John et al. (1985) applied the algorithmic model to a problem in human–computer interaction: the selection of an abbreviation technique for command names. Subjects performed a task in which a command appeared on the screen and the abbreviation was to be typed as quickly as possible. They performed under the following conditions: no abbreviation (i.e., abbreviations identical to the command stimuli; e.g., type the word *delete* in response to the command *delete*), rule based (vowel deletion, e.g., type *dlt* in response to *delete*, or special character, in which the abbreviation was the first letter of the command preceded by an arbitrary special character, e.g., */d* in response to *delete*), and nonsense (three-letter, meaningless nonsense syllables). For all except the no-abbreviation condition, a learning phase preceded the test phase. Both the initial-response time and execution time (the time between the initial response and typing of the last letter) were recorded. Initial response time was much shorter with no abbreviation (842 ms) than with rule-based abbreviation (1,457 ms), which was shorter than with the nonsense abbreviations (1,490 ms). For the rule-based abbreviations, initiation time was shorter for vowel deletion (1091 ms) than for special character (1,823 ms). For execution time, the special character condition was fastest (369 ms), followed by the nonsense abbreviations (866 ms), no-abbreviation (1,314 ms), and vowel deletion (1,394 ms).

John et al. (1985) developed algorithms for each task based on a priori task analyses and examination of subjects' reports and performance patterns in the experiment. A multiple regression analysis yielded a good fit to both the initiation- and execution-time data, with a mapping operator estimated to take about 60 ms, a retrieval operator 1,200 ms, and a motor operator 120 ms. John et al. noted that a qualitative analysis would have indicated only that the special character mapping is more complex than the vowel-deletion mapping. In contrast, the algorithmic theory

> allows us to go beyond that statement to say how much more complex the mapping is and explain the magnitude of the initial-response time differences. The model also allows the analysis to extend beyond the initial-response time to the execution time. (p. 219)

John et al. (1985) also indicated a limitation of the algorithmic theory: "At this point, the power of the theory is mainly explicative, because many of the algorithms were developed using empirical evidence" (p. 219).

John and Newell (1987) attempted to address this limitation by predicting the RTs a priori for two additional truncation abbreviation rules: a minimum-to-distinguish rule, in which the minimum number of letters to distinguish the commands in a set was used, and a two-letter truncation with exceptions to distinguish (e.g., *de* for *define* and *dl* for *delete*). John and Newell generated algorithms for these tasks through detailed task analyses and then used the values of the mapping, motor, perception, and retrieval operators determined in John et al.'s (1985) earlier study. Theoretical predictions for the initial-response times and execution times correlated highly with the observed times ($r^2 = .776$), indicating the value of the model for predictive purposes.

11.2.3 EVALUATION

Rosenbloom's (1986) algorithmic model of S-R compatibility perhaps has not received sufficient credit in the literature. It was the first formal model that allowed predictions for mean RT in different conditions to be derived. He integrated it with the chunking model of learning to provide a comprehensive model of the effects of S-R compatibility and practice on performance, which he implemented in a working computer program. The model clearly illustrates that spatial compatibility effects are a subset of the entire space of compatibility effects. It also emphasized the cognitive nature of the effects and that they are primarily a function of the task goals, as has been emphasized in more recent treatments of S-R compatibility. Finally, the work of John and Newell (1990) provided evidence that the algorithmic model could be used to make ballpark predictions about initiation time and execution time in a human–computer interaction task. These predictions were of sufficient accuracy to be of use for human factors engineers.

Despite these assets of the algorithmic model, it has not had widespread impact on current theorizing concerning S-R compatibility. There seem to be several reasons for this limited impact. One is that the model focuses exclusively on intentional response selection, or S-R translation, whereas research since the late 1980s has focused increasingly on automatic aspects of response selection. Not only does the model have nothing to say about phenomena that involve correspondence of irrelevant dimensions, such as the Simon effect, but it also is not applicable to set-level compatibility effects for different stimulus and response modalities or to crosstalk correspondence effects in dual-task performance. A second reason is that considerable subjectivity is involved in determination of the algorithms used to perform various tasks (Proctor & Dutta, 1992). After obtaining a set of empirical data, it is possible after the fact to derive one or more algorithms that will produce the differences in RT between the various conditions. John and Newell (1987) provided evidence that this problem can be dealt with to some extent by using previously determined parameter values, but the flexibility of the algorithmic model is a limitation when trying to derive precise predictions of the type needed for basic theory. In sum, the algorithmic model provides the most complete treatment of intentional response–selection processes to date, but it does not provide a comprehensive account of the full range of S-R compatibility effects.

11.3   SALIENT FEATURES CODING PERSPECTIVE

At approximately the same time as Rosenbloom (1986) was developing the algorithmic model of S-R compatibility, Proctor and Reeve (1985, 1986; Reeve & Proctor, 1990) proposed a perspective for intentional S-R translation that they called salient features coding (see Chapters 2, 5, and 8). As indicated in Chapter 2, the central idea behind the perspective is that "translation operations utilize the salient features of the stimuli and responses, with performance being best when there is a systematic correspondence between the salient stimulus features and salient response features" (Proctor & Reeve, 1985, p. 638). Because we have discussed the evidence relating to salient features accounts in detail in earlier chapters, we only summarize the major points here.

11.3.1 SPATIAL PRECUING EFFECTS

Proctor and Reeve (1985, 1986) developed the account initially to explain results obtained from four-choice tasks. When both the stimulus and response sets are four horizontal locations, the maximal precuing benefit occurs when the two left or two right locations are precued (Miller, 1982; Reeve & Proctor, 1984). Proctor and Reeve (1986) provided evidence that this precuing benefit could be attributed to the two halves of the arrays being salient, allowing the cued alternatives to be determined and prepared faster when the precue was consistent with this feature than when it was not. Although Proctor and Reeve (1986) did not develop a specific information-processing model of how feature correspondence speeded selection of the cued responses, it is easy to envisage models within Rosenbloom's (1986) algorithmic theory that would attribute the difference to additional operations that must be performed when the precue involves positions on each side of the display. This, in fact, is the essence of Adam, Hommel, and Umiltà's (2003) grouping model, described in Chapter 2, which attributes the smaller benefits obtained when locations on both sides are precued to a need to re-group the natural left–right groupings of the arrays.

11.3.2 TWO-DIMENSIONAL SYMBOLIC STIMULI MAPPED TO SPATIAL RESPONSES

Proctor and Reeve's (1985) primary reason for developing the salient features coding principle, though, was to explain why a similar benefit for the left and right locations is evident for tasks in which four two-dimensional symbolic stimuli are mapped to four horizontally arrayed response locations. When the stimuli are two letters (e.g., o, z) of large and small sizes, and letter identity is more salient than size, RT is shorter with a mapping for which the salient identity feature distinguishes the two left and two right locations (e.g., OozZ) than with a mapping for which it does not (e.g., OzoZ). Proctor and Reeve attributed this difference to correspondence of the salient letter-identity feature of the stimulus set with the salient left–right feature of the response set. Salient features correspondence has been able to predict S-R compatibility effects for a variety of two-dimensional stimulus and response sets in four-choice tasks (e.g., Proctor, Dutta, Kelly, & Weeks, 1994; see Chapter 5).

11.3.3 TWO-DIMENSIONAL SPATIAL S-R SETS

Vu and Proctor (2001a, 2002) provided a similar account in terms of relative salience of the two dimensions for the results of two-choice tasks in which the stimuli and responses are arrayed diagonally and can be coded with respect to both horizontal and vertical dimensions. The customary finding is that of right–left prevalence, where compatibility on the horizontal dimension is more important than compatibility on the vertical dimension (Nicoletti & Umiltà, 1984; see Chapter 5). Vu and Proctor (2001a, 2002) noted that the response environment of the typical experiment tends to make the horizontal dimension more salient (see also Hommel, 1996a). In a series of experiments, they showed that manipulations of the relative salience of the vertical and horizontal dimensions for both the stimulus and response sets systematically affect which dimension dominates the other, with a top–bottom prevalence effect evident for situations in which the vertical dimension is made salient. The strongest dominance of one dimension over the other was obtained when the relative salience for both the stimulus and response dimensions favored the same dimension.

Rubichi, Vu, Nicoletti, and Proctor (in press) recently reviewed the evidence concerning two-dimensional spatial compatibility effects. They concluded the following:

> Of the several explanations proposed to date for the prevalence effect in two-dimensional spatial coding, the only viable explanation is the relative salience account .... In general, the majority of the S-R set manipulations that have been studied so far provide support for the notion that the relative salience of the horizontal and vertical dimensions provided by how the features of the display–response configurations are grouped is the most important factor determining whether horizontal or vertical spatial codes have the strongest effect on responding. (pp. 34–35 of ms)

11.3.4 Orthogonal S-R Sets and Polarity Correspondence

The salient features coding perspective has also been applied to the orthogonal S-R compatibility effects for which two stimuli and two responses vary along orthogonal spatial dimensions. As described in Chapter 8, the central idea behind this application is that the stimuli and responses are coded asymmetrically, with one as salient, or positive polarity, and the other as nonsalient, or negative polarity. Weeks and Proctor (1990) attributed the up-right/down-left mapping advantage obtained in many situations to the fact that up and right tend to be coded as salient and down and left as nonsalient, and that mapping maintains correspondence between the salience, or polarities, of the codes. This account is the only workable explanation for the up-right/down-left advantage at present, and Cho and Proctor (2003) have amassed considerable evidence that also implicates such an account for the effects of orthogonal S-R compatibility that vary with response position and hand posture.

More generally, Proctor and Cho (in press) showed that asymmetric coding of the stimulus and response alternatives is prevalent in binary choice tasks. Correspondence of asymmetric stimulus and response codes is implicated as a major contributor to performance not only in orthogonal S-R compatibility tasks but also in word–picture verification tasks, parity judgment tasks, and the implicit associative test. Relational coding of the type that seems to underlie most orthogonal S-R compatibility effects is common for tasks in which one alternative can be represented as "salient," "positive polarity," or "figure" and the other as "nonsalient," "negative polarity," or "ground."

11.3.5 Evaluation

The salient features coding perspective has been applied primarily to intentional response selection, and therefore does not have much to say about automatic response activation. However, there is some evidence to suggest that the polarity coding effects may occur for irrelevant stimulus dimensions as well in at least some conditions

(Proctor & Cho, in press), and the perspective may also be applicable to the Simon effect in two-dimensional spatial tasks (e.g., Rubichi et al., in press). A major criticism of the salient features coding perspective is that explanations can be devised in a circular manner (e.g., Adam et al., 2003). However, the problem of circularity applies to coding accounts in general because one must know how the stimulus and response sets are coded in order to predict which mappings will be most compatible. There are several ways to deal with the problem of circularity (see, e.g., Proctor & Cho, in press), and it is not fatal to the salient features approach. In fact, the approach has been quite successful in providing coherent explanations for S-R compatibility effects that have proven difficult to explain in other manners.

11.4 A COMPUTATIONAL MODEL OF THE SIMON EFFECT

The algorithmic model of S-R compatibility and the salient features coding perspective focus primarily on intentional S-R translation processes and factors that influence their duration. With this emphasis, these approaches are primarily applicable to effects of S-R compatibility proper. Effects of correspondence of an irrelevant stimulus dimension with the response, as in the Simon effect, are usually attributed to a direct, or automatic, activation route, rather than to intentional translation. Zorzi and Umiltà (1995) developed a computational model of the Simon effect with this emphasis, introduced in Chapter 1.

11.4.1 THE ORIGINAL MODEL

Zorzi and Umiltà's (1995) model is implemented within a connectionist network of nodes connected by weighted excitatory and inhibitory links, which determine the activation values associated with the nodes. As shown in Figure 1.5, the model assumes that for the Simon task, the stimuli are represented by distinct position nodes, which code the left or right position of the stimulus, and feature nodes, which code the relevant stimulus dimension (e.g., red or green color). The responses are also represented as nodes signifying a left or right response. Both sets of stimulus nodes are linked to the response nodes. The left and right position nodes have fixed excitatory connections to the left and right response nodes, respectively. Activation of a position node not only primes the corresponding response, but it inhibits the noncorresponding response through an inhibitory link between the two responses. The relevant feature nodes are "learned" through presenting each of the four possible input patterns along with the required response to emulate what would occur when a subject is instructed about how to perform the task.

At stimulus onset, the input node for the irrelevant stimulus position is assigned a value of 1, and this activation then dissipates over time. Because discrimination of the relevant stimulus feature typically takes longer than discrimination of location, the input node for the relevant feature does not receive its activation value of 1 until several cycles later. After stimulus onset, the network continues to cycle until the activation of a response exceeds a threshold value.

Zorzi and Umiltà (1995) performed simulations showing that the model generated the Simon effect. The simulations showed both facilitation when stimulus location corresponded with the response and interference when it did not, and smaller effects when the relevant discrimination was more difficult than when it was easier. Their model also generates the common finding that the Simon effect decreases across the RT distribution.

11.4.2 THE MODIFIED MODEL

Tagliabue, Zorzi, Umiltà, and Bassignani (2000) modified Zorzi and Umiltà's (1995) model to examine how the Simon effect might be eliminated or reversed after practice with an incompatible spatial mapping, as described in Chapter 6. In the modified model, a distinction is made between short-term memory associations (STM links) and long-term memory associations (LTM links). STM links are established by instructions and are in effect for the specific task that is to be performed (e.g., pressing a right key to a red stimulus and a left key to a green stimulus), whereas LTM links are preexisting associations between stimulus and response locations that are relatively permanent. One modification of Zorzi and Umiltà's model was to place STM nodes between the input and response nodes to provide a temporary connection. As in the original model, activation of the position nodes is fast and automatic, and then decays, and the feature nodes do not become activated until after a delay. The STM nodes also introduce an additional delay in response activation produced by the relevant feature nodes.

The modified model simulated the basic Simon effect, with the overall effect having both facilitatory and inhibitory components. It also produced the customary pattern of the Simon effect decreasing as a function of time after stimulus onset. Tagliabue et al. (2000) compared whether the reduction of the Simon effect after practice with an incompatible spatial mapping could be attributed to changes in the LTM links or in formation of STM links between the stimulus position nodes and the response nodes. Their conclusion was that the practice effects were due to changes in the STM links. The modified model is thus able to account well for basic aspects of the Simon effect and the influence of prior practice with a spatial mapping on the Simon effect.

11.4.3 EVALUATION

The modified Zorzi and Umiltà (1995) model provides an adequate fit to many of the results obtained for the standard Simon effect for which location on the horizontal dimension is irrelevant. It also is sufficiently flexible to accommodate reversals in the Simon effect that are sometimes found after practice with an incompatible spatial mapping. However, because the model is limited to the Simon effect, it does not provide a comprehensive account of S-R compatibility effects.

11.5 THE DIMENSIONAL OVERLAP MODEL

The most influential model of the past two decades is Kornblum, Hasbroucq, and Osman's (1990) dimensional overlap model (see Figure 11.3). It is a dual-route model that includes both an automatic response–selection route and an intentional

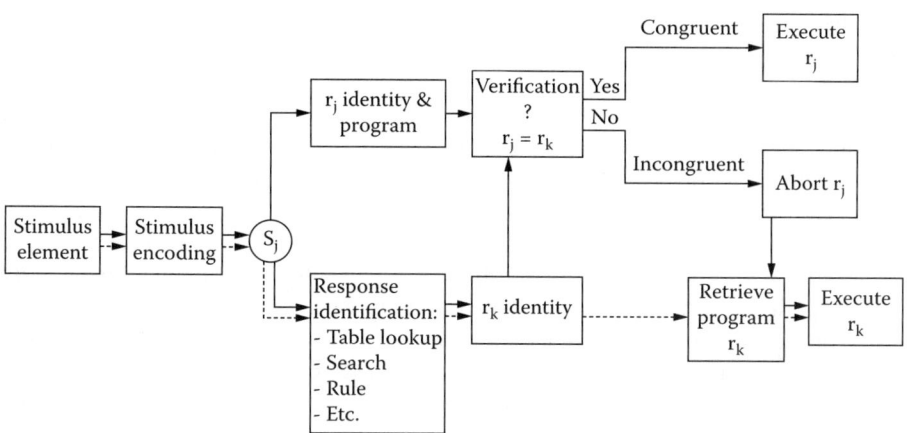

FIGURE 11.3 Kornblum et al.'s (1990) dimensional overlap model. From "Dimensional overlap: Cognitive basis for stimulus–response compatibility — A model and taxonomy," by S. Kornblum, T. Hasbroucq, and A. Osman, 1990, *Psychological Review, 97,* p. 257. Copyright 1990 by the American Psychological Association. Reprinted with permission.

translation route. Kornblum et al.'s model is intended to be a general model that can explain rule-based translation effects as well as Simon effects and effects of S-R compatibility proper.

11.5.1 THE ORIGINAL MODEL

Kornblum et al. (1990) noted,

> At the core of our model is the idea that when a particular S-R ensemble produces either high or low compatibility effects, it is because the stimulus and response sets in the ensemble have properties in common, and elements in the stimulus set automatically activate corresponding elements in the response set. (p. 253)

The statement "properties in common" refers to dimensional overlap, or similarity, between the stimulus and response sets. The dimensional overlap model is an exemplar of the category of dual-route models, with one route producing automatic activation of the corresponding response when sets have dimensional overlap (the top route in Figure 11.3) and the other producing intentional identification of the correct response (the bottom route in Figure 11.3).

According to the model, on presentation of a stimulus, the corresponding response is automatically activated and programmed regardless of the S-R mapping that is in effect for the task. The assigned response, which may or may not be the one that is automatically activated, is identified by way of the intentional response-identification route. Subsequent to identification, a verification process is performed to determine whether the activated response is the same as that identified as correct. If it is the same, then that response is executed; if not, then the automatically activated response must be aborted and the program for the identified response retrieved, and that response is then executed. Thus, automatic activation of the corresponding response contributes a benefit to RT when that is the correct response and a cost when it is not.

Dimensional overlap can also influence the duration of the response-identification process. When there is no dimensional overlap between the stimulus and response sets or the mapping for sets with overlap is random, the specific response to each stimulus must be determined by searching through a table or list of S-R pairs. However, when there is a systematic relation, response identification can benefit from application of a response–selection rule. The simplest rule is that of identity; that is, execute the response that corresponds to the stimulus. This identity rule is applicable when the mapping of stimuli to responses is compatible and leads to the quickest response identification. However, other rules, such as "mirror opposite," can also reduce identification time relative to searching for the specific assigned response, in agreement with results described in Chapters 1 and 3 (Fitts & Deininger, 1954; Morin & Grant, 1955). Thus, the intentional response-identification route provides an additional benefit for a compatible mapping compared with an incompatible mapping, as well as a benefit for a mapping that is not strictly compatible but has a systematic S-R relation compared with one that does not.

The dimensional overlap model attributes the Simon effect entirely to automatic activation of the corresponding response, consistent with most other explanations of the effect. A major difference is that it places the Simon effect at the level of response programming, whereas most other accounts place it at the level of abstract response codes. The model also attributes the effect of S-R compatibility proper in part to automatic activation. Because this activation occurs regardless of mapping, it should produce benefits for a compatible mapping and costs for an incompatible mapping. Finally, the model has an explicit inhibitory mechanism for aborting, or suppressing, the automatically activated response when it is incorrect.

11.5.2 EXAMINATIONS OF THE ORIGINAL MODEL

All of these aspects of the dimensional overlap model have been subjected to critical analysis. Hasbroucq, Guiard, and Ottomani (1990) questioned whether the automatic activation route of the model is necessary. Specifically, they cited Hasbroucq and Guiard's (1991) evidence, summarized in Chapter 6, that the Simon effect is due to competition in stimulus identification and not response selection. They noted that, in the dimensional overlap model, "the assumption of an automatic activation is only necessary to account for the Simon effect. Such an assumption appears to be pointless from the moment that … one ceases to take the Simon effect into consideration for modeling the mechanisms of SRC [S-R compatibility]" (p. 328). Consequently, Hasbroucq et al. focused on only the intentional response-identification route in what they called the list-rule model, which emphasizes the distinction between response selection based on rules versus searching through S-R pairings, as in Kornblum et al.'s (1990) model. However, as discussed in Chapter 6, the view that the Simon effect is due primarily to stimulus-identification processes has not stood up well to subsequent evidence, which creates difficulty for Hasbroucq et al.'s analysis.

Kornblum and Lee (1995) tested predictions of the dimensional overlap model in their Experiment 1 with stimulus and response sets that did and did not overlap

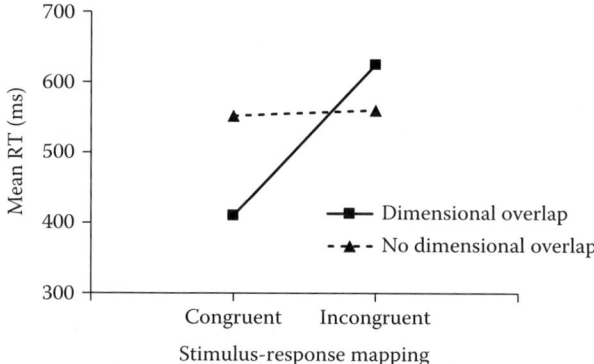

FIGURE 11.4 Results from Kornblum and Lee's (1995) tests of the dimensional overlap model. Reaction time (RT) in milliseconds as a function of S-R mapping for the S-R ensembles with dimensional overlap and those without.

on the relevant dimensions. They paired two stimulus sets (letters: B, J, Q, Y; locations on hand icons: index and middle fingers of left and right hands) with two response sets (vocal letter names: B, J, Q, Y; keypresses: index and middle fingers of each hand). Two combinations of S-R sets had dimensional overlap (letters to letter names and hand icons to keypresses), whereas two did not (hand icons to letter names and letters to keypresses). For the combinations with dimensional overlap, both compatible and incompatible mappings were used in different trial blocks, and for the combinations without dimensional overlap, two alternative mappings were used. As expected, there was no difference in RT for the two mappings of the nonoverlapping S-R combinations (see Figure 11.4). However, for the overlapping ones, RT was more than 200 ms shorter with the compatible mapping than with the incompatible mapping. Importantly, as illustrated in the figure, relative to the non-overlapping sets, the overlapping sets showed both facilitation with the compatible mapping and interference with the incompatible mapping, as predicted by the dimensional overlap model, although the facilitation was considerably larger than the interference. Kornblum (1992) reported similar results for a six-choice version of these tasks.

However, Proctor and Wang (1997a) provided evidence against the assumption that automatic activation of the corresponding response occurs in tasks examining S-R compatibility proper when the mapping is incompatible. Because dimensional overlap is higher when the stimulus and response sets have perceptual, as well as conceptual, overlap than when they do not, the benefit for compatible mappings and the cost for incompatible mappings should be larger for sets that have both types of overlap than for those that have only conceptual overlap. Proctor and Wang cited studies by Wang and Proctor (1996) and Proctor and Wang (1997b) that did not conform to this prediction (see Chapter 3). For example, Wang and Proctor had subjects perform two-choice left–right tasks with compatible and incompatible mappings for each combination of two stimulus sets and two response sets: Visual left and right locations or the visual words LEFT and RIGHT mapped to left and right keypresses or the vocal utterances "left" and "right." All of these sets have conceptual overlap, whereas only the spatial–manual and verbal–vocal relations also have perceptual overlap. Although the compatible mappings showed the predicted benefit for perceptual overlap, the incompatible mappings did not show the predicted cost. Several interpretations of these results are possible, but they suggest that automatic activation of the corresponding response does not occur for all mappings.

Shiu and Kornblum (1996) presented evidence for inhibition of the automatically activated, corresponding response when it is incorrect. They examined performance with an incompatible mapping in a variant of a negative priming task. Negative priming refers to the finding that RT is longer when the imperative stimulus on the current trial (the probe trial) was a distractor stimulus on the previous trial (the prime trial), compared with trials for which the probe stimulus is unrelated to the prime trial (Tipper & Milliken, 1996). In Shiu and Kornblum's experiment, four line-drawing pictures (bike, car, boat, or plane) and four written words corresponding to the pictures were used as stimuli, with each picture/word assigned to an incompatible vocal naming response from the same set (e.g., the picture and word *PLANE* were assigned the response "car"). Only a single stimulus was presented on each trial, with the idea being that a stimulus would activate its corresponding response, which would then have to be inhibited. Consistent with this hypothesis, responses were slower for trials on which the name of the prime trial stimulus (i.e., the corresponding response on the prime trial) was the correct response on the probe trial (e.g., respond "car" to picture or word *PLANE*, followed by "plane" to picture or word *BIKE*) than when there was no relation between the successive trials (e.g., respond "boat" to *PLANE*, followed by "car" to *BIKE*). Shiu and Kornblum interpreted this result as a negative priming effect resulting from inhibition of the congruent response for the prime stimulus.

Read and Proctor (2004) examined sequential effects of this type for a more standard spatial compatibility task in which four stimulus locations arranged in a row were mapped to a row of four keys, operated by the left middle, left index, right index, and right middle fingers. In Experiment 1, the S-R mapping was analogous to that used by Shiu and Kornblum (1996) in that the stimuli and their assigned responses were not related by a simple rule. This experiment yielded results similar to those of Shiu and Kornblum. However, when the mapping was changed to one that could be described by a simple rule (e.g., each stimulus was assigned to the mirror opposite response location), RT was shorter, rather than longer, when the incorrect compatible response to the prime trial stimulus became the correct response on the probe trial, compared with the neutral condition. The results of Read and Proctor's experiments suggest that if inhibition of an automatically activated corresponding response occurs, it is restricted to situations in which response selection must focus on the individual S-R pairings.

11.5.3 Kornblum's (1992) Expanded Model

Kornblum (1992, 1994) expanded the dimensional overlap model to include a stimulus-identification stage (see Figure 11.5; the complete model is shown in Figure 1.4). This stage is distinct from the response–selection stage. It sends a stimulus vector consisting of the attributes encoded for the stimulus to the response–selection

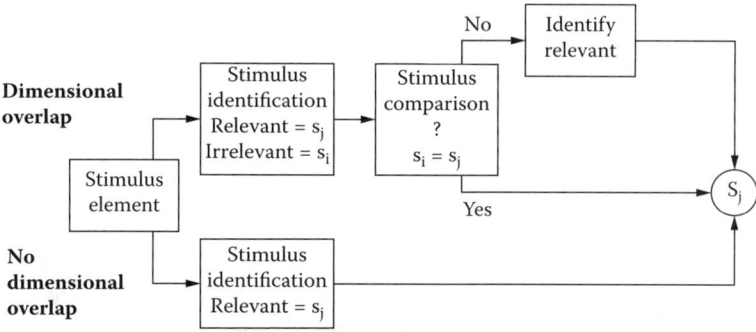

FIGURE 11.5 Kornblum's (1992) elaborated stimulus-identification module for the dimensional overlap model. From "Dimensional overlap and dimensional relevance in stimulus–response and stimulus–stimulus compatibility," by S. Kornblum, 1992. In G. E. Stelmach and J. Requin (Eds.), *Tutorials in motor behavior II* (p. 753). Copyright 1992 by Elsevier Science Publishers. Reprinted with permission.

module. When a stimulus has only a relevant dimension or the irrelevant and relevant dimensions do not overlap, then a single stimulus identification code is produced and passed on to the response–selection stage. When the relevant and irrelevant stimulus dimensions overlap, two identity codes are produced, one for each dimension. These stimulus codes are compared, and if they are the same, the identity code is passed on to the response–selection stage. However, if they are different, which of the codes is relevant must be determined before that code can be passed on to response selection. This extra step of identifying the relevant stimulus dimension lengthens the duration of the stimulus-identification stage, but only the single code for the relevant dimension is sent on to the response–selection stage.

The addition of stimulus–stimulus overlap allows the model to account for effects such as the Eriksen flanker effect and the Stroop effect, for which the relevant and irrelevant stimulus dimensions overlap (e.g., letter identity in the case of the standard flanker task and color in the case of the standard Stroop color-naming task). For cases in which the responses are manual and the relevant and irrelevant stimulus dimensions are nonspatial, as in the flanker task and keypress versions of the Stroop task (type 4 ensembles in Kornblum's, 1992, taxonomy; see Chapter 1), the correspondence effects are attributed entirely to the stimulus-identification stage. When the stimulus dimensions also overlap with the response dimension, as when the responses are spoken words from the same set as the relevant and irrelevant words (type 8 ensembles), the response–selection stage also contributes to the effects.

In Kornblum's (1992) taxonomy, type 3 ensembles (which include the Simon task) have overlap between the irrelevant stimulus dimension and the response, and type 4 ensembles (which include the flanker task) have overlap between the relevant and irrelevant stimulus dimensions. Because type 7 ensembles include both types of overlap, and their effects are attributed to different stages, response selection and stimulus identification, respectively, the model predicts that the correspondence effect for type 7 ensembles should be an additive combination of those for types 3 and 4 ensembles. Kornblum (1994) tested this prediction of the dimensional overlap model in an experiment in which left and right keypresses were made to the relevant stimulus dimension of red or green stimulus color. The stimuli were presented inside of an outline rectangle and could be accompanied by an irrelevant word (BLUE, GREEN, NOVEL, ELBOW). The position of the colors (left or right half, or upper or lower half) in the rectangle was also irrelevant. The irrelevant information could appear simultaneously with the relevant information or precede it by 200 ms. Subjects received four blocks of trials: pure S-S overlap (for which the irrelevant location word was congruent, incongruent, or neutral with respect to the relevant color); pure S-R overlap (for which the irrelevant stimulus location was congruent, incongruent, or neutral with respect to the correct response); mixed S-S and S-R overlap (for which the trial types for the two pure blocks were mixed); and composite S-S/S-R overlap (which included trials on which the irrelevant stimulus dimension was congruent or incongruent with both the relevant stimulus dimension and the correct response).

Similar results were obtained for the pure and mixed blocks: S-S overlap showed a nonsignificant 5-ms correspondence effect at the 0-ms lag but a 48-ms effect at the 200-ms lag, whereas S-R overlap showed significant correspondence effects of 40 ms at the 0-ms lag and 32 ms at the 200-ms lag. The composite S-S/S-R overlap condition likewise showed a nonsignificant S-S correspondence effect of 2 ms at the 0-ms lag but a 36-ms S-R correspondence effect. The results at the 200-ms lag were less consistent with the model's predictions, showing a 53-ms S-S correspondence effect and only a 17-ms S-R correspondence effect. On the whole, the results of Kornblum's (1994) experiment are generally consistent with the predicted independence of S-S and S-R correspondence effects.

Hommel (1997) performed additional tests of the prediction that S-S correspondence effects should be independent from S-R compatibility effects. He also used a type 7 ensemble, for which the relevant stimulus dimension was blue or green color and the irrelevant dimensions were the German words for BLUE and GREEN, and whether the stimulus appeared in a left or right location. An S-S correspondence effect of 20 ms and an S-R correspondence effect of 27 ms were obtained that did not interact. However, RT distribution analyses showed a pattern similar to that produced by Kornblum's (1994) SOA manipulation: The S-R correspondence effect was a decreasing function of RT quintile, but the S-S correspondence effect was an increasing function. In his Experiment 2, Hommel presented a target letter H or S to the left or right of fixation. It was flanked by four instances of the same (congruent) or opposite (incongruent) letter, two to the left and two to the right for half of the subjects, and two above and two below for the other half. S-S and S-R correspondence interacted, with the S-R correspondence effect being largest when the relevant and irrelevant stimulus dimensions corresponded than when they did not. This term also interacted with quintile in the RT distribution analysis: S-S and S-R correspondence effects had additive effects at the first and last quintile, but interacted underadditively at the intermediate ones. Hommel's third experiment used the global letter H or O composed from local letters H or O, which could correspond or not correspond with the global letter. In different sections of the task, the global letter or local letter was to be identified by a keypress. S-R correspondence interacted with S-S correspondence, again producing a larger S-R correspondence effect when there was S-S correspondence than when there was not.

Kornblum, Stevens, Whipple, and Requin (1999) focused on Kornblum's (1994) results suggesting that the time course of S-S correspondence effects is different from that for S-R correspondence effects and examined this issue in more detail. Their Experiment 1 used stimuli similar to those used by Kornblum, but examined only type 3 and type 4 ensembles at four intervals between onsets of the irrelevant and relevant information (0, 50, 100, and 200 ms). Consistent with Kornblum's previous results, the effect of S-S consistency increased across SOAs (being 2, 2, 29, and 53 ms at intervals of 0, 50, 100, and 200 ms, respectively), whereas the effect of S-R consistency decreased from 50 ms onward (being 34, 40, 34, and 15 ms, respectively). Experiment 2 used type 7 stimuli at intervals of 0, 100, 200, 400, and 800 ms. In agreement with Experiment 1, the S-R correspondence effect decreased monotonically as SOA increased, whereas the S-S correspondence effect increased up to 400 ms and then showed a decrease at the longest, 800-ms SOA. Kornblum et al. summarized the results of their Experiment 2, as indicating that "It is as if the overall RT at long delays was primarily determined by S-S consistency, just as at short delays it was influenced by S-R consistency, with a mixture of influences in between" (p. 693). In addition, the effects of S-S and S-R correspondence interacted marginally, with each correspondence effect being larger when the other relation was corresponding than when it was noncorresponding.

11.5.4 KORNBLUM ET AL.'S (1999) PARALLEL DISTRIBUTED PROCESSING MODEL

To account for the interaction between S-S and S-R correspondence and the different time courses of the effects, Kornblum et al. (1999) proposed a computational, parallel distributed processing model (which they called DO'97) based on the original dimensional overlap model. The key aspect of the model is that it postulates an inverted U-shaped irrelevant stimulus activation function. The architecture of the model is shown in Figure 11.6. The model consists of an input layer and an output layer. The input layer contains modules for the relevant and irrelevant stimulus dimensions; activation of a unit within a module represents activation of a stimulus feature. The relevant stimulus dimension is connected to the response module of the output layer by a controlled processing pathway, whereas when an irrelevant stimulus dimension overlaps with either the relevant stimulus dimension or the response, it is connected to the relevant stimulus unit or the response unit by an automatic processing connection. The automatic processing pathways consist of both excitatory connections to corresponding units and inhibitory connections to noncorresponding units, enabling the model to produce both facilitation and interference effects.

A distinctive feature of the model is that a threshold exists for the stimulus layer as well as for the response layer, with a constant positive value sent to the response layer when a relevant stimulus unit reaches threshold. No processing takes place in the response module until the stimulus threshold has been attained, meaning that automatic processes due to S-R dimensional overlap cannot influence performance until this time. The activation function for irrelevant stimulus units first increases and then decreases over time. When the irrelevant information is presented prior to the relevant information, the activation function for the irrelevant units will be shifted to earlier relative to that for the relevant unit. Consequently, as the asynchrony is

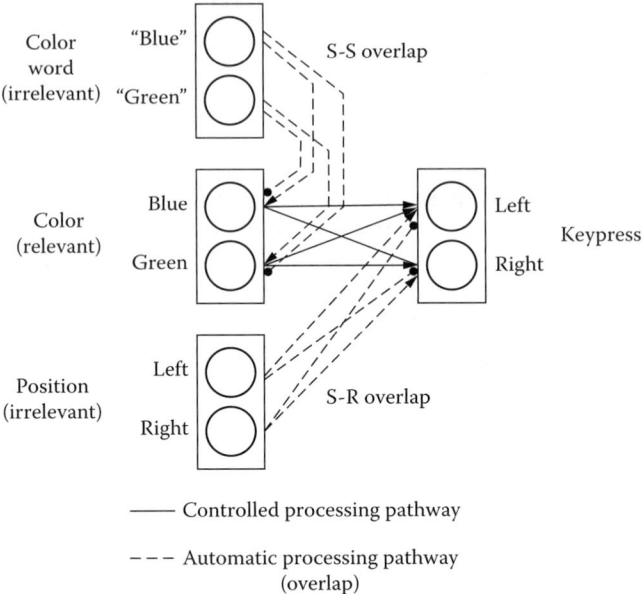

FIGURE 11.6 Kornblum et al.'s (1999) parallel distributed processing version of the dimensional overlap model. From "The effects of irrelevant stimuli: 1. The time course of stimulus–stimulus and stimulus–response consistency effects with Stroop-like stimuli, Simon-like tasks, and their factorial combinations," by S. Kornblum, G. T. Stevens, A. Whipple, and J. Requin, 1999, *Journal of Experimental Psychology: Human Perception and Performance, 25*, p. 695. Copyright 1999 by the American Psychological Association. Reprinted with permission

increased, the S-S correspondence effect will first increase and then decrease. Because activation from the irrelevant stimulus unit cannot begin to affect response selection until a relevant stimulus value has been passed to the response–selection stage, the activation function will most often be past its peak, resulting in a monotonically decreasing S-R correspondence effect.

Zhang, Zhang, and Kornblum (1999; Zhang & Kornblum, 1998) presented a slightly different parallel distributed processing model in which stimulus–stimulus overlap is modeled by convergence of the two input modules onto an intermediate module. They showed that the model could account for a variety of results for all eight ensemble types in Kornblum's (1992) dimensional overlap taxonomy. Kornblum and Stevens (2002) extended the dimensional overlap model to repetition effects, showing that the results from several experiments could be fit by assuming that the information requirements are less on repetition trials than on nonrepetition trials.

11.5.5 EVALUATION

Kornblum et al.'s (1990) dimensional overlap model was the first fully articulated dual-route model of response selection. Its emphasis on dimensional overlap, or similarity, as being central to S-R compatibility effects has had wide-ranging impact, as has the assumption that a variety of correspondence effects can be encompassed and related to each other within a single model. The model also includes a mechanism for inhibition of automatically activated responses, which provides a point of contact with the literature on negative priming effects. In its more recent incarnations, Kornblum and colleagues have shown that the general ideas can be incorporated into computational models that maintain many of the original assumptions while expanding the domain of the model to include time course and repetition effects. Despite the positive contribution of the model, some aspects of it have been questioned, including the assumption that a stimulus automatically activates its corresponding response in all situations and the model's emphasis on a unidirectional flow of information from stimulus to response.

11.6 THE THEORY OF EVENT CODING

Hommel, Müsseler, Aschersleben, and Prinz (2001) proposed a framework for perception and action planning, called the theory of event coding, that is intended to explain many aspects of perception–action relations, including S-R compatibility effects. In the framework, "perceptual events and action plans are coded in a common representational medium by feature codes with distal reference" (p. 849). The basic idea is that stimuli and responses are represented in the same system, or representational domain, on which perception, attention, intention, and action operate. These representations refer to external events in the environment and not to proximal sensory or motoric effects on the organism.

The theory of event coding places emphasis on structures called event codes, or event files (Hommel, 2004; Hommel et al., 2001). The rationale is an extension of one proposed by Kahneman, Treisman, and Gibbs (1992) to explain object-specific priming effects in letter-naming tasks. Kahneman et al. interpreted their results in terms of the concept of a temporary episodic representation called an object file, within which features of an object are linked and integrated. Hommel and colleagues proposed that actions can also be conceived of as being coded as linked and integrated features in action files. Because objects and actions are coded within a common representational system, the whole event, consisting of stimuli and responses, is integrated in an event file. Features that are bound in an event file are temporarily less available for other perceptions and actions.

The theory of event coding takes an ideomotor view of action (e.g., Koch, Keller, & Prinz, 2004), according to which intentional goals play a significant causal role in action. Actions are conceived of as being represented in terms of their anticipated sensory consequences, or response effects, with bidirectional associations existing between movements and the subsequent response effects. Consequently, the theory emphasizes that compatibility between responses and the effects they produce is important to performance.

11.6.1 BLINDNESS TO RESPONSE-COMPATIBLE STIMULI

Because of its emphasis on event codes in a common representational system, the theory of event coding predicts that action should influence perception. Müsseler and Hommel (1997a, 1997b) obtained evidence for such an influence with a phenomenon called blindness to response-compatible stimuli. In Müsseler and Hommel's (1997a) Experiment 1, each trial began with a left- or right-pointing arrow in the center of the screen. After determining the direction in which it pointed, the participants were to prepare to execute a double keypress with the two index fingers, after which they were to make as quickly as possible a speeded keypress response on the side compatible with the arrow direction. The double keypress initiated a brief presentation of a second arrow just to the right of screen center, followed by a pattern mask. About 1 s after the speeded response to the first arrow, the mask went off, and the subject was to make another left or right keypress, unspeeded, to indicate the direction in which the second arrow pointed. The primary finding was that identification of the second arrow was worse when it corresponded with the response being made to the first arrow (75.6% accuracy) than when it did not (83.6%). This is the phenomenon called blindness to response-compatible stimuli.

Müsseler and Hommel (1997a) obtained similar results when the second response was a verbal report of left or right (Experiment 2) and when the first stimulus was the written word LEFT or RIGHT instead of an arrow (Experiment 4), indicating that physical similarity of the stimuli or responses is not important. In their Experiment 5, the mapping of the first arrow to the speeded response was incompatible, and the blindness was to the stimulus congruent with the signaled response and not with the direction of the first stimulus. Müsseler and Hommel (1997b) showed that the blindness effect could be obtained as well for a detection task in which the subject was to indicate the presence or absence of the second arrow, regardless of which direction it pointed, by pressing a right or left key, respectively. Müsseler and Hommel (1997a, 1997b) attribute this blindness effect to inhibition of action-related codes produced when the response is executed, resulting in a refractory phase during which the individual is less sensitive to a stimulus sharing the same location code.

To examine the time course of the blindness effect, Wühr and Müsseler (2001) used a procedure in which three clicks, separated by 700 ms, signaled when to execute the responses. The first click was a warning, the second a signal to press both response keys (the buttons of a computer mouse, operated with the index and middle fingers of the right hand), and the third a signal to make the cued left or right button response. The second arrow was presented simultaneous with the onset of the signal to respond (the third click) or 280 or 980 ms before or after the signal. Contrary to Müsseler and Hommel's (1997a, 1997b) refractoriness account, poorer identification accuracy for a response-compatible second stimulus was evident when it preceded or was simultaneous with the signal to respond, but not when it occurred after that signal. Consequently, Wühr and Müsseler modified the account to state, "The perceptual processing of a stimulus feature is impaired as long as a shared perception–action feature code is integrated into the representation of the to-be-executed response" (p. 1260). Stevanovski, Oriet, and Jolicœur (2002) confirmed that the blindness effect is largest prior to execution of the prepared response using the original Müsseler and Hommel (1997a) double-keypress procedure, but with the second arrow presented at delays of 0, 50, 150, 250, 400, and 1,000 ms following the double keypress. The effect was much larger at the four shortest intervals, for which the stimulus typically occurred prior to execution of the cued response, than at the two longer intervals. However, there was a residual effect at these intervals, after response execution.

Müsseler and Hommel (1997a, Experiment 3) found no blindness effect when subjects were told to ignore the first arrow stimulus and only had to make the double-keypress response in the sequence. This result is consistent with the hypothesis that preparation and/or execution of the response is necessary to obtain the blindness effect. However, Stevanovski, Oriet, and Jolicœur (2003) reported four experiments showing a blindness effect for response-compatible stimuli in the absence of a response. They obtained an effect both when the initial arrow directed attention to a left or right location in which the second stimulus was to occur and when the arrow was completely uninformative. Stevanovski et al. concluded that the absence of a blindness effect when a response was not required in Müsseler and Hommel's study was due to their blank interval between offset of the first stimulus and onset of the second being too long (1,000 ms in Müsseler and Hommel's study compared with 200 ms in Stevanovski et al.'s). In a final experiment, Stevanovski et al. showed no blindness effect with their procedure when the interval was increased to 1,000 ms. Because their effects tended to be smaller than those in experiments requiring a response to the first stimulus, Stevanovski et al. decided that there are two components to the blindness effect: an action-related component and a symbolic component. They allowed that the feature binding account proposed by Hommel et al. (2001) could still explain their results if it is assumed that the symbolic component reflects binding of the perceptual feature in the object file and the action-related component feature binding in the action file.

11.6.2 INTERACTIONS OF PERCEPTION AND ACTION IN DYNAMIC TASKS

The theory of event coding emphasizes dynamic, as well as static, action effects. Research conducted by Schubo, Aschersleben, and Prinz (2001) illustrates perceptual–motor interactions in dynamic tasks of the type implied by the theory. Schubo et al. had subjects perform a serial overlapping response task in which a sequence of stimuli was presented, each requiring a response, but the response to be made on the current trial was the one corresponding to the stimulus on the previous trial. Thus, on each trial, the stimulus for that trial had to be encoded while the response designated by the stimulus on the previous trial was being executed. The stimulus on each trial was a dot that started on the left side of a display screen and followed a sinusoidal trajectory of low, medium, or high amplitude, with the cycle completed in 2 s. The response to each stimulus was a sinusoidal movement of the same magnitude, traced out with a stylus on a digitizer tablet, which was also to be completed in 2 s. The important finding was that the medium-amplitude response movement showed a contrast effect with the concurrent stimulus amplitude: The movement amplitude was highest when paired with the low amplitude stimulus, intermediate when paired with the medium-amplitude stimulus, and lowest when paired with the high-amplitude stimulus.

Schubo et al. (2001) also observed a contrast effect of the action on perception, with the amplitude of the subsequent movement to a stimulus influenced in a similar manner by the amplitude of the movement with which that stimulus was paired when it was being viewed. In a subsequent study, Schubo, Prinz, and Aschersleben (2004) showed that this contrast pattern was only evident with a short intertrial interval of 2 s, being absent at intertrial intervals of 4 and 6 s. Moreover, when the intertrial interval was 8 s, an assimilation effect was obtained such that a high-amplitude response movement paired with the stimulus encoding led to a higher amplitude movement on the next trial, and so on. The authors interpreted their results in terms of the theory of event coding, concluding that contrast between stimulus and response codes emerges when two S-R assignments compete with each other in perception. Assimilation in memory occurs after the perceptual competition ends.

11.6.3 RESPONSE–EFFECT COMPATIBILITY

Evidence that the relation between a response and the effect it produces is important to action has been demonstrated in studies of response–effect compatibility, that is, compatibility effects between a response and its effect on the environment. Kunde (2001, Experiment 1) had subjects perform a four-choice task in which they responded to a centered stimulus of one of four colors with one of four keypresses made with the index and middle fingers of the two hands. The response lit one box in a row of four on the lower part of the screen, with a corresponding mapping (for which the box corresponding to the key was lit) or a noncorresponding mapping (for which the two right keypresses were mapped to the two left boxes and the two left keypresses to the two right boxes). The major finding was that responses were 21 ms faster with the corresponding response–effect mapping than with the noncorresponding mapping.

Response–effect compatibility occurs for response features other than position. Kunde (2001) showed a similar response–effect compatibility effect for tasks in which a pressure-sensitive response key was pressed softly or forcefully, and these alternatives were mapped to response effects of quiet or loud tones. The compatibility effect was obtained for a standard two-choice task in which stimulus color signaled the response to be made (Experiment 2), as well as for a free response task in which subjects were free to make either response to a go signal (with the constraint that they make both responses about equally often). Kunde (2003) extended the results to short- and long-duration keypresses mapped to short- and long-duration tones. For both intensity and duration, response–effect compatibility has been shown to influence performance even when the response is cued with 100% certainty as long as 2 s before onset of the imperative stimulus, suggesting that such effects can influence response initiation. Koch and Kunde (2002) demonstrated that response–effect compatibility also occurs for spoken responses, with vocal color-name responses being faster when they correspond with produced visual colored stimuli or color words than when they do not correspond.

11.6.4 EVALUATION

The theory of event coding has been the source of considerable research over the past 5 to 10 years. It places stimulus–response compatibility effects within the context of a broader framework of perception and action, opening up research to a range of different, but related, phenomena and findings. The view that both percepts and actions are organized into event files has served to emphasize the close two-way interaction between perception and action, as well as to allow the theory to account for a range of phenomena, including sequential effects. At present though, the theory is best viewed as a general framework for directing research and interpreting results, and not as a predictive model.

11.7 THE ECOLOGICAL APPROACH

The ecological approach to S-R compatibility, based on the Gibsonian view of perception (Gibson, 1979), has been advocated most strongly by Michaels (1988, 1989) and colleagues. As applied to S-R compatibility effects, this approach implies that it is incorrect to think of differences in RT between various S-R mappings as being due to the durations of distinct information processes. Instead, one should look to physical characteristics of the environment with reference to an organism's possibilities for action. Emphasis is placed on the concept of affordances, which is that "the layout and motions of environmental surfaces and objects … afford certain actions, but not others" (Michaels & Stins, 1997, p. 334). According to Michaels and Stins, "Whether an affordance is realized (whether some action occurs) depends on the occasion (incl. the intentions of the actor) and on the availability of information that stabilizes and constrains the emerging action" (p. 333). The basic notion is that compatibility effects reflect the detection of affordances. Michaels and Stins indicate,

> We share a surprising amount with coding theory with respect to salience and flexibility, but disagree sharply, along fairly traditional ecological versus information-processing lines, in emphasizing information over internal representation, in the nature of the perception–action processes, on the importance of coordination, and on the generalizability of the results … to everyday perception and action. (p. 338)

A major difference of the affordance approach from the coding approach is that the action relations that are presumed to be primed automatically by the environment are themselves assumed to be very specific, rather than being at the level of more abstract codes.

11.7.1 DESTINATION AND ORTHOGONAL COMPATIBILITY EFFECTS

Michaels (1988) first applied the theory of affordances to S-R compatibility by testing "whether an apparently moving square whose optical properties specified a direction of motion and time-to-contact that would make it catchable more easily by one hand than by the other would show S-R compatibility and/or a Simon effect" (p. 233). As described in Chapter 5, her thought was that the apparent destination of the apparent-motion trajectory of a square on a display screen would afford a catching action (forward movement of a joystick) by the hand at the movement destination. However, the results of several experiments have provided little evidence that either the absolute position of the responding hands, the use of responses that mimic catching actions, or the apparent destination of the movement is important for the obtained effects (Michaels, 1993; Proctor et al., 1993; Proctor, Lu, Van Zandt, & Weeks, 1994). In sum, there is little evidence for an affordance account of the destination compatibility effect.

Michaels (1989) and Michaels and Schilder (1991) also extended the ecological approach to orthogonal S-R compatibility effects obtained with unimanual responses, as depicted in Chapter 8. In both studies, they demonstrated that the mapping effects varied in direction and magnitude as a function of which hand was used for responding and its position relative to body midline in the ipsilateral hemispace. They interpreted these findings as indicating that "the state of the action system 'sets up' perception" (Michaels, p. 263), determining the preferred hand movements to different stimuli. However, as described in detail in Chapter 8 and in Cho and Proctor (2003), subsequent results have been in much closer agreement with location coding accounts than with the view that the state of the action system sets up perception. As examples, similar effects are obtained when relative position for responses at body midline is varied and when keypresses are used instead of unimanual responses, and the effects of hand and hand posture seem to be due to the frame of reference provided by the hand.

11.7.2 SIMON EFFECTS FOR OBJECT PROPERTIES

Tucker and Ellis (1998) applied the concept of affordances to Simon effects obtained for object properties. They used the term *affordance* "to refer to the motor patterns whose representation visual objects and their properties give rise to, both during explicit goal-directed acts … as well as … before explicit intentions have been formed" (p. 833). Tucker and Ellis conducted experiments intended to show that the specific actions an object affords are "automatically potentiated" (p. 833). Their stimuli were black-and-white images of graspable household objects such as a teapot (see Figure 11.7). Subjects were to make a left or right button press with the respective hand according to whether the object shown was upright or inverted, and the left–right horizontal orientation of the object was irrelevant to the task. A Simon effect for location of the graspable part of the object was obtained, with RT 12 ms shorter when that part of the object corresponded with the correct response position than when it did not. In Experiment 2, subjects responded with the index and middle fingers of the right hand. In this case, no Simon effect was evident, leading the authors to conclude, "It is the relevance of this property for action that underlies the left–right coding" (p. 838). Based on these and other results, Tucker and Ellis proposed, "There is a continuum of response codes capable of being generated, ranging from the most abstract, relative level down to those more directly related to the actions made possible by the visual environment" (p. 843), all of which are capable of influencing performance.

In Tucker and Ellis's (1998) experiments, the affording object was relevant for the response, although the location of the handle was not. Phillips and Ward (2002) modified the procedure to make the affording object irrelevant as well. In their Experiment 1, the image of a skillet served as a priming display. The handle could be to the left or right side, oriented at an angle toward or away from the observer,

FIGURE 11.7 Example illustration of objects used by Tucker and Ellis (1998).

or located neutrally. The SOA between the prime skillet and target stimulus (one of two alternative arrangements of horizontal and vertical lines) was varied randomly between 0, 400, 800, and 1,200 ms. Responses were faster when the prime handle corresponded with the correct response than when it did not, but two aspects of the results suggested that this was not due to an automatic affordance. First, this correspondence effect developed across the 1,200-ms interval after prime onset, rather than decreasing. Second, the effect was just as large when the handle was away from the subject as when it was near, even though the away position should not provide as strong of an affordance for action from the corresponding hand.

In Experiment 2, Phillips and Ward (2002) had subjects perform with a crossed-hands placement such that the left hand operated the right key and the right hand the left key. In agreement with other studies of S-R compatibility and the Simon effect (see Chapters 3 and 4), the benefit was for correspondence of the handle direction with the correct response location, even though the response was made by the contralateral hand. This correspondence effect was stronger at the short SOAs than it was with a normal hand placement in Experiment 1, which Phillips and Ward attributed to the longer RT when hands are crossed. They provided evidence against the action specificity of the effect by showing in Experiment 3 that left- and right-foot responses yielded a result pattern almost identical to that obtained with hands in their Experiment 1. Lyons, Weeks, and Chua (2005) provided similar evidence about the lack of specificity to graspable stimuli, showing a "handle" correspondence effect for images of objects that are too large to be grasped (e.g., the trunk of an elephant). In short, these findings suggest that visual "affordances" evoke abstract spatial codes and not specific actions.

Although discrete press responses yield little evidence of affordance for specific motor actions, perhaps responses that differ in other respects will. Ellis and Tucker (2000) had subjects make power or precision grip responses to high- or low-pitch tones. This task was performed while viewing an object that could be grasped with a power or precision grip. Responses were made with a device held in the right hand that had a pressure switch grasped between the thumb and index finger, mimicking a precision grip, and a lever attached to another switch that was grasped between the surface of the palm and the other three fingers, mimicking a power grip. On each trial, the object appeared 700 ms prior to the tone, and subjects were told to remember it for a later recognition-memory test. Most important, precision grip responses were 4 ms faster in the presence of objects compatible with that grip and power grip responses were 15 ms faster in the presence of ones compatible with it. However, this finding was complicated by the fact that this object compatibility effect was evident for the mapping of high pitch to precision grip and low pitch to power grip but not for the opposite tone-grip mapping. Similarly, in Experiment 2, where the responses were clockwise and counterclockwise wrist rotations, clockwise responses were 10 ms faster than counterclockwise responses in the presence of clockwise compatible objects, but there was no difference in the presence of counterclockwise compatible objects. Again, however, this was qualified by an interaction with mapping, being evident only for the low-tone/clockwise mapping.

Tucker and Ellis (2004) evaluated the specificity of these action priming effects, examining whether concurrent visual input is necessary. They presented objects requiring precision or power grips for durations of 20, 30, 50, 150, or 300 ms, followed by a blank screen or a pattern mask. Stimulus duration had little influence on the grip correspondence effect, although the effect was more distinct when there was no mask. Thus, the visual stimuli do not need to be present at the time the response is being selected and executed. Experiment 2 showed similar grip correspondence effects when the stimuli were reduced in contrast by 90% or occluded behind a grid. Finally, in Experiment 3, the grasp compatibility effect was shown to be as large for words naming the objects as for pictures of the objects themselves. Tucker and Ellis concluded,

> The data force the conclusion that within an experimental set-up in which on-line reaching and grasping are not actually occurring the route through which these affordance effects are generated depends more upon stored knowledge of the object and its associated actions than upon the detailed physical parameters of the viewed object. (p. 199)

It should be apparent that such effects are affordance effects in only the loosest sense of the term and not in the sense intended by Gibson's ecological theory of perception.

11.7.3 EVALUATION

The ecological approach has had the positive effect of getting researchers to examine a variety of task environments that are closer to those encountered outside of the laboratory. However, the specific implications of affordance accounts have not fared well in empirical tests. For the most part, results of studies that were initially interpreted as more consistent with the ecological approach than with the information-processing approach have been shown to conform well with the views of S-R compatibility that have been developed from other lines of research. Although the ecological approach has not been widely adopted by researchers investigating S-R compatibility, the approach taken in the recently prominent theory of event coding is much closer to the ecological approach than those taken in previous models of compatibility effects.

11.8 CHAPTER SUMMARY

Much research on S-R compatibility has proceeded without formal theories. The first theory to move beyond a conceptual model was that of Rosenbloom (1986), which provided an algorithmic model of S-R compatibility effects that has been shown to be useful for applied problems and to fit within a unified theory of cognition (Newell, 1990). This model emphasizes intentional action of the type involved in effects of S-R compatibility proper, as does the salient features coding framework proposed by Proctor and Reeve (1985, 1986) that has been able to explain several recalcitrant phenomena. Zorzi and Umiltà's (1995) connectionist model provided an account of the more automatic aspects of S-R compatibility that are embodied in the Simon effect. Kornblum et al.'s (1990) dimensional overlap model provides an integrated explanation of S-R compatibility proper and the Simon effect by incorporating both intentional and automatic response–selection processes. In its elaborated form in which the stimulus-identification process is also specified, it offers explanations of the Stroop color-naming and Eriksen flanker effects as well (Kornblum, 1992; Kornblum & Lee, 1995).

The theory of event coding, with its emphasis on common coding for perception and action and a major role for anticipation of sensory effects in control of action, has opened up research to a variety of compatibility effects that have not been studied previously. The ecological approach, which also emphasizes perception–action relations but eschews information-processing analyses, has many of the same emphases as the theory of event coding, but its predictions concerning highly specific relations between the environment and the actions afforded by it have not fared well to date as explanations of compatibility effects. Though there is not complete agreement on how best to conceive of compatibility effects, the last 3 decades have seen an increasing sophistication in accounts of compatibility effects and in the range of phenomena encompassed by those accounts.

12 Incorporating S-R Compatibility in Design

12.1 INTRODUCTION

Performance of a person interacting with an interface is dependent on the location and arrangement of displays and controls, the mapping of display elements to controls, and the individual's prior knowledge, expectations, and mental representation of the task. The display and control arrangements should allow the user to detect and identify critical information and to select appropriate actions, with minimal effort. The previous chapters focused primarily on the empirical and theoretical aspects of S-R compatibility because of their importance to understanding the relationship between perception and action. Designers need to be aware of these fundamental aspects of S-R compatibility because of their relevance to practical and applied problems. An understanding of the nature of compatibility effects and the factors that are likely to influence them in different contexts can allow a designer to make informed decisions about display layouts and mappings. In the present chapter, we conclude the book by highlighting the practical significance of research on S-R compatibility.

Much of the research on S-R compatibility has used RT measures to illustrate the benefit for response selection of maintaining a compatible mapping of display and control elements. However, it is important to emphasize for purposes of application that compatibility effects often show up not only in the time to select actions, but also in the time to execute them (e.g., Worringham & Beringer, 1989) and in error rates (e.g., Mitchell & Vince, 1951). Kantowitz, Triggs, and Barnes (1990) noted that certain systems may introduce a delay between presentation of a relevant stimulus and the time at which the response to it is required, allowing operators longer to choose a response to the displayed information. In such cases, the influence of S-R compatibility on error rate is likely to be more essential than its influence on RT. As an example, Kantowitz et al. stated that if the delay for responding with a specific system is a few minutes, and

> poor S-R compatibility only influences the system by increasing RT by 500 ms, it may not merit any consideration in system design. But if it also increases operator error rates by 25%, no responsible system designer would fail to improve S-R compatibility within the system. (p. 369)

Errors caused by incompatibility can be illustrated by the confusion that many older voters experienced using a butterfly ballot (see Figure 12.1) in Palm Beach County, FL. This confusion caused controversy over the validity of the results of the 2000 U.S. presidential election. The butterfly ballot listed the presidential candidates in two columns, with the punch holes for voting placed linearly between

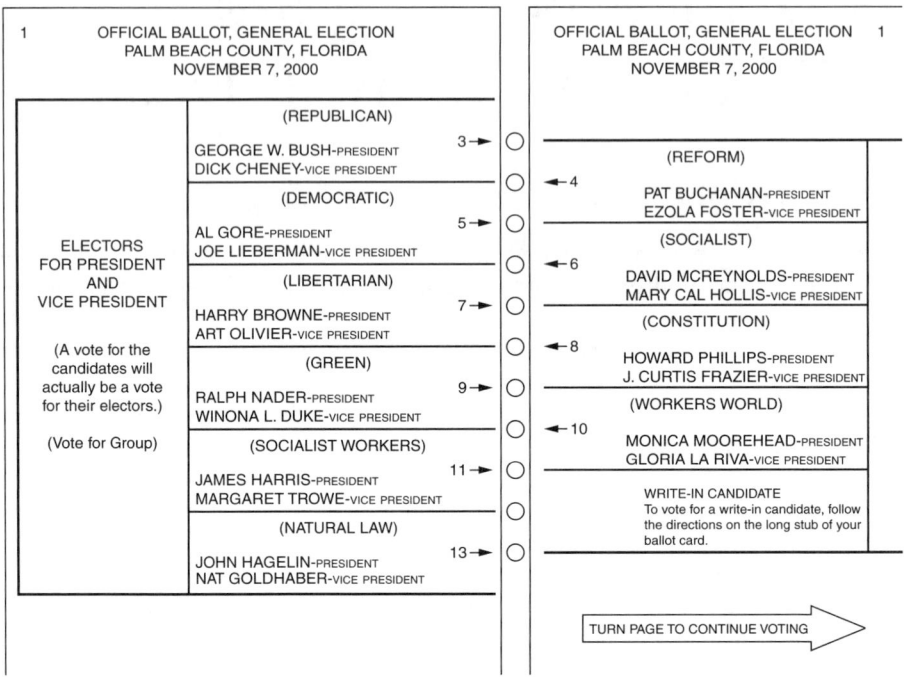

FIGURE 12.1   Reproduction of the butterfly ballot used in Palm Beach County, FL, for the 2000 U.S. presidential election.

those columns. An arrow pointed to the punch hole assigned to each candidate. To register a vote, the voter needed to insert a stylus into the punch hole for the candidate of choice among the centered column of holes.

With this ballot, some voters claimed that they selected the second punch hole on the list, intending to cast a vote for Al Gore, but registering instead a vote for Pat Buchanan. One reason this selection error occurred is that Gore was listed in the second position of the left-hand column, immediately below the other major party candidate, George W. Bush, creating a correspondence with the relative location of the second punch hole. The voters may also have had experience with election ballots in a single-column format, for which all candidates are listed on the left-hand side and their corresponding punch holes are displayed to the right in the same order. Such experience would result in an expectancy that the candidate location in the left column would correspond with the punch hole location.

Sinclair, Mark, Moore, Lavis, and Soldat (2000) conducted two experiments in Canada immediately after the 2000 U.S. presidential election to evaluate whether the butterfly ballot was more confusing and led to more errors than a single-column ballot. Both ballots listed the names of 10 politicians who were expected to run in an upcoming federal Canadian election. The arrangement of candidates on the ballot was designed to mimic the butterfly ballot used in the 2000 U.S. presidential election, with the candidates from the two dominant parties listed in the same positions as

George W. Bush and Al Gore, and the leader of a party that was expected to receive few votes listed in Pat Buchanan's position. College students were recruited as participants in Experiment 1. They received only one of the two ballot formats and were asked to vote for one of the candidates as Prime Minister by darkening the circle for the corresponding candidate. After casting their votes, the students were asked to write down the name of the candidate for whom they intended to vote and to rate, on a scale of 1 to 7, whether the ballot caused any confusion. The students rated the butterfly ballot to be more confusing than the single-column ballot but showed no difference in error rates between the two formats. Sinclair et al. attributed the lack of difference in error rates to the fact that college students are skilled at filling out complex scoring sheets.

Sinclair et al.'s (2000) Experiment 2 used participants recruited from a shopping center, with the intent of representing the voting public. These participants rated the butterfly ballot as more confusing than the single-column ballot, and they also made more errors on it. Analysis of the errors was consistent with the claim that voters intending to vote for the second candidate on the left, who was mapped to the third response location, accidentally voted for the first candidate on the right. Because consequences of incompatibility tend to be larger for older than younger adults (Proctor, Vu, & Pick, 2005), it is likely that older adults would make even more errors than this sample from the general population, as reports of the U.S. presidential election suggest.

12.2 PROXIMITY COMPATIBILITY AND STIMULUS–CENTRAL PROCESSING–RESPONSE COMPATIBILITY

In the field of human factors and ergonomics, many professionals are familiar with the principles of proximity compatibility and stimulus–central processing–response (S-C-R) compatibility, developed by Wickens and colleagues. The proximity compatibility principle (e.g., Wickens & Andre, 1990; Wickens & Carswell, 1995) states, "Displays relevant to a common task or mental operation (close task or mental proximity) should be rendered close together in perceptual space (close display proximity)" (Wickens & Carswell, p. 473). The idea is that displays of close proximity, in which multiple sources of information are integrated, will be more compatible with tasks that require information integration than will displays of far proximity. The concept of S-C-R compatibility, introduced in Chapter 3, elaborates S-R compatibility to include a central mediating component, emphasizing compatibility between stimulus and central processing and between central processing and response (Wickens, Sandry, & Vidulich, 1983). That is, some complex tasks require not only that the operator maintain S-R mappings, but also that stimulus input be incorporated into the operator's mental representations before selecting a response, and C designates these mediating representations.

12.2.1 Proximity Compatibility Principle

In many applied tasks, users interact with complex systems that provide information through multiple sources, and this information needs to be integrated by the user.

The principle of proximity compatibility emphasizes that there is a relation between different types of displays and mental workload. Tasks that require integration of information from different displays or display components will benefit from display elements that are arrayed in close proximity, whereas tasks that require attention to be focused on one display or display component will benefit when the elements are perceived separately. According to Wickens and colleagues (e.g., Wickens & Carswell, 1995), the perceptual proximity in a display — the perceptual similarity that exists between the display components — should match the processing proximity required by a task, whether two or more sources of information must be integrated or processed separately.

Perceptual proximity can refer to color (e.g., similar colors tend to be grouped together; a target that is different in color from the background will be easy to detect because it will "pop out"), distance (e.g., spatial proximity between elements; displays placed in closed proximity to each other are likely to be grouped together), perceptual codes (e.g., verbal codes are distinct from manual codes), and so on. Processing proximity is more abstract, but it is based on the same premise that when two or more sources of information are used for the same task, they should have close proximity when the information needs to be integrated and distant proximity if it does not. Degrees of processing proximity from high to low can be characterized as follows (Wickens & Carswell, 1995):

(a) *Integrative processing* entails active integration of components or sources of information. Common integrative processes include computational integration, or the adding or subtracting of information, and Boolean integration, or the requirement of the information to meet specific conditions (e.g., use of "and" or "or" in the specification).

(b) *Nonintegrative processing* requires processing of similar components, such as monitoring multiple displays that yield related information. Nonintegrative processing can involve displays that share the same metrics (metric similarity), displays that covary over time such that an increase in one causes a systematic increase or decrease in the other (covariance similarity), displays that are functionally similar or processed in a similar manner (functional similarity), and those to which response must be made in the same time frame (temporal similarity).

(c) *Independent processing* involves no interaction or similarity between information sources or processing mechanisms. Although information from two or more displays can be processed together, processing of information from each is independent.

In some cases, when multiple display elements or multiple displays are aligned or grouped in a particular manner, certain features of the display will "emerge." The emergent feature can reduce the attentional demands required for a user to monitor multiple display elements and increase monitoring performance. This is because the user can monitor the single emergent global property (orientation) or object (form), rather than having to monitor each component separately before integrating the

Memory set = '2 2 1 2 2 1'

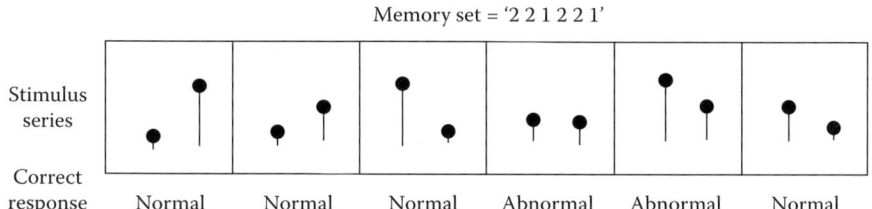

Stimulus series						
Correct response	Normal	Normal	Normal	Abnormal	Abnormal	Normal

FIGURE 12.2 Example of memory set, homogeneous, non-object stimulus series, and correct response in Carswell and Wickens's (1996) Experiment 1. From "Mixing and matching lower-level codes for object displays: Evidence for two sources of proximity compatibility," by C. M. Carswell and C. D. Wickens, 1996, *Human Factors, 38*, p. 5. Copyright 1996 by the Human Factors Ergonomics Society. Reprinted with permission.

information provided by each. Bennett and Flach (1992) noted that emergent displays can be classified into integral displays, for which the emergent feature is so salient that the individual components are "lost," and configural displays, for which the emergent feature is present but the individual components still maintain their individual perceptual properties. Integral displays benefit integrative processing and produce a large cost for selective processing of display elements. Configural displays are less effective than integral displays for integrative tasks, but produce a smaller cost for selective tasks. Bennett and Flach noted that performance for integrated tasks improves more from the presence of emergent features, or configurality of the display, than from the use of object displays per se.

Carswell and Wickens (1996) performed two experiments that examined performance on two integration tasks using object displays and separable configurations of displays. On each trial, subjects were shown a series of six displays, each with two sources of information (see Figure 12.2). Both sources needed to be processed and compared to determine whether each of the displays was "normal" or "abnormal" relative to a predesignated indicator as to which display element should have a greater value than the other. This predesignation was accomplished by providing subjects with a memory set of six numbers prior to the series of displays. For example, if the memory set was {2, 2, 1, 2, 2, 1}, then the second source of information should be greater than the first on Trials 1 and 2, and the first source of information should be greater than the second source on Trial 3, and so on. If the sources of information displayed were consistent with the memory set for a trial, then subjects would respond "normal." If the sources were inconsistent with the memory set for a trial, then subjects would respond "abnormal." The values for the two sources of information were presented in a homogeneous format (which was assumed to promote emergent features) or heterogeneous format (which was not), and some displays of each format could form an object. Performance was better with homogeneous than heterogeneous formats, and the benefit for object displays was mainly evident for homogeneous formats. These findings confirm Bennett and Flach's (1992) hypothesis that the emergent property is more important than the formation of an object.

In Carswell and Wickens's (1996) Experiment 2, subjects were asked to perform a conjunction task instead of a comparison task. They were given a memory set that specified the particular levels of the two sources of information in each of three displays (e.g., 5-1, 3-5, 2-2) and were to respond "normal" if the displayed information was consistent with the levels indicated by the memory set and "abnormal" if the information was inconsistent with those levels. As in Experiment 1, there was a benefit for using homogeneous formats to display the two sources of information. Furthermore, there was a benefit for object displays using both formats. Thus, Carswell and Wickens concluded that proximity compatibility benefits performance on integrated tasks through emergent features, which are more likely to be produced with homogeneous formats.

Hong, Thong, and Tam (2004) applied the proximity compatibility principle to the design of product listings on E-commerce web sites. They varied whether the products were listed in text only or with an image and text, and whether the product listings were formatted in a list (one product on each row) or an array (more than one product on a row). Users were asked to shop, at their own pace, for a product from a particular category specified by the experimental instructions. Performance measures included information search time and effectiveness (operationally defined in terms of recognition of previously seen images, if presented, and the brand names). Because the users were expected to compare different products within a category prior to making a purchase, this task could be considered to be an integrative task, and as a result should benefit from close spatial proximity of product images, brand names, and information. Thus, the proximity compatibility principle predicts better performance with the list format than the array format because similar types of information (e.g., product name and price) are presented in closer spatial proximity (see Figure 12.3).

Hong et al. (2004) found that users, on average, made about 7.5 clicks on the product-listing page to access more detailed information about the product. This

List format:

Image 1	Product description 1	Price 1
Image 2	Product description 2	Price 2
Image 3	Product description 3	Price 3
Image 4	Product description 4	Price 4
Image 5	Product description 5	Price 5
Image 6	Product description 6	Price 6

Array format:

Image 1 Price 1 Product description 1	Image 2 Price 2 Product description 2	Image 3 Price 3 Product description 3
Image 4 Price 4 Product description 4	Image 5 Price 5 Product description 5	Image 6 Price 6 Product description 6

FIGURE 12.3   Illustration of the list and array formats used by Hong et al. (2004).

finding indicates that the users were indeed examining and comparing products prior to making a purchase. Users spent less time searching for information when images of the products were available and when the product information was presented in a list format. The image–text presentation led to better recognition of the image but not of the brand name of the product. Recognition of both the image and brand name was better with the list presentation than with the array presentation. On a questionnaire filled out at the end of the shopping trip, users indicated a more positive attitude for image–text mode of presentation than for the text-only mode, and their perception did not depend on whether the presentation was in the list or array format. Thus, E-commerce sites should make use of both image and textual information when listing products and should present the products in a list format that allows easier comparison of related information due to the information being presented in close spatial proximity.

Marino and Mahan (2005) applied the proximity compatibility principle to the area of label design. They compared food labels that provided nutritional information in the form of an object display (a polar coordinate display that connected the individual nutritional components into a polygon) and separable display (listing of nutrition facts; see Figure 12.4a). The area of the polygon is an emergent feature of the object display that can be used to make comparisons about the nutritional quality of different items, with a smaller area representing higher nutritional quality. When subjects were asked to decide which of two items being presented had a higher nutritional quality (or that both items were equal), object displays led to shorter decision times than separable displays. However, there was no difference in the proportion of correct responses for the two display formats. When subjects had to decide how many servings of a food item they would need to consume in order to achieve close to 100% of the daily recommended value of a target nutrient, separable displays led to shorter decision times than the object displays. Based on these findings, Marino and Mahan concluded that task–display matching can enhance the effectiveness of label designs by facilitating the ease with which users are able to make judgments about nutrition information.

However, Bennett and Fritz (2005) noted that the polar coordinate object display used in Marino and Mahan's (2005) study may have been a poor design choice because "there is a poor mapping between the numerous emergent features produced by the polar coordinate display and the domain property they are intended to represent" (p. 132). Bennett and Fritz showed that with the polar coordinate object display, the same nutrients and daily recommended values can produce polygons of drastically different areas (in their example, 2,000 vs. 6,050 square units) by varying the position of the different nutrients that do not have equal daily recommended values. Because the mapping between the area and nutritional quality is not direct (i.e., smaller area does not equal higher quality in all cases, as Marino and Mahan assumed), nutritional decisions made on the basis of the emergent area feature may not always be accurate. Bennett and Fritz also noted that use of polar coordinate displays does not allow easy comparison between categories of nutrients and recommended use of a bar graph display to represent the nutritional information (see Figure 12.4b). This graphical display format provides redundant emergent features

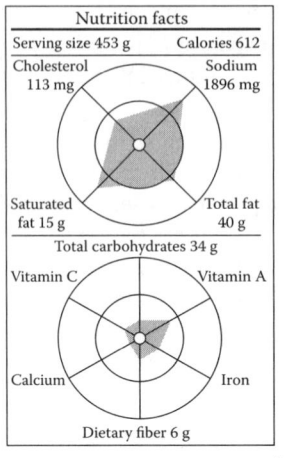

(a)

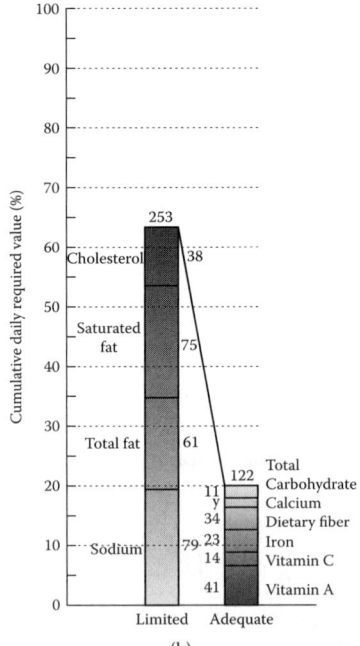

(b)

FIGURE 12.4 Illustration of the polar coordinate object display and list display used in Marino and Mahan's (2005) study (a); (b) is the bar graph display recommended by Bennett and Fritz (2005). Figure 12.4a is from "Configural displays can improve nutrition-related decisions: An application of the proximity compatibility principle," by C. J. Marino and R. P. Mahan, 2005, *Human Factors, 47,* p. 125. Copyright 2005 by the Human Factors Ergonomics Society. Reprinted with permission; Figure 12.4b is from "Objects and mappings: Incompatible principles of display design – A critique of Marino and Mahan," by K. B. Bennett and H. I. Fritz, 2005, *Human Factors, 47,* p. 135. Copyright 2005 by the Human Factors Ergonomics Society. Reprinted with permission.

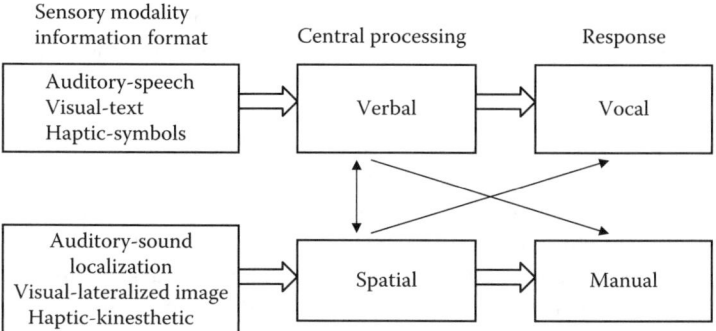

FIGURE 12.5 Illustration of S-C-R compatibility for verbal and spatial information as a function of sensory modality, central processing, and response code.

about nutritional quality (the height of the bar graphs and the orientation of the line connecting the two graphs) and specific daily recommended values that can facilitate decisions about individual nutrients as well.

12.2.2 S-C-R Compatibility

The code representation for a task is important because once the nature of the code is determined, the task can be designed to maximize compatibility between the stimulus code and central processes based on the person's mental model of the system, and between the central processes and the response code. According to Wickens (1984), there are four input code representations based on the auditory and visual modalities and the verbal and spatial perceptual codes (see Figure 12.5). Both speech and text inputs utilize verbal codes, but speech involves the auditory modality and text the visual modality. Similarly, both sound and analog pictorial input use spatial codes, with the former being auditory and the latter visual. S-C compatibility is demonstrated by the fact that tasks that require verbal processing (e.g., word meaning) will benefit more from stimuli that are verbal in nature (speech stimuli) than spatial in nature (visual stimuli placed in distinct physical locations), whereas those that require spatial processing codes (e.g., tracking of a visual target's location) are more compatible with analog pictorial or spatial stimuli than word stimuli. For manual (spatial) and vocal (verbal) responses, C-R compatibility is demonstrated in that performance is better when a manual response is made to a spatial task and vocal response to a verbal task than with the alternative pairings.

Evidence for S-C-R compatibility was obtained in a dual-task study by Wickens et al. (1983) and Vidulich and Wickens (1985); the same data were reported as Experiment 1 in both studies. Subjects performed a tracking task (spatial codes) by moving a joystick for one task. For the other task, subjects performed a memory-search task (verbal codes) in which they had to hold a set of three letters, presented visually or auditorily, in memory and indicate whether target letters presented in the same modality were in the memory set by pressing a YES or NO button or saying "yes" or "no." As predicted by S-C-R compatibility, performance on the memory-search task was better when the stimuli for it were presented auditorily and responses were made vocally than with the other modality combinations.

In Wickens, Vidulich, and Sandry-Garza's (1984) Experiment 1, dual tasks similar to those used by Wickens et al. (1983) were performed, but the stimulus displays for the two tasks were both visual and could appear on the left or right side, and the responses were both manual and could be made by moving a joystick or pressing keys placed to the left or right side. Because spatial tasks tend to be processed by the right hemisphere, which controls movement of the left hand, and verbal tasks by the left hemisphere, which controls movement of the right hand, S-C-R compatibility predicts better performance when the tracking-task display and response are located in the left visual field and the memory-search display and response in the right visual field than when the locations of the tasks are the opposite. The pattern of results was consistent with this prediction of the S-C-R compatibility principle.

Wickens et al. (1983, Experiment 2) and Vidulich and Wickens (1985, Experiment 2) reported the data from a study in which subjects performed a primary task of flying an F-18 simulator mock-up through a three-dimensional tunnel presented on a head-up display. This primary task is more demanding than the tracking task described earlier. In addition to the primary task, subjects performed one of two secondary tasks: verbal communication or spatial target localization. For the verbal communication task, subjects responded to visually displayed or auditorily presented commands by making manual or vocal responses. For the spatial target-localization task, subjects had to identify targets (e.g., houses or hangers) that were on the "ground" by specifying their names manually or vocally.

Results showed that secondary task type interacted with both input and output modality. For the verbal communication task, auditory input and speech output yielded better performance than visual input and manual output. For the spatial target-identification task, visual input and manual output yielded better performance than auditory input and speech output. These findings are consistent with S-C-R compatibility predictions. When the verbal communication secondary task was performed with the primary task, which was spatial in nature, timesharing was more efficient than when the input modality for the communication task was auditory. However, when the secondary task was spatial target localization, there was no difference in performance for auditory and visual input. There was also an interaction of output modality and task: Unexpectedly, performance was superior with manual responses for the secondary task, rather than vocal responses, even though the primary task also involved manual responses. This finding demonstrates that there can be efficient timesharing of two C-R compatible conditions even when they involve the same output modality.

Wickens et al.'s (1984) Experiment 2 and Vidulich and Wickens's (1985) Experiment 3 report data from an experiment in which subjects performed a threat evaluation task and a fault detection task in which the stimuli were presented visually or auditorily (sequence of tones) and the responses were manual or vocal. For the threat evaluation task, subjects were to indicate the likelihood that another aircraft would turn to a collision course with their aircraft by responding (manually or vocally) with a response of high, medium, or low. Because the threat evaluation task is spatial, it has high compatibility with visual stimuli and manual responses. For the fault detection task, subjects were asked to proceed through a sequence of verbal commands to diagnose whether the system and its components were faulty. The computer responded to the commands with visual or speech stimuli with a "yes" or "no" feedback. The "no" feedback meant that the system was not functioning correctly, and subjects had to give verbal commands to check all of the parts of the system. After checking all components of the system, the subjects were to report the failed components. One group performed the diagnosis with only manual responses and the other with both manual and vocal responses. Because the fault detection task is verbal, it has high compatibility with auditory stimuli and vocal responses.

Comparison of dual-task performance for the two most compatible conditions (visual–manual for the threat evaluation task and auditory–speech for the fault detection task) with the two least compatible conditions (auditory–speech for the threat evaluation task and visual–manual for the fault detection task) showed that there was a large benefit of maintaining S-C-R compatibility for both tasks. The influence of S-C-R compatibility in the dual-task context can also be seen in that the dual-task decrement increased when the display compatibility for the other task was low and when the response compatibility was low.

Xun, Guo, and Zhang (1998)* subsequently conducted a study that also exam ined performance as a function of the secondary task's input modality when the primary task was spatial in nature. Subjects performed two computer-simulated flight operation tasks. The primary task involved a two-dimensional tracking task of a target that moved in a rectangle display at the top part of the monitor. The secondary task was a spatial memory task in which the stimulus was presented visually (a series of arrow movements) or auditorily (speech stimuli of movement direction). After hearing or seeing the sequence of movements, subjects were to report the sequence by pressing response keys that corresponded to the movements. Single-task performance for the secondary tasks was also examined. For primary-task performance, tracking was better when the secondary task employed auditory input rather than visual input. For performance of the secondary task in isolation, S-C-R compatibility effects were evident: Performance on the spatial memory task was better with visual (spatial) inputs than verbal (speech) input. However, in dual-task conditions, performance on the secondary task deteriorated more when the input was visual than when it was auditory. Based on these findings, Xun et al. concluded that S-C-R compatibility predicted performance in single-task conditions but not in dual-task conditions. In the latter conditions, Xun et al. suggested that minimizing resource competition by presenting the primary and secondary task stimuli in dif-ferent modalities is more important than maintaining S-C-R compatibility.

In sum, the combinations of verbal–vocal and visuospatial–manual are more compatible than the alternative combinations at both the S-C and C-R levels. Although these findings are indicative of set-level compatibility, Eberts and Posey

---

* The original article was written in Chinese. We thank Huijun Wang for translating this article into English.

(1990) argued that the concept of S-C-R compatibility is more important in complex tasks of the type often encountered in real life because these tasks require many cognitive intermediaries between the stimulus and the response. If the task involves processing spatial codes, then analog pictorial stimuli should be used for presentation of information and manual responses should be required; if the task involves processing verbal codes, the stimuli should be speech-based and responses vocal. However, one weakness of S-C-R compatibility is that for tasks that require a combination of verbal and spatial processing, the theory does not make any predictions. Furthermore, for dual-task contexts, there is some evidence that minimizing resource competition may be more beneficial than maintaining S-C-R compatibility.

Eberts and Posey (1990) expanded S-C-R compatibility to include the concept of mental model because the code of representation in central processing is affected by the mental model formed by the person when the task is performed. A mental model refers to an individual's representation of some aspect of the world (e.g., a system) based on his or her understanding of it (see Fischer, 1991). Eberts and Posey stated that the structure of the mental model must be considered in addition to code representation in order to present stimulus information that is compatible with the structures and determine the responses to be made to them. With an image-based spatial mental model, the person would think about the system and its performance through forming an image of the system and visualizing how the system would function under different scenarios. Another way of characterizing image-based spatial mental models is as simulations of the systems in the person's head. The main point for understanding the mental model is that when the designer knows how the person represents the information or problem, the task can be designed to be compatible with those representations. Incorporation of mental models in the S-C-R compatibility theory allows the theory to accommodate changes in representation as a function of learning and practice.

12.3 GUIDELINES FOR COMPATIBILITY EFFECTS IN INTERFACE DESIGN

Since the publication of Fitts and Seeger's (1953) study, it has been widely accepted that performance is best when displays and their corresponding controls are configured in similar arrangements, with each display mapped to the spatially corresponding control. Maintaining compatibility is particularly important when stress or mental workload is high, because performance deteriorates more with incompatible mappings than with compatible mappings (see Fitts, Bahrick, Briggs, & Noble, 1959; Loveless, 1962). Consequently, many guidelines for interface designs underscore the need to maintain S-R compatibility (e.g., Andre & Wickens, 1990). Yet, maintaining compatibility is not simple because, as emphasized in this book, compatibility effects depend on the task goals, task set, frames of reference provided by the task environment, and prior experience, among other things.

A vivid illustration of this point comes from studies in which operator orientation was varied, as in applied tasks such as heavy equipment operation in which an operator's orientation relative to the display and controls may change. Worringham

Visual field	Compatible	Compatible	Incompatible	Incompatible
Control display	Compatible	Incompatible	Compatible	Incompatible
Body position	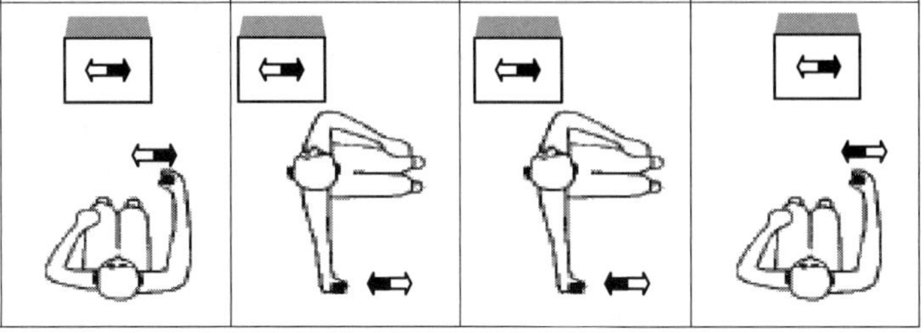			

FIGURE 12.6 Display and control positions relative to the subject in Worringham and Beringer's (1989) study.

and Beringer (1989) had subjects perform a task in which they had to operate a joystick with the right hand to move a cursor from a home dot to a target box that occurred in one of several locations around the home dot. In addition to a situation in which both the display and joystick were to the front of the subject and moved in parallel, situations were examined in which the joystick was located to the operator's right side and the display was to the operator's left side (with the head turned to the left), in various combinations (see Figure 12.6). This allowed the authors to dissociate effects of control–display compatibility (congruence between the direction of control and display movements), visual field compatibility (congruence between the direction of the control, as viewed by the subject, if he or she were looking at it, and display movements), and visual–trunk compatibility (control movement in the same direction relative to the operator's trunk as the movement of the visual display).

Worringham and Beringer (1989) found visual–trunk compatibility to be unimportant, which is not too surprising. However, visual field compatibility was much more important than control–display compatibility, which is surprising. This can be understood by considering the situation in which the observer's head is turned to the left to view the display and the control is on the opposite side, to the right of the trunk. In this case, control–display compatibility is maintained when the joystick moves forward to move the cursor forward and backward to move it backward, and visual field compatibility is maintained when the mapping of joystick to cursor movements is opposite. Yet, the condition with visual field compatibility yielded better performance than the one with control–display compatibility. Worringham and Beringer (1998) replicated this result more recently, ruling out an alternative possibility that they called muscle synergy compatibility, and Chua, Weeks, Ricker, and Poon (2001) obtained similar results for a more standard two-choice–reaction task and a task in which subjects had to synchronize movements with an oscillating visual display.

Another example of an applied problem involving changes of operator orientation for which recent findings are relevant is that of unmanned air vehicle (UAV) control. McCarley and Wickens (2005) describe that for external pilots controlling a UAV through a joystick control, many issues of motion compatibility can arise as functions of the relation between the headings of the UAV and pilot. If their headings are the same, then movement of the joystick to the left or right results in corresponding movement of the UAV relative to the pilot. However, if the headings of the UAV and pilot are not the same, then the mapping of the joystick movement and movement of the UAV relative to the pilot will not necessarily correspond, being completely incompatible in extreme cases. This spatial incompatibility can lead to mishaps and accidents. The research on mixed mappings suggests that the increased errors may be due in part to pilots taking longer to respond because the S-R mapping changes depending on the heading of the plane.

12.3.1 COMPATIBILITY PRINCIPLES

Many compatibility principles exist that can assist a designer in predicting which display–control arrangements and mappings will be most compatible.

12.3.1.1 Element-Level Compatibility Principles

- *Spatial Compatibility:* When displays and controls are arranged in similar layouts, each display should be mapped to the control that spatially corresponds with the location of the display.
  Example: Elevator buttons should be arrayed vertically; the top button should be pressed to move the elevator up and bottom button to move the elevator down.
- *Parallel vs. Orthogonal Mappings*: Displays and controls that vary along parallel dimensions and are compatibly mapped (e.g., left–right display locations mapped to left–right controls) are to be preferred to those that vary along orthogonal dimensions (e.g., top–bottom display locations mapped to left–right controls). Alternative mappings of orthogonal dimensions also yield performance differences.
  Example: For a digital display of station frequency, it is better for the control buttons to be arrayed vertically, with the up button mapped to increasing the frequency and the down button to decreasing the frequency, than for the buttons to be arranged horizontally.
- *Structural Compatibility*: When the elements of stimulus and response sets have an ordered relation or other structural feature, the S-R mapping should maintain structural correspondence.
  Example: When conducting a survey for which a question has five alternative answers ranging from least to most, and the answer is to be designated by a number from 1 to 5, then the numbers should be assigned in ascending order from least to most.
- *Spatial Correspondence:* Correspondence effects occur for stimulus and response locations when stimulus location is irrelevant to the task and for other irrelevant dimensions that overlap with the response dimension.
  Example: A collision avoidance signal occurring to the right of the driver may induce the driver to respond with a turn toward the signal rather than away from it.

- *Cross-Task Correspondence:* When stimuli or responses for one task overlap with the stimuli or responses for another task, cross-task correspondence effects often occur. Responding is faster and more accurate when the elements of the two tasks correspond than when they do not.

  Example: When pressing a control mounted on the left side of a vehicle steering wheel to operate some function of the vehicle's radio, if an unexpected event occurs on the road to which a rapid steering action is required, the action will likely be initiated more quickly for a maneuver to the left than for a maneuver to the right.

- *Proximity Compatibility:* Performance is best when display proximity matches the task proximity. The elemental components of a display should be arranged in close perceptual proximity to facilitate performance on tasks that require integration of display elements, but separate displays should be used for tasks that require selective attention to an individual component of the display.

  Example: The gas gauge and speedometer of an automobile are typically separated because the speed of the vehicle needs to be determined separately from the amount of gas remaining. Some newer cars include a high-proximity display that integrates the two sources of information so that the driver can easily determine how far the car can be driven at the current speed before running out of gas.

- *Movement Compatibility:* The direction of movement for a control should be consistent with the direction of movement of the indicator or system to which it is mapped, or with population stereotypes for control–display relations.

  Example: To increase the temperature using a thermostat that has a vertical scale and a slide control that moves the target indicator up and down, upward movement of the control should produce a corresponding upward movement of the indicator (to higher temperature) and a downward movement of the control should produce a downward movement of the indicator (to lower temperature). For a thermostat with a vertical scale and a rotary control located underneath it, a clockwise rotation of the control should result in an increase in the target temperature. (See Chapter 9 for population stereotypes such as clockwise-to-right-or-up, Warrick's principle, clockwise-to-increase, and clockwise-away.)

- *Prevalence Effects with Two-Dimensional Compatibility:* When the display–control configuration can be coded along two dimensions at once, a prevalence effect may occur in which coding with respect to one dimension dominates the other. The dominant dimension, often horizontal, will be the one made salient by the display and control environments.

  Example: A display–control configuration that has response buttons on both sides of the display (e.g., an automatic teller machine) will make the horizontal dimension more salient than the vertical dimension, so options that are aligned horizontally may be responded to faster than those that are aligned vertically.

12.3.1.2 Set-Level Compatibility Principles

- *Ideomotor Compatibility:* Performance benefits when the response given to a stimulus corresponds to the sensory feedback provided by the stimulus.

  Example: When listening to an automated menu on the telephone, it is more natural to say the menu option in response to the message than to enter one or more numbers on the keypad.

- *S-C-R Compatibility:* The nature of task representation needs to be taken into account to determine the input and output modes for the stimuli and responses. For verbal codes, auditory stimuli and speech responses are more compatible. For spatial codes, visual stimuli and manual responses are more compatible.

  Example: In a flying task, the nature of the task is visual–spatial. Thus, when a spatial action such as movement of the aircraft to the left or right is desired, the warning stimulus would be more effective if it was presented visually than if it was presented with speech stimuli.

12.3.2 FACTORS CONTRIBUTING TO THE SIZE OF COMPATIBILITY EFFECTS

Dimensional Overlap: Compatibility effects occur for any situation in which the stimulus dimension overlaps with the response dimension. This overlap may be conceptual (categorized along the same dimension), perceptual (physically similar), or structural (maintaining an ordered relationship) in nature. Dimensional overlap on the relevant or irrelevant stimulus dimension can affect performance. For S-R configurations with high dimensional overlap, responding is facilitated if the mapping is compatible, and interference occurs if the mapping is incompatible.

Intentions and Task Goals: Compatibility effects arise from the processes that mediate perception and action. Thus, they are not an automatic consequence of physical relations but depend on the operators' intentions and the task goals. Differences in performance can be obtained with the same S-R configuration and mapping assignments depending on how the relations are expressed to the individual.

Salience of Stimulus and Response Features or S-R sets: For multidimensional stimulus and response environments, different emphases can be placed on the individual dimensions. Compatibility effects are usually evident for the individual dimensions, but depending on the relative salience of a particular dimension, that dimension may exert a greater or lesser influence on performance. Changes in subjective judgments can also be obtained depending on the dimension that is made salient. It is particularly important to maintain a consistent mapping of salient features of the stimulus set with salient features of the response set.

Relative Location and Frames of Reference: Compatibility effects occur with respect to relative location, not just absolute location. Thus, it is important to maintain the relative location of displays and their associated controls. Relative location coding can occur with respect to many frames of reference. These frames of reference include body midline, direction of attention, and location relative to environmental cues and objects. When multiple displays or controls are organized into groups, locations can be coded at both the global and local levels. Ambiguous display–control relations can be avoided by adding distinct reference frames.

Pure vs. Mixed Mappings: Performance is better when the same mapping is used for all display–control pairs than when different mappings are used. The most ideal layout is one in which all display–control mappings are compatible. When mappings are mixed, performance often suffers more for the compatible relation than for the incompatible relation.

Rule Consistency: When all display elements cannot be mapped compatibly to their spatially corresponding controls, a mapping that produces a systematic relation (e.g., opposite) is better than a random mapping. For example, performance is much better with a mirror-opposite relation between displays and controls than with a random one.

Practice: Practice improves performance on almost any task. Practice with an incompatible display–control configuration improves performance, but performance typically is worse than it would be with the same amount of practice if the configuration were compatible. In other words, the cost of incompatibility cannot simply be overcome by practice. Moreover, under conditions of high psychological stress, people may revert back to their natural response tendencies.

Learning and Transfer: If an operator has previous experience with a related display–control configuration, the previously learned relationship may transfer to the current situation. This transfer can facilitate performance if the display–control relations conform to the operator's experience, but interfere if they do not.

Correlated S-R Dimensions. When certain S-R pairings occur more often than others, subjects develop expectancies (through either explicit or implicit learning of the relation). Performance is enhanced when the S-R relation on the current trial is consistent with the subject's expectancy and suffers when it is not.

Stimulus–Response Repetition: Performance of responses in a sequence is affected by immediately preceding events. Sequential effects analyses can reveal to interface designers whether certain display–control arrangements lead to better performance than others on sequential tasks.

Dual-Task Performance: Performance of the second of two tasks in close succession usually shows a decrement. This dual-task interference is reduced when the same mapping rule is used for both tasks, when both tasks are ideomotor compatible, and by practice.

12.4 NAÏVE AND EXPERIENCED JUDGMENTS

One reason designers should have a thorough understanding of the data and theory that underlie the topic of S-R compatibility is that the benefit in performance with certain display–control configurations over others may not always be intuitive. This section reviews the literature on judgments of compatibility with naïve and experienced users.

12.4.1 STIMULUS–RESPONSE COMPATIBILITY

Payne (1995) conducted a study to examine whether naïve subjects were accurate in their ratings of performance for diagrams that illustrated four-choice tasks in which the inner or outer S-R pair was mapped compatibly or incompatibly (see

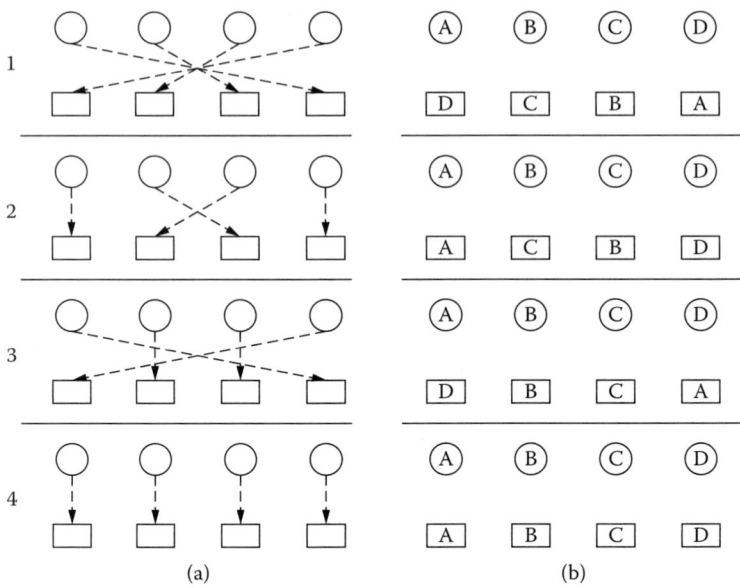

(a)   (b)

FIGURE 12.7 Diagram of mapping conditions in Payne's (1995) and Vu and Proctor's (2001b) studies. Panel a uses arrows to depict the mapping relations and panel b uses letters. From "Stimulus–response compatibility in interface design" by K.-P. L. Vu, and R. W. Proctor, 2001. In M. J. Smith, G. Salvendy, D. Harris, and R. J. Koubek (Eds.), *Usability evaluation and interface design: Cognitive engineering, intelligent agents, and virtual reality* (Vol. 1, p. 1369). Copyright 2001 by Lawrence Erlbaum Associates. Reprinted with permission.

Figure 12.7a). Results from studies of mixed mappings with four-choice tasks, described in Chapter 7, indicate that performance should be best for the condition in which both the inner and outer pairs are mapped compatibly, intermediate when they are both mapped incompatibly, and worst when one pair is mapped compatibly and the other incompatibly (see, e.g., Duncan, 1977a). Payne found that subjects correctly predicted that performance would be best when both S-R pairs were mapped compatibly, but misjudged the benefit for maintaining a consistent mapping for the S-R pairs. That is, subjects incorrectly predicted that performance would be better in the mixed-mapping conditions than in the condition in which both pairs were mapped incompatibly.

Payne (1995) suggested that the inaccurate performance rankings may have been due to subjects decomposing the configuration into individual pairings and using a summing heuristic to determine the overall performance. For example, because it is intuitive that performance is better with a compatible than incompatible S-R mapping, the overall performance can be estimated by assigning each compatible S-R pairing a rank of +1 and each incompatible pairing a rank of –1, and taking the sum. In this case, higher values would lead to predictions of better performance. However, this strategy for estimating performance does not take into consideration that for the condition in which both pairs are incompatible, there is only one mapping rule in effect, and subjects can adopt a "respond to the mirror opposite" rule for all S-R pairs. In contrast, for the mixed conditions, subjects need to select the appropriate mapping before responding (Duncan, 1977a), which slows performance. Thus, Payne summarized the two critical aspects that need to be taken into account for estimating performance as (a) the number of mapping rules (the lower the better), and (b) the nature of the mapping rule (responding with the "same" rule is easier than with the "opposite" rule).

Vu and Proctor (2001b) evaluated whether inaccurate naïve judgments were based on the perceived complexity of the diagrams used to depict the mapping relations. That is, by glancing at the diagrams displaying the S-R mappings (see Figure 12.7a), the condition in which both pairs are mapped incompatibly looks more complex than the two mixed mapping conditions. There is a possibility, then, that subjects were not attending to the individual S-R pairings but to the perceived complexity of the diagram. To evaluate this possibility, Vu and Proctor compared naïve judgments of performance with the four mapping conditions using a diagram similar to one used in Payne's (1995) study, along with diagrams in which each mapping condition was presented on a separate page, and each S-R pairing on a separate row. The logic was that display complexity would not be a factor if each pairing were presented individually. Moreover, Vu and Proctor used letters (A, B, C, and D) to indicate the stimulus and its assigned response (see Figure 12.7b), to reduce any perceived perceptual complexity associated with the arrows and to make the structural relationship more evident.

Vu and Proctor (2001b) found no difference between using letters and arrows to indicate the individual S-R mappings. However, presenting each condition and S-R pairing separately tended to reduce the RT estimates for the incompatible condition to the level of the mixed conditions. Thus, the findings of Vu and Proctor are in agreement with Payne's (1995) findings of inaccurate naïve judgments of compatibility. Vu and Proctor also evaluated whether actual performance with the various mapping relations leads to better subjective estimates of overall performance for the different mapping conditions. Subjects were given diagrams depicting the four different mapping conditions, with arrows depicting the mapping relation. After subjects gave their performance estimates for all four conditions, they performed 40 trials with each condition, and after each condition gave new estimates of performance for the different conditions. After experiencing the task with the different mappings, subjects adjusted their judgments of performance to more accurately match actual performance. Thus, minimal experience with alternative display–control mapping relations was sufficient to improve relative estimates of performance.

Vu and Proctor (2001b, 2003) also evaluated naïve and experienced judgments for two-choice tasks when only one mapping (compatible or incompatible) was described or when compatible and incompatible mappings were mixed within a trial block. The stimulus mode (physical locations, arrow directions, and locations words) and response mode (manual or verbal) were varied to determine whether subjects were aware of set-level compatibility effects (see Chapter 1). Subjects were accurate at predicting that performance would be better with a compatible than incompatible spatial mapping when only one mapping was in effect for a task (see also Tlauka, 2004). However, they were not very accurate at predicting set-level compatibility effects or the change in the magnitude of the compatibility effect when compatible and incompatible mappings were mixed. Performance with 40 trials of the different mapping conditions and tasks did improve subjects' estimates of set-level compatibility effects for pure tasks, but subjects were still unable to predict changes in the compatibility effect with mixed mappings.

Vu and Proctor (2003, Experiment 3) increased the number of practice trials to 100 for mixed mappings of physical locations to keypresses and found that with the extra practice, subjects were able to correctly predict the elimination of the S-R compatibility effect. Overall, the findings from Vu and Proctor's (2001b, 2003) studies indicate that the relation between the mapping of display and response elements on performance may not be intuitive for many people. Thus, when a designer is confronted with an interface design of display and control mappings for which he or she has not had prior experience, actually performing the task with the various combinations of possible mappings can provide insight regarding which is the most natural mapping. It should be noted, that for the tasks mentioned above, it takes approximately 10 minutes to perform 100 trials for each mapping condition.

Tlauka (2004) also examined judgments of compatibility for two-choice tasks when the stimuli were physical locations, arrow directions, and location words. He found that subjects were accurate in predicting the more compatible S-R mappings when the display and response configuration yielded an unambiguous spatial relation between the individual stimuli and responses than when it did not. Additionally, Tlauka examined compatibility judgments for normal, adjacent-hands placements and crossed-hands placements, and found that the judgments of relative compatibility followed the spatial locations of the stimuli and responses and not the hands. Moreover, when both response locations were placed to the left or right of the body midline and one stimulus location was to the left and the other to the right of midline, compatibility judgments were made with respect to relative location of the stimuli and responses. Tlauka also showed that subjects' judgments of compatibility for numerical relations were not as robust as for spatial relations. When the stimuli were centered numbers, there was an overall tendency to judge the right response as being preferred for both small and large numbers. The SNARC effect, mentioned in Chapter 8, shows, though, that the mapping of smaller numbers to the left and larger numbers to the right is more compatible than the alternative mapping. Thus, Tlauka's findings are in agreement with Vu and Proctor's (2001b, 2003) findings in showing that naïve judgments of compatibility are relatively accurate for S-R relations that have direct spatial correspondence but not for others.

12.4.2 Display–Control Layouts

12.4.2.1 Numeric Keypads and Keyboards

The numeric keypad is extremely useful for data entry of numbers. The 10-key pad is used on telephones and automatic teller machines for entry of phone numbers and personal identification numbers (PINs). Some of these keypads also contain letters of the alphabet on keys 2 through 9 to allow for use of alphanumeric strings that can be remembered easily (i.e., 1-800-GET-HELP is easier to remember than 1-800-438-4357; the PIN of ANIMAL is easier to remember than 264625). Today, the alphanumeric 10-key pad is used on cellular phones to allow for popular features such as text messaging. Originally, the 10-key pad did not include the letters Q and Z, which are letters that could be used in words that are used to form phone numbers, PINs, and text messages. The Q and Z letters could be placed on the 1 or 0 keys because no other letters reside on those keys, or they could be placed in alphabetical order on the 7 and 9 keys, respectively.

Blanchard, Lewis, Ross, and Cataldo (1993) examined user performance and preference for the position of the Q and Z keys on a 10-key pad. In their laboratory experiment, subjects used the original telephone keys that did not have the letters Q and Z on them, half of the time with the 1 key as input for both Q and Z and half with the 7 and 9 keys as input for Q and Z, respectively. Performance was measured by total keying time and the number of errors made in dialing alphanumeric strings of names, 1-800 numbers, and four-letter passwords. Half of the strings of each type contained either the letter Q or Z. After subjects used each of the two mappings for the Q and Z keys, preference ratings for the mappings were obtained. Results showed no difference in keying time or errors for the two mappings, but the Q-to-7 and Z-to-9 mapping received higher preference ratings than the alternative Q/Z-to-1 mapping. This preference for the Q-to-7 and Z-to-9 mapping was also replicated in a field study for which 83% of a larger sample of 400 people preferred the Q-to-7 and Z-to-9 mapping. This preference has been incorporated into the design of many 10-key numeric pads, including those on the authors' telephones and cellular phones.

Smith and Cronin (1993) conducted a study examining performance and preference for a standard keyboard and a Kinesis split keyboard, designed to reduce stress on the hands and produce more comfort for the user. Experienced typists (at least 45 words/minute) who had several years' experience with the traditional keyboard used both types of keyboards to type random strings of letters. The typists practiced for 7 hours with the Kinesis keyboard prior to the test session. A subset of the subjects was also measured for electrical activity of the muscle, which was used to indicate postural deviation (less deviation is better for the user). The Kinesis keyboard yielded lower postural deviation scores than the traditional keyboard. However, there was no difference in typing speed or errors for the two keyboards. For the preference data, more subjects preferred the Kinesis keyboard for comfort and usability, but more subjects preferred the traditional keyboard for performance, even though it showed no better performance than the Kinesis keyboard. Smith and Cronin's (1993) study, as well as that of Blanchard et al. (1993), shows that subjects may prefer a particular mapping even when it yields no better performance than an alternative mapping.

12.4.2.2 Stovetop Configurations

Chapanis and Lindenbaum (1959) had subjects perform with different stove configurations that varied with respect to the mapping of burner to control knob (see Figure 12.8). They found that performance was best with the control–burner mapping of Condition 2 (see also Ray & Ray, 1979). Shinar and Acton (1978) evaluated whether subjects would predict better performance with different burner–control mappings. They presented subjects with a diagram of the four-burner arrangement used in

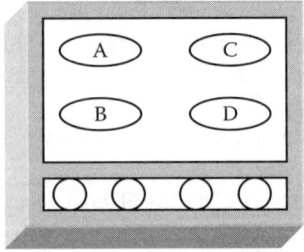

Condition	Control-burner mapping (left to right)			
2	A	B	D	C
3	A	B	C	D
4	B	A	D	C
5	B	A	C	D

FIGURE 12.8 Diagram of the burner-control arrangement used by Chapanis and Linden-baum (1959; top), and the specific control-burner mappings for the different conditions (bottom).

Chapanis and Lindenbaum's study and asked subjects to label the four control knobs with the letter of the burner that the knob should control. More subjects responded that they would prefer the burner–control mapping of Condition 3 (31%) than Condition 2 (25%); see Figure 12.8. This finding suggests that subjects are not very accurate at predicting the mapping that will lead to the best performance (see also Tlauka, 2004).

Hsu and Peng (1993) made the observation that subjects could have been biased to choose the knob assignment of Condition 3 in Shinar and Acton's (1978) study because that assignment maintains the ordered relation implied by the labels of A, B, C, and D. To evaluate whether the labeling had a biasing effect on subjects' mapping preferences, Hsu and Peng conducted a study in which the burners were designated by symbols (see Figure 12.9a) and a diagram in which the control knobs were numbered from 1 to 4 (from left to right; see Figure 12.9b), and Chinese subjects had to indicate which burner the control would turn on. In addition to using a paper-and-pencil test to obtain subjects' preferences, Hsu and Peng also conducted a computer simulation of the task in which subjects had to respond to a light on a burner with an assigned keypress to measure the subjects' response times and error rates.

Hsu and Peng (1993) found that for the paper-and-pencil test, subjects indicated a preference for the response mapping of Condition 3, replicating Shinar and Acton's (1978) finding. However, preference for Condition 3 was significantly stronger with the letter labels (49%) than with the symbolic labels (38%). Thus, there does seem to be a bias for the response assignment that maintains the structural relationship

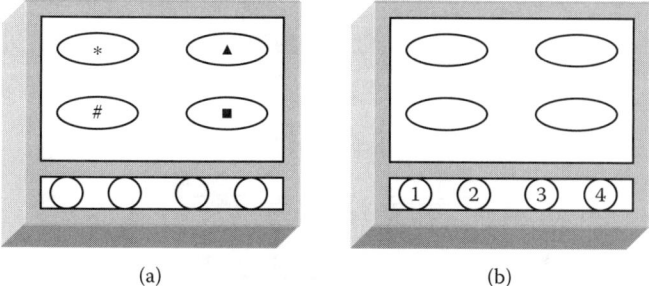

(a)   (b)

FIGURE 12.9 Illustration of the control-burner arrangement used by Hsu and Peng (1993). In panel a, the burners were designated by symbols and subjects indicated which control should operate the individual burner. In panel b, the controls were labeled by numbers and subjects indicated to which burner each control knob should be mapped.

between the labels. Data from the computer-simulation task showed that Chinese students produced similar RTs for the Condition 2 and 3 mappings, but had a much lower error rate for Condition 3 than Condition 2 (4.33% vs. 10.50%). Thus, Hsu and Peng's data suggest that Chinese subjects were able to accurately identify the burner–control mapping that yields better performance. Hsu and Peng noted that, for American subjects, the discrepancy between Chapanis and Lindenbaum's (1959) performance data and Shinar and Acton's preference data can be reconciled if the response bias from the letter labels in Shinar and Acton's study were removed.

Payne (1995) also examined naïve judgments of compatibility for ambiguous mappings of stove burners to control knobs that were used in Chapanis and Lindenbaum's (1959) study. Subjects were given diagrams of three burner–control mapping configurations in which the control knob for each burner was designated by a symbol given to the knob and burner (see Figure 12.10). Arbitrary symbols were used instead of the letters A, B, C, and D to avoid ordered relationships implied by the labels. The first diagram on the page gave a baseline of 100 errors out of a large number of trials, and subjects were asked to indicate the number of errors that would be committed for the other two configurations. Results showed that subjects had difficulty judging which stove configuration would lead to fewer errors, even though the performance data obtained by Chapanis and Lindenbaum showed marked differences (10.75% and 9.67% error rate for configurations 4 and 3, respectively, and a 6.33% error rate for design 2). Only 30% of subjects rated configuration 2 as yielding the best performance, which is what would have been expected by chance alone. Thus, the results of this experiment indicate that naïve judgments of performance with various display–control configurations may not be very accurate, which is consistent with Shinar and Acton's (1978) findings.

Wu (1997) evaluated performance and preference of burner–knob mappings by having Chinese subjects perform the task with the various mappings on a wooden model of a four-burner stove as well as with a paper-and-pencil test. For the paper-and-pencil test, Wu used a variant of the four-burner stove mapping diagrams (see Figure 12.11) used by Shinar and Acton (1978) and Hsu and Peng (1993). There were four versions of the paper-and-pencil test, two that used letters to designate

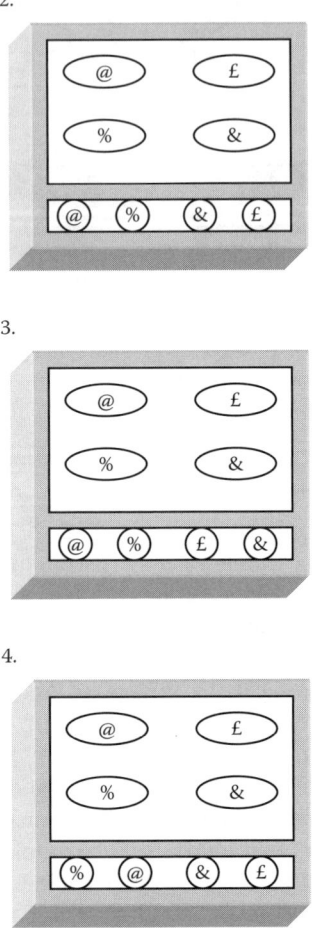

FIGURE 12.10 Illustration of the control-burner arrangement and mappings used by Payne (1995). The conditions, 2 through 4, corresponded to the same conditions in Chapanis and Lindenbaum's (1959) study, with the difference being that symbols were used to designate the burner and control mappings rather than letters.

the burner or control and two that used symbols (versions a vs. s). For each type, one version of the paper-and-pencil test had the letters designated on the burners and the other on the control knobs (versions a and s vs. a′ and s′; see Figure 12.11). For each version of the paper-and-pencil test, subjects wrote the letter or symbol on the control or burner to designate the preferred burner-to-control mappings. For the performance test, 30 Chinese students were asked to push one of four vertically arranged control knobs to turn off a lit burner of a wooden model of a four-burner stove. A computer was hooked up to the model stove and could light up any one of the four burners. The computer also controlled the burner–control mapping.

Wu (1997) found that the designation labels used for the burners and controls did affect preference judgments. When the alphabetical designators A, B, C, and D

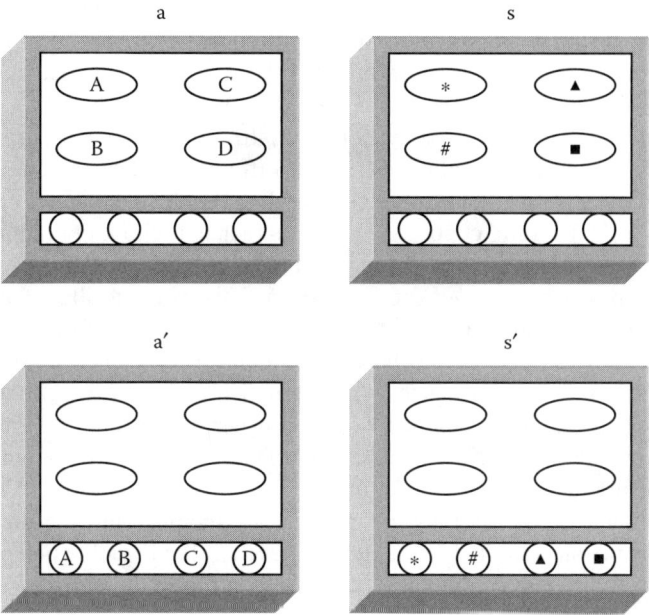

FIGURE 12.11 Diagram of the control-burner arrangements used in Wu's (1997) study. In panels a and a′, letters were used to designate the burner or control, and in panels s and s′, symbols were used. The burners were labeled for the top arrangements and controls were labeled in the bottom arrangements.

were used, more subjects indicated that they preferred to perform with mapping Condition 3, because the knobs maintained the ordered relationship between the letters. However, although there was an overall preference for the Condition 3 mapping, performance with the actual burner showed better performance with the Condition 5 mapping. Comparison of Hsu and Peng's (1993) results with those of Wu showed that Chinese subjects prefer the Condition 3 mapping. Although Hsu and Peng found that the Chinese subjects also showed better performance with the Condition 3 mapping with a computer simulation, Wu found better performance with the Condition 5 mapping using a physical model of a wooden stove. The difference between performances with the same burner–control mapping is likely due to use of the computer simulation versus an actual model of the stove.

These studies illustrate that subjects' stated preferences for specific display–control mappings does not mean that performance with those mappings will be better than with others. The dissociation between preference and performance measures has been demonstrated with many applied tasks (see Andre & Wickens, 1995; Bailey, 1993). For example, Camacho, Steiner, and Berson (1990, Experiment 2) had subjects use a touch-screen to select items on a menu. The items were represented by icons or alphanumeric designators. Subjects preferred the use of icons over alphanumeric text. Although use of icons led to faster response times than did use of alphanumeric stimuli, they also resulted in 22% more errors. Expert preferences for certain system designs also do not guarantee good human performance with the system. Keyson and Parsons (1990, Experiment 1) had professional ergonomists compare five different computer system prototypes. The ergonomists gave high preference ratings to the system that yielded the worst performance and the lowest preference rating to the system that yielded the best performance.

Furthermore, data patterns may not generalize across response apparatuses or cultures. The varied results obtained in the different studies reviewed reflect the ambiguous nature of the mapping relations when the display cannot be mapped to its spatially corresponding control. The cost associated with more ambiguous mappings for stove configurations can be reduced by connecting each burner to its control with sensor lines (Chapanis & Yoblick, 2001). This example illustrates that when spatial compatibility cannot be maintained, the potential costs of incompatibility can be remedied by clearly marking the display–control mappings. In summary, it is probably wise to test performance with alternative display–control configurations. However, time limitations in applied work settings may restrict the amount of testing that can be conducted, and for complex interfaces, the numerous combinations of display–control mappings may make the testing of each implausible. Consequently, an understanding of the various compatibility principles and theories can allow designers to generate a limited set of alternatives that should yield good performance. A detailed usability test can then be conducted on those alternatives.

12.5 CHAPTER SUMMARY

We began the book by noting that the concept of S-R compatibility introduced by Paul Fitts and colleagues in the 1950s is one of the great ideas in the history of human performance research. Since its introduction, compatibility effects of a variety of types have been the subject of an ever-increasing amount of theoretical and applied interest. Although the general design principle of maintaining spatial compatibility is important, it should be clear from the material covered in the book as a whole and in the present chapter that S-R compatibility encompasses much more than this general principle. Throughout the book we have emphasized that there are a variety of compatibility effects that occur for both relevant and irrelevant information when there is dimensional overlap among stimulus and response sets or overlearned S-R relations. Many factors come into play in determining what actions will be most compatible with the stimulus conditions in any particular situation.

The guidelines described in this chapter incorporate current knowledge about compatibility effects, but they are by no means exhaustive. In many complex tasks, the compatibility relations may be hard to predict due to interactions among the many task components and because "in general, people are unaware of the many information processing mechanisms that are responsible for the speed and accuracy of human–machine interaction" (Andre & Wickens, 1995, p.10). Thus, compatibility principles should be used as a starting point in the design phase, but they cannot replace testing to verify the efficiency with which users can perform using a given interface or product within a particular task environment.

References

Adam, J. J. (2000). The additivity of stimulus–response compatibility with perceptual and motor factors in a visual choice–reaction time task. *Acta Psychologica, 105*, 1–7.

Adam, J. J., Boon, B., Paas, F. G. W. C., & Umiltà, C. (1998). The up-right/down-left advantage for vertically oriented stimuli and horizontally oriented responses: A dual-strategy hypothesis. *Journal of Experimental Psychology: Human Perception and Performance, 24*, 1582–1595.

Adam, J. J., Hommel, B., & Umiltà, C. (2003). Preparing for perception and action (I): The role of grouping in the response-cuing paradigm. *Cognitive Psychology, 46*, 302–358.

Adam, J. J., Paas, F. G. W. C., Teeken, J. C., van Loon, E. M., van Boxtel, M. P. J., Houx, P. J., & Jolles, J. (1998). Effects of age on performance in a finger-precuing task. *Journal of Experimental Psychology: Human Perception and Performance, 24*, 870–883.

Aglioti, S., Tassinari, G., & Berlucchi, G. (1996). Spatial stimulus–response compatibility in callosotomy patients and subjects with callosal agenesis. *Neuroscience and Biobehavioral Reviews, 20*, 623–629.

Allport, A., Styles, E. A., & Hsieh, H. (1994). Shifting intentional set: Exploring the dynamic control of tasks. In C. Umiltà & M. Moscovitch (Eds.), *Attention and performance XV* (pp. 421–452). Cambridge, MA: MIT Press.

Allport, A., & Wylie, G. (2000). Task-switching, stimulus–response bindings, and negative priming. In S. Monsell & J. Driver (Eds.), *Attention and performance XVIII: Control of cognitive processes* (pp. 35–70). Cambridge, MA: MIT Press.

Alluisi, E. A., Strain, G. S., & Thurmond, J. B. (1964). Stimulus–response compatibility and the rate of gain of information. *Psychonomic Science, 1*, 111–112.

Anderson, J. R., Taatgen, N. A., & Byrne, M. D. (2005). Learning to achieve perfect time-sharing: Architectural implications of Hazeltine, Teague, and Ivry (2002). *Journal of Experimental Psychology: Human Perception and Performance, 31*, 749–761.

Andre, A. D., Haskell, I., & Wickens, C. D. (1991). S-R compatibility effects with orthogonal stimulus and response dimensions. In *Proceedings of the Human Factors Society 35th Annual Meeting* (pp. 1546–1550). Santa Monica, CA: Human Factors Society.

Andre, A. D., & Wickens, C. D. (1990). Display–control compatibility in the cockpit: Guidelines for display layout analysis. *Technical Report: NASA Ames Research Center.* Moffett Field, CA.

Andre, A. D., & Wickens, C. D. (1995, October). When users want what's not best for them. *Ergonomics in Design*, 10–14.

Ansorge, U. (2003). Influences of response-activating stimuli and passage of time on the Simon effect. *Psychological Research, 67*, 174–183.

Ansorge, U., & Wühr, P. (2004). A response-discrimination account for the Simon effect. *Journal of Experimental Psychology: Human Perception and Performance, 30*, 365–377.

Anzola, G. P., Bertolini, G., Buchtel, H. A., & Rizzolatti, G. (1977). Spatial compatibility and anatomical factors in simple and choice–reaction time. *Neuropsychologia, 15*, 295–302.

Arend, U., & Wandmacher, J. (1987). On the generality of logical recoding in spatial interference tasks. *Acta Psychologica, 65*, 193–210.

Atkins, P. (2003). *Galileo's finger: The ten great ideas of science*. London: Oxford University Press.

Azuma, R., Prinz, W., & Koch, I. (2004). Dual-task slowing and the effects of cross-task compatibility. *Quarterly Journal of Experimental Psychology, 57A*, 693–713.

Bailey, R.W. (1993). Performance versus preference. *Proceedings of the Human Factors and Ergonomics Society 37th Annual Meeting* (pp. 282–286). Santa Monica, CA: Human Factors and Ergonomics Society.

Bächtold, D., Baumüller, M., & Brugger, P. (1998). Stimulus–response compatibility in representational space. *Neuropsychologia, 36*, 731–735.

Baldo, J. V., Shimamura, A. P., & Prinzmetal, W. (1998). Mapping symbols to response modalities: Interference effects on Stroop-like tasks. *Perception & Psychophysics, 60*, 427–437.

Barber, P., & O'Leary, M. (1997). The relevance of salience: Towards an activation account of irrelevant stimulus–response compatibility effects. In B. Hommel & W. Prinz (Eds.), *Theoretical issues in stimulus–response compatibility* (pp. 135–172). Amsterdam: North-Holland.

Barber, P. J., O'Leary, M. J., & Simon, J. R. (1994). Defining stimulus congruity: A rejoinder to Guiard, Hasbroucq, and Possamai (1994). *Psychological Research, 56*, 213–215.

Bauer, D. W., & Miller, J. (1982). Stimulus–response compatibility and the motor system. *Quarterly Journal of Experimental Psychology, 34A*, 367–380.

Beh, H. C., Roberts, R. D., and Prichard-Levy, A. (1994). The relationship between intelligence and choice–reaction time within the framework of an extended model of Hick's law: A preliminary report. *Personality and Individual Differences, 16*, 891–897.

Beller, H. K. (1975). Naming, reading, and executing directions. *Journal of Experimental Psychology: Human Perception and Performance, 1*, 154–160.

Bennett, K. B., & Flach, J. M. (1992). Graphical displays: Implications for divided attention, focused attention, and problem-solving. *Human Factors, 34*, 513–533.

Bennett, K. B., & Fritz, H. I. (2005). Objects and mappings: Incompatible principles of display design – A critique of Marino and Mahan. *Human Factors, 47*, 131–137.

Bergum, B. O., & Bergum, J. E. (1981). Population stereotypes: An attempt to measure and define. In *Proceedings of the Human Factors Society 25th Annual Meeting* (pp. 662–665). Santa Monica, CA: Human Factors Society.

Bertelson, P. (1961). Sequential redundancy and speed in a serial two-choice responding task. *Quarterly Journal of Experimental Psychology, 13*, 90–102.

Bertelson, P. (1963). S-R relationships and reaction times to new versus repeated signals in a serial task. *Journal of Experimental Psychology, 65*, 478–484.

Beyak, B., Weeks, D., & Chua, R. (1999). *Salience of response features in the spatial precuing task*. Unpublished manuscript.

Biel, G. A., & Carswell, C. M. (1993). Musical notation for the keyboard: An examination of stimulus–response compatibility. *Applied Cognitive Psychology, 7*, 433–452.

Blanchard, H. E., Lewis, S. H., Ross, D., & Cataldo, G. (1993). User performance and preference for alphabetic entry from 10-key pads: Where to put Q and Z? *Proceedings of the Human Factors and Ergonomics Society 37th Annual Meeting* (pp. 225–229). Santa Monica, CA: Human Factors and Ergonomics Society.

Bosbach, S., Prinz, W., & Kerzel, D. (2004). A Simon effect with stationary moving stimuli. *Journal of Experimental Psychology: Human Perception and Performance, 30*, 39–55.

Brass, M., Bekkering, H., & Prinz, W. (2001). Movement observation affects movement execution in a simple response task. *Acta Psychologica, 106*, 3–22.

Brebner, J. (1973). S-R compatibility and changes in RT with practice. *Acta Psychologica, 37*, 93–106.

Brebner, J. (1979). The compatibility of spatial and nonspatial relationships. *Acta Psychologica, 43,* 23–32.

Brebner, J., & Sandow, B. (1976). The effect of scale side on population stereotype. *Ergonomics, 19,* 571–580.

Brebner, J., Shephard, M., & Cairney, P. (1972). Spatial relationships and S-R compatibility. *Acta Psychologica, 36,* 1–15.

Broadbent, D. E. (1958). *Perception and communication.* London: Pergamon Press.

Broadbent, D. E., & Gregory, M. (1962). Donders' B- and C-reactions and S-R compatibility. *Journal of Experimental Psychology, 63,* 575–578.

Camacho, M. J., Steiner, B. A., & Berson, B. L. (1990). Icons vs. alphanumerics in pilot-vehicle interfaces. *Proceedings of the Human Factors Society 34th Annual Meeting* (pp. 11–15). Santa Monica, CA: Human Factors Society.

Campbell, K. C., & Proctor, R. W. (1993). Repetition effects with categorizable stimulus and response sets. *Journal of Experimental Psychology: Learning, Memory, and Cognition, 19,* 1345–1362.

Card, S. K., Moran, T. P., & Newell, A. (1983). *The psychology of human–computer interaction.* Hillsdale, NJ: Erlbaum.

Carswell, C. M., & Wickens, C. D. (1996). Mixing and matching lower-level codes for object displays: Evidence for two sources of proximity compatibility. *Human Factors, 38,* 1–22.

Chan, A. H. S., & Courtney, A. J. (2001). Color associations for Hong Kong Chinese. *International Journal of Industrial Ergonomics, 28,* 165–170.

Chan, A. H. S., Courtney, A. J., & So, K. W. Y. (2000). Circular displays with thumbwheels: Hong Kong Chinese preferences. *Human Factors and Ergonomics in Manufacturing, 10,* 453–463.

Chan, A. H. S., Shum, V. W. Y., Law, H. W., & Hui, I. K. (2003). Precise effects of control position, indicator type, and scale side on human performance. *International Journal of Advanced Manufacturing Technology, 22,* 380–386.

Chapanis, A., & Lindenbaum, L. E. (1959). A reaction time study of four control–display linkages. *Human Factors, 1,* 1–7.

Chapanis, A., & Yoblick, D. A. (2001). Another test of sensor lines on control panels. *Ergonomics, 44,* 1302–1311.

Chase, W. G., & Clark, H. H. (1971). Semantics in the perception of verticality. *British Journal of Psychology, 62,* 311–326.

Cho, Y. S., & Proctor, R. W. (2001). Effect of an initiating action on the up-right/down-left advantage for vertically arrayed stimuli and horizontally arrayed responses. *Journal of Experimental Psychology: Human Perception and Performance, 27,* 472–484.

Cho, Y. S., & Proctor, R. W. (2002). Influences of hand posture and hand position on compatibility effects for up–down stimuli mapped to left–right responses: Evidence for a hand referent hypothesis. *Perception & Psychophysics, 64,* 1301–1315.

Cho, Y. S., & Proctor, R. W. (2003). Stimulus and response representations underlying orthogonal stimulus–response compatibility effects. *Psychonomic Bulletin & Review, 10,* 45–73.

Cho, Y. S., & Proctor, R. W. (2004a). Effects of stimulus-set location on orthogonal stimulus–response compatibility. *Psychological Research, 69,* 106–114.

Cho, Y. S., & Proctor, R. W. (2004b). Influences of multiple spatial stimulus and response codes on orthogonal stimulus–response compatibility. *Perception & Psychophysics, 66,* 1003–1017.

Cho, Y. S., & Proctor, R. W. (2005). Representing response position relative to display location: Influence on orthogonal stimulus–response compatibility. *Quarterly Journal of Experimental Psychology, 58A,* 839–864.

Christensen, C. A., Ivkovich, D., & Drake, K. J. (2001). Late positive ERP peaks observed in stimulus–response compatibility tasks tested under speed-accuracy instructions. *Psychophysiology, 38,* 404–416.

Chua, R., Pollock, B. J., Elliott, D., Swanson, L. R., & Carnahan, H. (1995). The influence of age on manual asymmetries in movement preparation and execution. *Developmental Neuropsychology, 11,* 129–137.

Chua, R., Weeks, D. J., Ricker, K. L., & Poon, P. (2001). Influence of operator orientation on relative organizational mapping and spatial compatibility. *Ergonomics, 44,* 751–765.

Cotton, B., Tzeng, O., & Hardyck, C. (1977). A response instruction by visual-field interaction: S-R compatibility effect or? *Bulletin of the Psychonomic Society, 10,* 475–477.

Courtney, A. J. (1986). Chinese population stereotypes: Color associations. *Human Factors, 28,* 97–99.

Courtney, A. J. (1988). Chinese response preferences for display–control relationships. *Human Factors, 30,* 367–372.

Courtney, A. J. (1992). Control–display stereotypes for multicultural user systems. *IEEE Transactions on Systems, Man, and Cybernetics, 22,* 681–687.

Courtney, A. J. (1994a). The effect of scale side, indicator type, and control plane on direction-of-turn stereotypes for Hong Kong Chinese subjects. *Ergonomics, 37,* 865–877.

Courtney, A. J. (1994b). Hong Kong Chinese direction-of-motion stereotypes. *Ergonomics, 37,* 417–426.

Craft, J. L., & Simon, J. R. (1970). Processing symbolic information from a visual display: Interference from an irrelevant directional cue. *Journal of Experimental Psychology, 83,* 415–420.

Davis, R., Moray, N., & Treisman, A. (1961). Imitative responses and the rate of gain of information. *Quarterly Journal of Experimental Psychology, 13,* 78–89.

De Houwer, J. (1998). The semantic Simon effect. *Quarterly Journal of Experimental Psychology, 51A,* 683–688.

De Houwer, J. (2004). Spatial Simon effects with nonspatial responses. *Psychonomic Bulletin & Review, 11,* 49–53.

De Houwer, J. Beckers, T., Vandorpe, S., & Custers, R. (2005). Further evidence for the role of mode-independent short-term associations in spatial Simon effects. *Perception & Psychophysics, 67,* 659–666.

De Houwer, J., Crombez, G., Baeyens, F., & Hermans, D. (2001). On the generality of the affective Simon effect. *Cognition & Emotion, 15,* 189–206.

De Houwer, J., & Eelen, P. (1998). An affective variant of the Simon paradigm. *Cognition & Emotion, 12,* 45–61.

De Jong, R. (1995). Strategical determinants of compatibility effects with task uncertainty. *Acta Psychologica, 88,* 187–207.

De Jong, R., Liang, C.-C., & Lauber, E. (1994). Conditional and unconditional automaticity: A dual-process model of effects of spatial stimulus–response correspondence. *Journal of Experimental Psychology: Human Perception and Performance, 20,* 731–750.

De Jong, R., Wierda, M., Mulder, G., & Mulder, L. J. M. (1988). The use of partial information in response processing. *Journal of Experimental Psychology: Human Perception and Performance, 14,* 682–692.

Dehaene, S., Bossini, S., & Giraux, P. (1993). The mental representation of parity and number magnitude. *Journal of Experimental Psychology: General, 122,* 371–396.

Deininger, R. L., & Fitts, P. M. (1955). Stimulus–response compatibility, information theory, and perceptual-motor performance. In H. Quastler (Ed.), *Information theory in psychology; problems and methods* (pp. 316–341). Glencoe, IL: Free Press.

Delaney, P. F., Reder, L. M., Staszewski, J. J., & Ritter, F. E. (1998). The strategy-specific nature of improvement: The power law applies by strategy within task. *Psychological Science, 9*, 1–7.

Destrebecqz, A., & Cleeremans, A. (2001). Can sequence learning be implicit? New evidence with the process dissociation procedure. *Psychonomic Bulletin & Review, 8*, 343–350.

Donders, F. C. (1868/1969). On the speed of mental processes. In W. G. Koster (Ed.), *Acta Psychologica, 30, Attention and Performance II* (pp. 412–431). Amsterdam: North-Holland.

Duncan, J. (1977a). Response selection errors in spatial choice–reaction tasks. *Quarterly Journal of Experimental Psychology, 29*, 415–423.

Duncan, J. (1977b). Response selection rules in spatial choice–reaction tasks. In S. Dornic (Ed.), *Attention and Performance VI* (pp. 49–61). Hillsdale, NJ: Erlbaum.

Duncan, J. (1978). Response selection in spatial choice reaction: Further evidence against associative models. *Quarterly Journal of Experimental Psychology, 30*, 429–440.

Duncan, J. (1979). Divided attention: The whole is more than the sum of its parts. *Journal of Experimental Psychology: Human Perception and Performance, 5*, 216–228.

Dutta, A., & Proctor, R. W. (1992). Persistence of stimulus–response compatibility effects with extended practice. *Journal of Experimental Psychology: Learning, Memory, and Cognition, 18*, 801–809.

Dutta, A., & Proctor, R. W. (1993). The role of feedback in learning spatially incompatible stimulus–response assignments: Does it have one? *Proceedings of the 37th Annual Meeting of the Human Factors and Ergonomics Society* (pp. 1320–1324). Santa Monica, CA: Human Factors and Ergonomics Society.

Eberts, R. E., & Posey, J. W. (1990). The mental model in stimulus–response compatibility. In R. W. Proctor & T. G. Reeve (Eds.), *Stimulus–response compatibility: An integrated perspective* (pp. 389–425). Amsterdam: North Holland.

Ehrenstein, A., & Proctor, R. W. (1998). Selecting mapping rules and responses in mixed compatibility four-choice tasks. *Psychological Research, 61*, 231–248.

Eimer, M. (1998). The lateralized readiness potential as an on-line measure of central response activation processes. *Behavior Research Methods, Instruments, & Computers, 30*, 146–156.

Eimer, M., & Schlaghecken, F. (1998). Effects of masked stimuli on motor activation: Behavioral and electrophysiological evidence. *Journal of Experimental Psychology: Human Perception and Performance, 24*, 1737–1747.

Eimer, M., & Schlaghecken, F. (2001). Response facilitation and inhibition in manual, vocal, and oculomotor performance: Evidence for a modality-unspecific mechanism. *Journal of Motor Behavior, 33*, 16–26.

Eimer, M., & Schlaghecken, F. (2002). Links between conscious awareness and response inhibition: Evidence from masked priming. *Psychonomic Bulletin & Review, 9*, 514–520.

Eimer, M., & Schlaghecken, F. (2003). Response facilitation and inhibition in subliminal priming. *Biological Psychology, 64*, 7–26.

Eimer, M., Schubö, A., & Schlaghecken, F. (2002). Locus of inhibition in the masked priming of response alternatives. *Journal of Motor Behavior, 34*, 3–10.

Elithorn, A., & Lawrence, C. (1955). Central inhibition — Some refractory observations. *Quarterly Journal of Experimental Psychology, 7*, 116–127.

Ellis, R., & Tucker, M. (2000). Micro-affordance: The potentiation of components of action by seen objects. *British Journal of Psychology, 91*, 451–471.

Eriksen, B. A., & Eriksen, C. W. (1974). Effects of noise letters upon the identification of a target letter in a nonsearch task. *Perception & Psychophysics, 16*, 143–149.

Fischer, G. (1991). The importance of models in making complex systems comprehensible. In M. J. Tauber & D. Ackermann (Eds.), *Mental models and human–computer interaction 2* (pp. 3–36). Amsterdam: North-Holland.

Fischer, G. W., & Hawkins, S. A. (1983). Strategy compatibility, scale compatibility, and the prominence effect. *Journal of Experimental Psychology: Human Perception and Performance, 19,* 580–597.

Fitts, P. M. (1964). Perceptual-motor skill learning. In A. W. Melton (Ed.), *Categories of human learning* (pp. 243–285). New York: Academic Press.

Fitts, P. M., Bahrick, H. P., Briggs, G. E., & Noble, M. E. (1959). *Skilled performance: Part II. Task variables and performance.* Final report. Columbus: The Ohio State University Research Foundation.

Fitts, P. M., & Biederman, I. (1965). S-R compatibility and information reduction. *Journal of Experimental Psychology, 69,* 408–412.

Fitts, P. M., & Deininger, R. L. (1954). S-R compatibility: Correspondence among paired elements within stimulus and response codes. *Journal of Experimental Psychology, 48,* 483–492.

Fitts, P. M., & Seeger, C. M. (1953). S-R compatibility: Spatial characteristics of stimulus and response codes. *Journal of Experimental Psychology, 46,* 199–210.

Fitts, P. M., & Simon, C. W. (1952). The arrangements of instruments, the distance between instruments, and the position of instrument pointers as determinants of performance in an eye-hand co-ordination task. USAF Air Materiel Command Report No. 5832.

Fitts, P. M., & Switzer, G. (1962). Cognitive aspects of information processing: 1. The familiarity of S-R sets and subsets. *Journal of Experimental Psychology, 63,* 321–329.

Forrin, B. (1975). Naming latencies to mixed sequences of letters and digits. In P. M. A. Rabbitt & S. Dornic (Eds.), *Attention and performance V* (pp. 345–356). New York: Academic Press.

Forrin, B., & Morin, R. E. (1967). Effects of context on reaction time to optimally coded signals. *Acta Psychologica, 27,* 188–196.

Freudenrich, C. C. (2005). How air traffic control works. Retrieved July 16, 2005 from http://electronics.howstuffworks.com/air-traffic-control1.htm.

Frowein, H. W., & Sanders, A. F. (1978). Effects of visual stimulus degradation, S-R compatibility, and foreperiod duration on choice–reaction time and movement time. *Bulletin of the Psychonomic Society, 12,* 106–108.

Garvey, W. D. (1957). The effects of "task-induced stress" on man-machine system performance. U.S. NRL Rep. No. 5015.

Garvey, W. D., & Knowles, W. B. (1954). Response time patterns associated with various display–control relationships. *Journal of Experimental Psychology, 47,* 315–322.

Garvey, W. D., & Mitnick, L. L. (1955). Effect of additional spatial references on display–control efficiency. *Journal of Experimental Psychology, 50,* 276–282.

Gibbs, C. B. (1949). Progress report on the first year's work. MRC APU Rep. No. 113.

Gibson, J. J. (1979). *The ecological approach to visual perception.* Boston: Houghton Mifflin.

Gignac, G. E., & Vernon, P. A. (2004). Reaction time and the dominant and nondominant hands: An extension of Hick's law. *Personality and Individual Differences, 16,* 733–739.

Goodman, D., & Kelso, J. (1980). Are movements prepared in parts? Not under compatible (naturalized) conditions. *Journal of Experimental Psychology: General, 109,* 475–495.

Gordon, P. C. (1990). Perceptual-motor processing in speech. In R. W. Proctor & T. G. Reeve (Eds.), *Stimulus–response compatibility: An integrated perspective* (pp. 343–362). Amsterdam: North-Holland.

Gordon, P. C., & Meyer, D. E. (1984). Perceptual-motor processing of phonetic features in speech. *Journal of Experimental Psychology: Human Perception and Performance, 10*, 153–178.

Gottsdanker, R., & Stelmach, G. E. (1971). The persistence of psychological refractoriness. *Journal of Motor Behavior, 3*, 301–312.

Graham, N. E. (1952). Manual tracking on a horizontal scale and in the four quadrants of a circular scale. *British Journal of Psychology, 43*, 70–77.

Greenwald, A. G. (1970a). A choice–reaction time test of ideomotor theory. *Journal of Experimental Psychology, 86*, 20–25.

Greenwald, A. G. (1970b). Sensory feedback mechanisms in performance control: With special reference to the ideomotor mechanism. *Psychological Review, 77*, 73–99.

Greenwald, A. G. (1972). On doing two things at once: Timesharing as a function of ideomotor compatibility. *Journal of Experimental Psychology, 94*, 52–57.

Greenwald, A. G. (2003). On doing two things at once: III. Confirmation of perfect timesharing when simultaneous tasks are ideomotor compatible. *Journal of Experimental Psychology: Human Perception and Performance, 29*, 859–868.

Greenwald, A. G. (2005). A reminder about procedures needed to reliably produce perfect timesharing: Comment on Lien, McCann, Ruthruff, and Proctor (2005). *Journal of Experimental Psychology: Human Perception and Performance, 31*, 221–225.

Greenwald, A. G., McGhee, D. E., & Schwartz, J. L. K. (1998). Measuring individual differences in implicit cognition: The implicit association test. *Journal of Personality and Social Psychology, 74*, 1464–1480.

Greenwald, A. G., & Shulman, H. G. (1973). On doing two things at once: II. Elimination of the psychological refractory period effect. *Journal of Experimental Psychology, 101*, 70–76.

Grice, G. R., Canham, L., & Boroughs, J. M. (1984). Combination rule for redundant information in reaction time tasks with divided attention. *Perception & Psychophysics, 35*, 451–463.

Grice, G. R., Canham, L., & Gwynne, J. M. (1984). Absence of a redundant-signals effect in a reaction time task with divided attention. *Perception & Psychophysics, 36*, 565–570.

Griew, S. (1964). Age, information transmission and the positional relationship between signals and responses in the performance of a choice task. *Ergonomics, 7*, 267–277.

Grosjean, M., & Mordkoff, J. T. (2002). Postresponse stimulation and the Simon effect: Further evidence of action-effect integration. *Visual Cognition, 9*, 528–539.

Guiard, Y. (1983). The lateral coding of rotations: A study of the Simon effect with wheel-rotation responses. *Journal of Motor Behavior, 15*, 331–342.

Guiard, Y., Hasbroucq, T., & Possamai, C.-A. (1994). Stimulus congruity, irrelevant spatial SR correspondence, and display–control arrangement correspondence: A reply to O'Leary, Barber, and Simon (1994). *Psychological Research, 56*, 210–212.

Hasbroucq, T., & Guiard, Y. (1986). Response determination in tactile motor tasks: Body- vs. device-centered cues. *Cahiers de Psychologie Cognitive, 6*, 367–377.

Hasbroucq, T., & Guiard, Y. (1991). Stimulus–response compatibility and the Simon effect: Toward a conceptual clarification. *Journal of Experimental Psychology: Human Perception and Performance, 17*, 246–266.

Hasbroucq, T., & Guiard, Y. (1992). The effects of intensity and irrelevant location of a tactile stimulation in a choice–reaction time task. *Neuropsychologia, 30*, 91–94.

Hasbroucq, T., Guiard, Y., & Ottomani, L. (1990). Principles of response determination: The list-rule model of SR compatibility. *Bulletin of the Psychonomic Society, 28*, 327–330.

Hazeltine, E., Teague, D., & Ivry, R. B. (2002). Simultaneous dual-task performance reveals parallel response selection after practice. *Journal of Experimental Psychology: Human Perception and Performance, 28*, 527–545.

Healy, A. F., Wohldmann, E. L., & Bourne, L. E., Jr. (2005). The procedural reinstatement principle: Studies on training, retention, and transfer. In A. F. Healy (Ed.), *Experimental cognitive psychology and its applications* (pp. 59–71). Washington, DC: American Psychological Association.

Heathcote, A., Brown, S., & Mewhort, D. J. K. (2000). The power law repealed: The case for an exponential law of practice. *Psychonomic Bulletin & Review, 7*, 185–207.

Hedge, A., & Marsh, N. W. A. (1975). The effect of irrelevant spatial correspondences on two-choice response time. *Acta Psychologica, 39*, 427–439.

Heister, G., Ehrenstein, W. H., & Schroeder-Heister, P. (1987). Spatial S-R compatibility with unimanual two-finger choice reactions: Effects of irrelevant stimulus location. *Perception & Psychophysics, 42*, 195–201.

Heister, G., Ehrenstein, W. H., & Schroeder-Heister, P. (1988). Spatial S-R compatibility effects with unimanual two-finger choice reactions for prone and supine hand positions. *Perception & Psychophysics, 40*, 271–278.

Heister, G., & Schroeder-Heister, P. (1994). Spatial S-R compatibility: Positional instructions vs. compatibility instruction. *Acta Psychologica, 85*, 15–24.

Heister, G., Schroeder-Heister, P., & Ehrenstein, W. H. (1990). Spatial coding and spatio-anatomical mapping: Evidence for a hierarchical model of spatial stimulus–response compatibility. In R. W. Proctor & T. G. Reeve (Eds.), *Stimulus–response compatibility: An integrated perspective* (pp. 117–143). Amsterdam: North-Holland.

Heyes, C., & Ray, E. (2004). Spatial S-R compatibility effects in an intentional imitation task. *Bulletin of the Psychonomic Society, 11*, 703–708.

Hick, W. E. (1952). On the rate of gain of information. *Quarterly Journal of Experimental Psychology, 4*, 11–26.

Hoffmann, E. R. (1997). Strength of component principles determining direction of turn stereotypes — linear displays with rotary controls. *Ergonomics, 40*, 199–222.

Hoffmann, J., & Koch, I. (1997). Stimulus–response compatibility and sequential learning in the serial reaction time task. *Psychological Research, 60*, 87–97.

Holding, D. H. (1957). Direction of motion relationships between controls and displays moving in different planes. *Journal of Applied Psychology, 41*, 93–97.

Hommel, B. (1993a). Inverting the Simon effect by intention: Determinants of direction and extent of effects of irrelevant spatial information. *Psychological Research, 55*, 270–279.

Hommel, B. (1993b). The relationship between stimulus processing and response selection in the Simon task: Evidence for a temporal overlap. *Psychological Research, 55*, 280–290.

Hommel, B. (1993c). The role of attention for the Simon effect. *Psychological Research, 55*, 208–222.

Hommel, B. (1994). Spontaneous decay of response-code activation. *Psychological Research, 56*, 261–268.

Hommel, B. (1995). Stimulus–response compatibility and the Simon effect: Toward an empirical clarification. *Journal of Experimental Psychology: Human Perception and Performance, 21*, 764–775.

Hommel, B. (1996a). No prevalence of right–left over top–bottom spatial codes. *Perception & Psychophysics, 43*, 102–110.

Hommel, B. (1996b). The cognitive representation of action: Automatic integration of perceived action effects. *Psychological Research, 59*, 176–186.

Hommel, B. (1997). Interactions between stimulus–stimulus congruence and stimulus–response compatibility. *Psychological Research, 59*, 248–260.

Hommel, B. (1998a). Automatic stimulus–response translation in dual-task performance. *Journal of Experimental Psychology: Human Perception and Performance, 24*, 1368–1384.

Hommel, B. (1998b). Event files: Evidence for automatic integration of stimulus–response episodes. *Visual Cognition, 5*, 183–216.

Hommel, B. (2004). Event files: Feature binding in and across perception and action. *Trends in Cognitive Science, 11*, 494–500.

Hommel, B., & Lippa, Y. (1995). S-R compatibility effects due to context-dependent spatial stimulus coding. *Psychonomic Bulletin & Review, 2*, 370–374.

Hommel, B., Müsseler, J., Aschersleben, G., & Prinz, W. (2001). The theory of event-coding (TEC): A framework for perception and action planning. *Behavioral and Brain Sciences, 24*, 849–878.

Hommel, B., & Prinz, W. (Eds.) (1997). *Theoretical issues in stimulus–response compatibility.* Amsterdam: North-Holland.

Hommel, B., Proctor, R. W., & Vu, K.-P. L. (2004). A feature-integration account of sequential effects in the Simon task. *Psychological Research, 68*, 1–17.

Hong, W., Thong, J. Y. L., & Tam, K. Y. (2004). Designing product listing pages on e-commerce websites: An examination of presentation mode and information format. *International Journal of Human–Computer Studies, 61*, 481–503.

Hsu, S. H., & Peng, Y. (1993). Control/display relationship of the four-burner stove. A reexamination. *Human Factors, 35*, 745–749.

Humphries, M., & Shephard, A. H. (1955). Performance on several control–display arrangements as a function of age. *Canadian Journal of Psychology, 9*, 231–238.

Humphries, M., & Shephard, A. H. (1959). Age and training in the development of a perceptual-motor skill. *Perceptual and Motor Skills, 9*, 3–11.

Hyman, R. (1953). Stimulus information as a determinant of reaction time. *Journal of Experimental Psychology, 45*, 188–196.

Hyman, R., & Umiltà, C. (1969). The information hypothesis and nonrepetitions. *Acta Psychologica, 30*, 37–53.

Iacoboni, M., Woods, R. P., & Mazziotta, J. C. (1996). Brain-behavior relationships: Evidence from practice effects in spatial stimulus–response compatibility. *Journal of Neurophysiology, 76*, 321–331.

Inhoff, A. W., Rosenbaum, D. A., Gordon, A. W., & Campbell, J. A. (1984). Stimulus–response compatibility and motor programming of manual response sequences. *Journal of Experimental Psychology: Human Perception and Performance, 10*, 724–733.

Ivry, R. B., Franz, E. A., Kingstone, A., & Johnston, J. C. (1998). The psychological refractory period effect following callostomy: Uncoupling of lateralized response codes. *Journal of Experimental Psychology: Human Perception and Performance, 24*, 463–480.

Ivry, R. B., & Hazeltine, E. (2000). Task-switching in a callosotomy patient and in normal participants: Evidence for response-related sources of interference. In S. Monsell & J. Driver (Eds.), *Control of cognitive processes: Attention and performance XVIII* (pp. 401–424). Cambridge, MA: MIT Press.

Jennings, J. R., Van der Molen, M. W., Van der Veen, F. M., & Debski, K. B. (2002). Influence of preparatory schema on the speed of responses to spatially compatible and incompatible stimuli. *Psychophysiology, 39*, 496–504.

Jensen, A. R. (1980). Chronometric analysis of intelligence. *Journal of Social Biological Structure, 3*, 103–122.

Jentsch, I., & Sommer, W. (2002). Functional localization and mechanisms of sequential effects in serial reaction time tasks. *Perception & Psychophysics, 64,* 1169–1188.

John, B. E., & Newell, A. (1987). Predicting the time to recall computer command abbreviations. In *CHI & GI 1987 Conference Proceedings: Human Factors in Computing Systems and Graphics Interface* (pp. 33–40). New York: ACM.

John, B. E., & Newell, A. (1990). Toward an engineering model of stimulus–response compatibility. In R. W. Proctor & T. G. Reeve (Eds.), *Stimulus–response compatibility: An integrated perspective* (pp. 427–479). Amsterdam: North-Holland.

John, B. E., Rosenbloom, P. S., & Newell, A. (1985). A theory of stimulus–response compatibility applied to human–computer interaction. In *CHI '85 Conference Proceedings: Human Factors in Computing Systems* (pp. 213–219). New York: ACM.

Kahneman, D., Treisman, A., & Gibbs, B. J. (1992). The reviewing of object files: Object-specific integration of information. *Cognitive Psychology, 24,* 175–219.

Kantowitz, B. H., Triggs, T. J., & Barnes, V. E. (1990). Stimulus–response compatibility and human factors. In R. W. Proctor & T. G. Reeve (Eds.), *Stimulus–response compatibility: An integrated perspective* (pp. 365–388). Amsterdam: North-Holland.

Katz, A. N. (1981). Spatial compatibility effects with hemifield presentation in a unimanual two-finger task. *Canadian Journal of Psychology, 35,* 63–68.

Kerzel, D., & Bekkering, H. (2000). Motor activation from visible speech: Evidence from stimulus–response compatibility. *Journal of Experimental Psychology: Human Perception and Performance, 26,* 634–647.

Keyson, D. K., & Parsons, K. C. (1990). Designing the user interface using rapid prototyping. *Applied Ergonomics, 21,* 207–211.

Kinoshita, S., & Peek-O'Leary, M. (2005). Does the compatibility effect in the race implicit association test reflect familiarity or affect? *Psychonomic Bulletin & Review, 12,* 442–452.

Klapp, S. T., Greim, D. M., Mendicino, C. M., & Koenig, S. (1979). Anatomic and environmental dimensions of stimulus–response compatibility: Implication for theories of memory coding. *Acta Psychologica, 43,* 367–379.

Klapp, S. T., & Haas, B. W. (2005). Nonconscious influence of masked stimuli on response selection is limited to concrete stimulus–response associations. *Journal of Experimental Psychology: Human Perception and Performance, 31,* 193–209.

Klapp, S. T., & Hinkley, L. B. (2002). The negative compatibility effect: Unconscious inhibition influences reaction time and response selection. *Journal of Experimental Psychology: General, 131,* 255–269.

Kleinsorge, T. (1999). Die Kodierungsabhaengigkeit orthogonaler Reiz-Reaktions-Kompatibilitaet. [Coding specificity of orthogonal S-R compatibility]. *Zeitshrift für Experimentelle Psychologie, 46,* 249–264.

Knowles, W. B, Garvey, W. D., & Newlin, E. P. (1953). The effect of speed and load on display–control relationships. *Journal of Experimental Psychology, 46,* 65–75.

Koch, I., Keller, P., & Prinz, W. (2004). The ideomotor approach to action control: Implications for skilled performance. *International Journal of Sport & Exercise Psychology, 2,* 362–375.

Koch, I., & Kunde, W. (2002). Verbal response–effect compatibility. *Memory & Cognition, 30,* 1297–1303.

Koch, I., Metin, B., & Schuch, S. (2003). The role of temporal unpredictability for process interference and code overlap in perception–action dual tasks. *Psychological Research, 67,* 244–252.

Koch, I., & Prinz, W. (2002). Process interference and code overlap in dual-task performance. *Journal of Experimental Psychology: Human Perception and Performance, 28,* 192–201.

Kornblum, S. (1965). Response competition and/or inhibition in two-choice–reaction time. *Psychonomic Science, 2*, 55–56.

Kornblum, S. (1967). Choice–reaction time for repetitions and nonrepetitions. *Acta Psychologica, 27*, 178–187.

Kornblum, S. (1968). Serial-choice–reaction time: Inadequacies of the information hypothesis. *Science, 159*, 432–434.

Kornblum, S. (1991). Stimulus–response coding in four classes of stimulus–response ensembles. In J. Requin & G. E. Stelmach (Eds.), *Tutorials in motor neuroscience* (pp. 3–15). Dordrecht: Kluwer Academic Publishers.

Kornblum, S. (1992). Dimensional overlap and dimensional relevance in stimulus–response and stimulus–stimulus compatibility. In G. E. Stelmach & J. Requin (Eds.), *Tutorials in motor behavior II* (pp. 743–777). Amsterdam: North-Holland.

Kornblum, S. (1994). The way irrelevant dimensions are processed depends on what they overlap with: The case of Stroop- and Simon-like stimuli. *Psychological Research, 56*, 130–135.

Kornblum, S., Hasbroucq, T., & Osman, A. (1990). Dimensional overlap: Cognitive basis for stimulus–response compatibility — A model and taxonomy. *Psychological Review, 97*, 253–270.

Kornblum, S., & Lee, J.-W. (1995). Stimulus–response compatibility with relevant and irrelevant stimulus dimensions that do and do not overlap with the response. *Journal of Experimental Psychology: Human Perception and Performance, 21*, 855–875.

Kornblum, S., & Stevens, G. (2002). Sequential effects of dimensional overlap: Findings and issues. In W. Prinz & B. Hommel (Eds.), *Attention and performance IX* (pp. 9–49). Oxford, UK: Oxford University Press.

Kornblum, S., Stevens, G. T., Whipple, A., & Requin, J. (1999). The effects of irrelevant stimuli: 1. The time course of stimulus–stimulus and stimulus–response consistency effects with Stroop-like stimuli, Simon-like tasks, and their factorial combinations. *Journal of Experimental Psychology: Human Perception and Performance, 25*, 688–714.

Kosslyn, S. M. (1994). *Image and brain: The resolution of the imagery debate.* Cambridge, MA: MIT Press.

Kunde, W. (2001). Response–effect compatibility in manual choice–reaction tasks. *Journal of Experimental Psychology: Human Perception and Performance, 27*, 387–394.

Kunde, W. (2003). Temporal response–effect compatibility. *Psychological Research, 67*, 153–159.

Kunde, W., & Stöcker, C. (2002). A Simon effect for stimulus–response duration. *Quarterly Journal of Experimental Psychology, 55A*, 581–592.

Kvälseth, T. O. (1980). An alternative to the Hick–Hyman's and Sternberg's law. *Perceptual and Motor Skills, 50*, 1281–1282.

Kvälseth, T. O. (1989). Longstreth et al.'s reaction time model: Some comments. *Bulletin of the Psychonomic Society, 27*, 358–360.

Ladavas, E. (1987). Influence of handedness on spatial compatibility effects with perpendicular arrangement of stimuli and responses. *Acta Psychologica, 64*, 13–23.

Ladavas, E., & Moscovitch, M. (1984). Must egocentric and environmental frames of reference be aligned to produce spatial S compatibility effects? *Journal of Experimental Psychology: Human Perception and Performance, 10*, 205–215.

Lamberts, K., Tavernier, G., & d'Ydewalle, G. (1992). Effects of multiple reference points in spatial stimulus–response compatibility. *Acta Psychologica, 79*, 115–130.

Larish, D. D. (1986). Influences of stimulus–response translations on response programming: Examining the relationship of arm, direction, and extent of movement. *Acta Psychologica, 61*, 53–70.

Larish, D. D., & Frekany, G. A. (1985). Planning and preparing expected and unexpected movements: Reexamining the relationships of arm, direction, and extent of movement. *Journal of Motor Behavior, 14*, 322–340.

Larish, D. D., & Stelmach, G. E. (1982). Preprogramming, programming, and reprogramming of aimed hand movements as a function of age. *Journal of Motor Behavior, 14*, 322–340.

Learmount, D., & Norris, G. (1990). Lessons to be learned. *Flight International,* 31 October – 6 November, 24–26.

Leonard, J. A. (1955). Factors which influence channel capacity. In H. Quastler (Ed.), *Information theory in psychology; problems and methods* (pp. 306–315). Glencoe, IL: Free Press.

Leonard, J. A. (1958). Partial advance information in a choice–reaction task. *British Journal of Psychology, 49*, 89–96.

Leonard, J. A. (1959). Tactile choice reactions. *Quarterly Journal of Experimental Psychology, 11*, 76–83.

Leuthold, H., Sommer, W., & Ulrich, R. (1996). Partial advance information and response preparation: Inferences from the lateralized readiness potential. *Journal of Experimental Psychology: General, 125*, 307–323.

Lien, M.-C., McCann, R. S., Ruthruff, E., & Proctor, R. W. (2005a). Confirming and disconfirming theories about ideomotor compatibility in dual-task performance: A reply to Greenwald (2005). *Journal of Experimental Psychology: Human Perception and Performance, 31*, 226–229.

Lien, M.-C., McCann, R. S., Ruthruff, E., & Proctor, R. W. (2005b). Dual-task performance with ideomotor compatible tasks: Is the central processing bottleneck intact, bypassed, or shifted in locus? *Journal of Experimental Psychology: Human Perception and Performance, 31*, 122–144.

Lien, M.-C., & Proctor, R. W. (2000). Multiple spatial correspondence effects on dual-task performance. *Journal of Experimental Psychology: Human Perception and Performance, 26*, 1260–1280.

Lien, M.-C., & Proctor, R. W. (2002). Stimulus–response compatibility and psychological refractory period effects: Implications for response selection. *Psychonomic Bulletin & Review, 9*, 212–238.

Lien, M.-C., Proctor, R. W., & Allen, P. A. (2002). Ideomotor compatibility in the psychological refractory period effect: 29 years of oversimplification. *Journal of Experimental Psychology: Human Perception and Performance, 28*, 396–409.

Lien, M.-C., Proctor, R. W., & Ruthruff, E. (2003). Still no evidence for perfect timesharing with two ideomotor-compatible tasks: A reply to Greenwald (2003). *Journal of Experimental Psychology: Human Perception and Performance, 29*, 1267–1272.

Lien, M.-C., Schweickert, R., & Proctor, R. W. (2003). Task switching and response correspondence in the psychological refractory period paradigm. *Journal of Experimental Psychology: Human Perception and Performance, 29*, 692–712.

Lindley, R. H., Bathhurst, K., Smith, W. R., & Wilson, S. M. (1993). Hick's law, IQ, and singularity or specificity of mind: A psychometric analysis. *Personality and Individual Differences, 15*, 129–135.

Lippa, Y. (1996). A referential-coding explanation for compatibility effects of physically orthogonal stimulus and response dimensions. *Quarterly Journal of Experimental Psychology, 49A*, 950–971.

Lippa, Y., & Adam, J. J. (2001). Orthogonal stimulus–response compatibility resulting from spatial transformations. *Perception & Psychophysics, 63,* 156–174.

Lleras, A., Moore, C. M., & Mordkoff, J. T. (2004). Looking for the source of the Simon effect: Evidence of multiple codes. *American Journal of Psychology, 117*, 531–542.

Logan, G. D. (1994). Spatial attention and the apprehension of spatial relations. *Journal of Experimental Psychology: Human Perception and Performance, 20*, 1015–1036.

Logan, G. D. (2003). Simon-type effects: Chronometric evidence for keypress schemata in typewriting. *Journal of Experimental Psychology: Human Perception and Performance, 29*, 741–757.

Logan, G. D., & Gordon, R. D. (2001). Executive control of visual attention in dual-task situations. *Psychological Review, 108*, 393–434.

Logan, G. D., & Schulkind, M. D. (2000). Parallel memory retrieval in dual-task situations: I. Semantic memory. *Journal of Experimental Psychology: Human Perception and Performance, 26*, 1260–1280.

Logan, G., & Zbrodoff, N. J. (1979). When it helps to be misled: Facilitative effects of increasing the frequency of conflicting stimulus in a Stroop-like task. *Memory & Cognition, 7*, 166–174.

Longstreth, L. E. (1988). Hick's law: Its limit is 3 bits. *Bulletin of the Psychonomic Society, 26*, 8–10.

Longstreth, L. E., El-Zahhar, N., & Alcorn, M. B. (1985). Exceptions to Hick's law: Explorations with a response duration measure. *Journal of Experimental Psychology: General, 114*, 417–434.

Los, S. A. (1996). On the origin of mixing costs: Exploring information processing in pure and mixed blocks of trials. *Acta Psychologica, 94*, 145–188.

Loveless, N. E. (1956). Display–control relationships on circular and linear scales. *British Journal of Psychology, 47*, 271–282.

Loveless, N. E. (1959). The effect of the relative position of control and display upon their direction-of-motion relationship. *Ergonomics, 2*, 381–385.

Loveless, N. E. (1962). Direction-of-motion stereotypes: A review. *Ergonomics, 5*, 357–383.

Lu, C.-H., & Proctor, R. W. (1994). Processing of an irrelevant location dimension as a function of the relevant stimulus dimension. *Journal of Experimental Psychology: Human Perception and Performance, 20*, 286–298.

Lu, C.-H., & Proctor, R. W. (1995). The influence of irrelevant location information on performance: A review of the Simon and spatial Stroop effects. *Psychonomic Bulletin & Review, 2*, 174–207.

Lu, C.-H., & Proctor, R. W. (1998, October). *Mapping effects for orthogonally oriented stimulus and response sets.* Poster presented at the 42nd Annual Meeting of the Human Factors and Ergonomics Society, Chicago.

Lu, C.-H., & Proctor, R. W. (2001). Influence of irrelevant information on human performance: Effects of S-R association strength and relative timing. *Quarterly Journal of Experimental Psychology, 54*, 95–136.

Lyons, J., Weeks, D. J., & Chua, R. (2005). *The influence of object orientation on speed of object identification: Affordance facilitation or cognitive coding.* Manuscript submitted for publication.

Marble, J. G., & Proctor, R. W. (2000). Mixing location-relevant and location-irrelevant choice–reaction tasks: Influences of location mapping on the Simon effect. *Journal of Experimental Psychology: Human Perception and Performance, 26*, 1515–1533.

Marino, C. J., & Mahan, R. P. (2005). Configural displays can improve nutrition-related decisions: An application of the proximity compatibility principle. *Human Factors, 47*, 121–130.

Masaki, H., Wild-Wall, N., Sangals, J., & Sommer, W. (2004). The functional locus of the lateralized readiness potential. *Psychophysiology, 41,* 220–230.

Mattes, S., Leuthold, H., & Ulrich, R. (2002). Stimulus–response compatibility in intensity–force relations. *Quarterly Journal of Experimental Psychology, 55A,* 1175–1191.

McCann, R. S., & Johnston, J. C. (1992). Locus of the single-channel bottleneck in dual-task interference. *Journal of Experimental Psychology: Human Perception and Performance, 18,* 471–484.

McCarley, J. S., & Wickens, C. D. (2005). *Human factors implications of UAVs in the National Airspace.* Technical Report AHFD-05-05/FAA-05-01. Atlantic City International Airport, NJ: Federal Aviation Administration.

Meiran, N. (1996). Reconfiguration of processing mode prior to task performance. *Journal of Experimental Psychology: Learning, Memory, and Cognition, 22,* 1423–1442.

Meiran, N., Chorev, Z., & Sapir, A. (2002). Component processes in task switching. *Cognitive Psychology, 41,* 211–253.

Merkel, J. (1885). Die zeitliche Verhaltnisse de Willenstatigkeit [The temporal relations of activities of the will]. *Philosophische Studien, 2,* 73–127.

Merz, F., Kalveram, K.-T., & Huber, K. (1981). Der Einfluss Kognitiver Faktoren auf Steuerleistungen [The influence of cognitive factors on control performance]. In L. Tent (Ed.), *Erkennen – Wollen – Handeln.* (pp. 327–335). Göttingen: Hogrefe.

Meyer, D. E., & Kieras, D. E. (1997). A computational theory of executive cognitive processes and multiple-task performance: I. Basic mechanisms. *Psychological Review, 104,* 3–65.

Michaels, C. F. (1988). S-R compatibility between response position and destination of apparent motion: Evidence of the detection of affordances. *Journal of Experimental Psychology: Human Perception and Performance, 14,* 231–240.

Michaels, C. F. (1989). S-R compatibilities depend on eccentricity of responding hand. *Quarterly Journal of Experimental Psychology, 41A,* 262–272.

Michaels, C. F. (1993). Destination compatibility, affordances, and coding rules: A reply to Proctor, Van Zandt, Lu, & Weeks (1993). *Journal of Experimental Psychology: Human Perception and Performance, 19,* 1121–1127.

Michaels, C. F., & Schilder, S. (1991). Stimulus–response compatibilities between vertically oriented stimuli and horizontally oriented response: The effects of hand position and posture. *Perception & Psychophysics, 49,* 342–348.

Michaels, C. F., & Stins, J. F. (1997). An ecological approach to stimulus–response compatibility. In B. Hommel & W. Prinz (Eds.), *Theoretical issues in stimulus–response compatibility* (pp. 333–360). Amsterdam: North-Holland.

Miller, J. (1982). Discrete versus continuous stage models of human information processing: In search of partial output. *Journal of Experimental Psychology: Human Perception and Performance, 8,* 273–296.

Miller, J. (2006). Simon congruency effects based on stimulus and response numerosity. *Quarterly Journal of Experimental Psychology, 59,* 387–396.

Miller, J. O., Atkins, S. G., & Van Nes, F. (2005). Compatibility effects based on stimulus and response numerosity. *Psychonomic Bulletin & Review, 12,* 265–270.

Mitchell, M. J. H. (1947). Direction of movement of machine controls. III. Ministry of Supply, S.M. 10018(S).

Mitchell, M. J. H. (1948). Direction of movement of machine controls. IV. Medical Research Council. A.P.U. 48/371.

Mitchell, M. J. H., & Vince, M. A. (1951). The direction of movement of machine controls. *Quarterly Journal of Experimental Psychology, 3,* 24–35.

Monsell, S. (2003). Task switching. *Trends in Cognitive Sciences, 7,* 134–140.

Mordkoff, T. (1998). The gating of irrelevant information in selective-attention tasks [Abstract]. *Abstracts of the Psychonomic Society, 3,* 193.

Morin, R. E., & Forrin, B. (1962). Mixing of two types of S-R associations in a choice–reaction time task. *Journal of Experimental Psychology, 64,* 137–141.

Morin, R. W., Forrin, B., & Archer, W. (1961). Information processing behavior: The role of irrelevant stimulus information. *Journal of Experimental Psychology, 61,* 89–96.

Morin, R. E., & Grant, D. A. (1955). Learning and performance on a keypressing task as a function of the degree of spatial stimulus–response correspondence. *Journal of Experimental Psychology, 49,* 39–47.

Mowbray, G. H., & Rhodes, M. V. (1959). On the reduction of choice–reaction time with practice. *Quarterly Journal of Experimental Psychology, 11,* 16–23.

Murrell, K. F. H. (1952). Direction of motion relationships between valves and gauges when mounted in panels. Results from a printed test. *Naval Motion Study Unit Rep. No. 51.*

Müsseler, J., & Hommel, B. (1997a). Blindness to response-compatible stimuli. *Journal of Experimental Psychology: Human Perception and Performance, 23,* 861–872.

Müsseler, J., & Hommel, B. (1997b). Detecting and identifying response-compatible stimuli. *Psychonomic Bulletin & Review, 4,* 125–129.

Newell, A. (1990). *Unified theories of cognition.* Cambridge, MA: Harvard University Press.

Newell, A., & Rosenbloom, P. S. (1981). Mechanisms of skill acquisition and the law of practice. In J. R. Anderson (Ed.), *Cognitive skills and their acquisitions* (pp. 1–55). Hillsdale, NJ: Erlbaum.

Nicoletti, R., Anzola, G. P., Luppino, G., Rizzolatti, G., & Umiltà, C. (1982). Spatial com patibility effects on the same side of the body midline. *Journal of Experimental Psychology: Human Perception and Performance, 8,* 664–673.

Nicoletti, R., & Umiltà, C. (1984). Right–left prevalence in spatial compatibility. *Perception & Psychophysics, 35,* 333–343.

Nicoletti, R., & Umiltà, C. (1985). Responding with hand and foot: The right/left prevalence in spatial compatibility is still present. *Perception & Psychophysics, 38,* 211–216.

Nicoletti, R., & Umiltà, C. (1989). Splitting visual space with attention. *Journal of Experimental Psychology: Human Perception and Performance, 15,* 164–169.

Nicoletti, R., & Umiltà, C. (1994). Attention shifts produce spatial stimulus codes. *Psychological Research, 56,* 144–150.

Nicoletti, R., Umiltà, C., & Ladavas, E. (1984). Compatibility due to the coding of the relative position of the effectors. *Acta Psychologica, 57,* 133–143.

Nicoletti, R., Umiltà, C., Tressoldi, E. P., & Marzi, C. A. (1988). Why are right–left spatial codes easier to form than above-below ones? *Perception & Psychophysics, 43,* 287–292.

Nissen, M. J., & Bullemer, P. (1987). Attentional requirements of learning: Evidence from performance measures. *Cognitive Psychology, 19,* 1–32.

Notebaert, W., Soetens, E., & Melis, A. (2001). Sequential analysis of a Simon task-evidence for an attention-shift account. *Psychological Research, 65,* 170–184.

Nuerk, H.-C., Iverson, W., & Willmes, K. (2004). Notational modulation of the SNARC and the MARC (linguistic markedness of response codes) effect. *Quarterly Journal of Experimental Psychology, 57,* 835–863.

O'Leary, M. J., & Barber, P. J. (1993). Interference effects in the Stroop and Simon paradigms. *Journal of Experimental Psychology: Human Perception and Performance, 19,* 830–844.

O'Leary, M. J., Barber, P. J., & Simon, J. R. (1994). Does stimulus correspondence account for the Simon effect? Comments on Hasbroucq and Guiard (1991). *Psychological Research, 56,* 203–209.

Olson, G. M., & Laxar, K. (1973). Asymmetries in processing the terms "right" and "left." *Journal of Experimental Psychology, 100*, 284–290.

Osman, A., Moore, C. M., & Ulrich, R. (1995). Bisecting RT with lateralized readiness potentials: Precue effects after LRP onset. *Acta Psychologica, 90*, 111–127.

Pachella, R. G. (1974). The interpretation of reaction time in information-processing research. In B. H. Kantowitz (Ed.), *Human information processing: Tutorials in performance and cognition* (pp. 41–82). Hillsdale, NJ: Erlbaum.

Pashler, H. (1984). Processing stages in overlapping tasks: Evidence for a central bottleneck. *Journal of Experimental Psychology: Human Perception and Performance, 10*, 358–377.

Pashler, H. (1994). Dual-task interference in simple tasks: Data and theory. *Psychological Bulletin, 16*, 220–224.

Pashler, H., & Baylis, G. (1991). Procedural learning: 2. Intertrial repetition effects in speeded choice task. *Journal of Experimental Psychology: Learning, Memory, and Cognition, 17*, 33–48.

Pashler, H., & Johnston, J. C. (1989). Chronometric evidence for central postponement in temporally overlapping tasks. *Quarterly Journal of Experimental Psychology, 41A*, 19–45.

Pashler, H., & Johnston, J. C. (1998). Attentional limitations in dual-task performance. In H. Pashler (Ed.), *Attention* (pp. 155–189). Hove, UK: Psychology Press.

Payne, S. J. (1995). Naïve judgments of stimulus–response compatibility. *Human Factors, 37*, 495–506.

Petropoulos, H., & Brebner, J. (1981). Stereotypes for direction-of-movement of rotary controls associated with linear displays: The effects of scale presence and position, of pointer direction, and distances between the control and display. *Ergonomics, 24*, 143–151.

Phillips, J. C., & Ward, R. (2002). S-R correspondence effects of irrelevant visual affordance: Time course and specificity of response activation. *Visual Cognition, 9*, 540–558.

Pierce, J. R., & Karlin, J. E. (1957). Reading rates and the information rate of a human channel. *Bell Systems Technical Journal, 36*, 497–516.

Possamaï, C.-A., Burle, B., Osman, A., & Hasbroucq, T. (2002). Partial advance information, number of alternatives, and motor processes: An electromyographic study. *Acta Psychologica, 111*, 125–139.

Proctor, R. W., & Cho, Y. S. (2001). The up-right/down-left advantage occurs for both participant-paced and computer-paced conditions: An observation on Adam, Boon, Paas, & Umiltà (1998). *Journal of Experimental Psychology: Human Perception and Performance, 27*, 466–471.

Proctor, R. W., & Cho, Y. S. (2003). Effects of response eccentricity and relative position on orthogonal stimulus–response compatibility with joystick and keypress responses. *Quarterly Journal of Experimental Psychology, 56A*, 309–327.

Proctor, R. W., & Cho, Y. S. (in press). Polarity correspondence: A general principle for performance of speeded binary classification tasks. *Psychological Bulletin*.

Proctor, R. W., & Dutta, A. (1992). Unified theories must explain the codependencies among perception, cognition, and action. *Behavioral and Brain Sciences, 15*, 453–454.

Proctor, R. W., & Dutta, A. (1993). Do the same stimulus–response relations influence choice reactions initially and after practice? *Journal of Experimental Psychology: Learning, Memory, and Cognition, 19*, 922–930.

Proctor, R. W., Dutta, A., Kelly, P. L., & Weeks, D. J. (1994). Cross-modal compatibility effects with visual/spatial and auditory/verbal stimulus and response sets. *Perception & Psychophysics, 55*, 42–47.

Proctor, R. W., Koch, I., & Vu, K.-P. L. (in press). The role of preparation in the right–left prevalence effect. *Memory & Cognition*.

Proctor, R. W., & Lu, C.-H. (1994). Referential coding and attention-shifting accounts of the Simon effect. *Psychological Research, 56*, 185–195.

Proctor, R. W., & Lu, C.-H. (1999). Processing irrelevant location information: Practice and transfer effects in choice–reaction tasks. *Memory & Cognition, 27*, 63–77.

Proctor, R. W., Lu, C.-H., & Van Zandt, T. (1992). Enhancement of the Simon effect by response precuing. *Acta Psychologica, 81*, 53–74.

Proctor, R. W., Lu, C.-H., Van Zandt, T., & Weeks, D. J. (1994). Affordances, codes, and decision processes: A response to Michaels (1993). *Journal of Experimental Psychology: Human Perception and Performance, 20*, 452–455.

Proctor, R. W., Marble, J. G., & Vu, K.-P. L. (2000). Mixing incompatibly mapped location-relevant trials with location-irrelevant trials: Effects of stimulus mode on the reverse Simon effect. *Psychological Research, 64*, 11–24.

Proctor, R. W., & Pick, D. F. (1998). Lateralized warning tones produce typical irrelevant-location effects on choice reactions. *Psychonomic Bulletin & Review, 5*, 124–129.

Proctor, R. W., & Pick, D. F. (1999). Deconstructing Marilyn: Robust effects of face contexts on stimulus–response compatibility. *Memory & Cognition, 27*, 986–995.

Proctor, R. W., & Pick, D. F. (2003). Display–control arrangement correspondence and logical recoding in the Hedge and Marsh reversal of the Simon effect. *Acta Psychologica, 112*, 259–278.

Proctor, R. W., Pick, D. F., Vu, K.-P. L., & Anderson, R. E. (2005). The enhanced Simon effect for older adults is reduced when the irrelevant location information is conveyed by an accessory stimulus. *Acta Psychologica, 119*, 21–40.

Proctor, R. W., & Reeve, T. G. (1985). Compatibility effects in the assignment of symbolic stimuli to discrete finger response. *Journal of Experimental Psychology: Human Perception and Performance, 11*, 623–639.

Proctor, R. W., & Reeve, T. G. (1986). Salient-feature coding operations in spatial precuing tasks. *Journal of Experimental Psychology: Human Perception and Performance, 12*, 277–285.

Proctor, R. W., & Reeve, T. G. (1988). The acquisition of task-specific productions and modification of declarative representations in spatial-precuing tasks. *Journal of Experimental Psychology: General, 117*, 182–196.

Proctor, R. W., & Reeve, T. G. (Eds.) (1990). *Stimulus–response compatibility: An integrated perspective*. Amsterdam: North-Holland.

Proctor, R. W., Reeve, T. G., Weeks, D. J., Campbell, K. C., & Dornier, L. (1997). Translating between orthogonally oriented stimulus and response arrays in four-choice–reaction tasks. *Canadian Journal of Experimental Psychology, 51*, 85–97.

Proctor, R. W., Van Zandt, T., Lu, C.-H., & Weeks, D. J. (1993). Stimulus–response compatibility for moving stimuli: Perception of affordances or directional coding? *Journal of Experimental Psychology: Human Perception and Performance, 19*, 81–91.

Proctor, R. W., & Vu, K.-P. L. (2002a). Eliminating, magnifying, and reversing spatial compatibility effects with mixed location-relevant and irrelevant trials. In W. Prinz & B. Hommel (Eds.), *Common mechanisms in perception and action: Attention and performance, Vol. XIX* (pp. 443–473). Oxford: Oxford University Press.

Proctor, R. W., & Vu, K.-P. L. (2002b). Mixing location irrelevant and relevant trials: Influence of stimulus mode on spatial compatibility effects. *Memory & Cognition, 30*, 281–294.

Proctor, R. W., & Vu, K.-P. L. (2002c, August). *Violations of the spatial compatibility principle: Theoretical and applied implications*. Paper presented at the 2002 Annual Convention of the American Psychological Association, Chicago.

Proctor, R. W., & Vu, K.-P. L. (2003). Human information processing: An overview for human– computer interaction. In J. A. Jacko & A. Sears (Eds.), *The human–computer interaction handbook: Fundamentals, evolving technologies, and emerging applications* (pp. 35–51). Mahwah, NJ: Erlbaum.

Proctor, R. W., Vu, K.-P. L., & Marble, J. G. (2003). Eliminating spatial compatibility effects for location-relevant trials by intermixing location-irrelevant trials. *Visual Cognition, 10,* 15–50.

Proctor, R. W., Vu, K.-P. L., & Nicoletti, R. (2003). Does right–left prevalence occur for the Simon effect? *Perception & Psychophysics, 65,* 1318–1329.

Proctor, R. W., Vu, K.-P. L., & Pick, D. F. (2005). Aging and response selection in spatial choice tasks. *Human Factors, 47,* 250–270.

Proctor, R. W., Vu, K.-P. L., & Pick, D. F. (in press). A deficit in effortful selection of cued responses for older adults. *Journal of Motor Behavior.*

Proctor, R. W., & Wang, H. (1997a). Differentiating types of set-level compatibility. In B. Hommel & W. Prinz (Eds.), *Theoretical issues in stimulus–response compatibility* (pp. 11–37). Amsterdam: North-Holland.

Proctor, R. W., & Wang, H. (1997b). Set- and element-level stimulus–response compatibility effects for different manual response sets. *Journal of Motor Behavior, 29,* 351–365.

Proctor, R. W., Wang, D.-Y. D., & Pick, D. F. (2004). Stimulus–response compatibility with wheel-rotation responses: Will an incompatible response coding be used when a compatible coding is possible? *Psychonomic Bulletin & Review, 11,* 811–847.

Proctor, R. W., Wang, H., & Vu, K.-P. L. (2002) Influences of conceptual, physical, and structural similarity on stimulus–response compatibility. *Quarterly Journal of Experimental Psychology, 55A,* 59–74.

Ragot, R., Cave, C., & Fano, M. (1988). Reciprocal effects of visual and auditory stimuli in a spatial compatibility situation. *Bulletin of the Psychonomic Society, 26,* 350–352.

Ragot, R., & Fiori, N. (1994). Mental processing during reactions toward and away from a stimulus: An ERP analysis of auditory congruence and S-R compatibility. *Psychophysiology, 31,* 439–446.

Ragot, R., & Guiard, Y. (1992). Stimulus congruence and stimulus–response compatibility: Two variables disentangled in an auditory reaction time task. *European Journal of Cognitive Psychology, 4,* 219–232.

Ratcliff, R., & Smith, P. L. (2004). A comparison of sequential sampling models for two-choice–reaction time. *Psychological Review, 111,* 333–367.

Ray, R. D., & Ray, W. D. (1979). An analysis of domestic cooker control design. *Ergonomics, 22,* 1243–1248.

Read, L. E., & Proctor, R. W. (2004). Spatial stimulus–response compatibility and negative priming. *Psychonomic Bulletin & Review, 11,* 41–48.

Reeve, T. G., & Proctor, R. W. (1984). On the advance preparation of discrete finger responses. *Journal of Experimental Psychology: Human Perception and Performance, 10,* 541–553.

Reeve, T. G., & Proctor, R. W. (1990). The salient-features coding principle for spatial- and symbolic-compatibility effects. In R. W. Proctor & T. G. Reeve (Eds.), *Stimulus–response compatibility: An integrated perspective* (pp. 163–180). Amsterdam: North-Holland.

Reeve, T. G., Proctor, R. W., Weeks, D. J., & Dornier, L. (1992). Salience of stimulus and response features in choice–reaction tasks. *Perception & Psychophysics, 52,* 453–460.

Riggio, L., Gawryszewski, L. G., & Umiltà, C. (1986). What is crossed in crossed-hand effects? *Acta Psychologica, 62,* 89–100.

Rogers, R.D., & Monsell, S. (1995). Cost of a predictable switch between simple cognitive tasks. *Journal of Experimental Psychology: General, 124,* 207–231.

Romaiguère, P., Hasbroucq, T., Possamaï, C.-A., & Seal, J. (1993). Intensity to force translation: A new effect of stimulus–response compatibility revealed by analysis of response time and electromyographic activity of a prime mover. *Cognitive Brain Research, 1,* 197–201.

Rosenbaum, D. A. (1980). Human movement initiation: Specification of arm, direction, and extent. *Journal of Experimental Psychology: General, 109,* 444–474.

Rosenbaum, D. A., Gordon, A. M., Stillings, N. A., & Feinstein, M. H. (1987). Stimulus–response compatibility in the programming of speech. *Memory & Cognition, 15,* 217–224.

Rosenbloom, P. S. (1986). The chunking of goal hierarchies: A model of stimulus–response compatibility and practice. In J. Laird, P. Rosenbloom, & A. Newell, *Universal subgoaling and chunking: The automatic generation and learning of goal hierarchies* (pp. 133–282). Boston: Kluwer.

Rosenbloom, P. S., & Newell, A. (1987). An integrated computational model of stimulus–response compatibility and practice. In G. H. Bower (Ed.), *The psychology of learning and motivation* (Vol. 21, pp. 1–52). San Diego: Academic Press.

Ross, S., Shepp, B. E., & Andrews, T. G. (1955). Response preferences in display–control relationships. *Journal of Applied Psychology, 39,* 425–428.

Roswarski, T. E., & Proctor, R. W. (1996). Multiple spatial codes and temporal overlap in choice–reaction tasks. *Psychological Research, 59,* 196–211.

Roswarski, T. E., & Proctor, R. W. (2000). Auditory stimulus–response compatibility: Is there a contribution of stimulus-hand correspondence? *Psychological Research, 63,* 148–158.

Roswarski, T. E., & Proctor, R. W. (2003). The role of instructions, practice, and stimulus-hand correspondence on the Simon effect. *Psychological Research, 67,* 43–55.

Rothermund, K., & Wentura, D. (2004). Underlying processes in the Implicit Association Test: Dissociating salience from associations. *Journal of Experimental Psychology: General, 133,* 139–165.

Rubichi, S., Nicoletti, R., Iani, C., & Umiltà, C. (1997). The Simon effect occurs relative to the direction of an attention shift. *Journal of Experimental Psychology: Human Perception and Performance, 23,* 1353–1364.

Rubichi, S., Nicoletti, R., Pelosi, A., & Umiltà, C. (2004). Right–left prevalence effect with horizontal and vertical effectors. *Perception & Psychophysics, 66,* 255–263.

Rubichi, S., Nicoletti, R., & Umiltà, C. (2005). Right–left prevalence with task irrelevant spatial codes. *Psychological Research, 69,* 167–178.

Rubichi, S., Vu, K.-P. L., Nicoletti, R., & Proctor, R. W. (in press). Spatial coding in two dimensions. *Psychonomic Bulletin & Review.*

Ruthruff, E., Johnston, J. C., & Van Selst, M. (2001). Why practice reduces dual-task interference. *Journal of Experimental Psychology: Human Perception and Performance, 27,* 3–21.

Ruthruff, E., Johnston, J. C., Van Selst, M., Whitsell, S., & Remington, R. (2003). Vanishing dual-task interference after practice: Has the bottleneck been eliminated or is it merely latent? *Journal of Experimental Psychology: Human Perception and Performance, 29,* 280–289.

Ruthruff, E., Van Selst, M., Johnston, J. C., & Remington, R. (in press). How does practice reduce dual-task interference: Integration, automation, or just stage-shortening? *Psychological Research.*

Sanders, A. F. (1998). *Elements of human performance*. Mahwah, NJ: Erlbaum.

Schlaghecken, F., & Eimer, M. (2004). Masked prime stimuli can bias "free" choices between response alternatives. *Psychonomic Bulletin & Review, 11*, 463–468.

Schroeder-Heister, P., Heister, G., & Ehrenstein, W. H. (1988). Spatial S-R compatibility under head tilt. *Acta Psychologica, 69*, 35–44.

Schubo, A., Aschersleben, G., & Prinz, W. (2001). Interactions between perception and action in a reaction task with overlapping S-R assignments. *Psychological Research, 65*, 145–157.

Schubo, A., Prinz, W., & Aschersleben, G. (2004). Perceiving while acting: Action affects perception. *Psychological Research, 68*, 208–215.

Schumacher, E. H., & D'Esposito, M. (2002). Neural implementation of response selection in humans as revealed by localized effects of stimulus–response compatibility on brain activation. *Human Brain Mapping, 17*, 193–201.

Schumacher, E. H., Lauber, E. J., Glass, J. M., Zurbriggen, E. L., Gmeindl, L., Kieras, D. E., & Meyer, D. E. (1999). Concurrent response–selection processes in dual-task performance: Evidence for adaptive executive control for task scheduling. *Journal of Experimental Psychology: Human Perception and Performance, 25*, 791–814.

Schumacher, E. H., Seymour, T. L., Glass, J. M., Fencsik, D. E., Lauber, E. J., Kieras, D. E., & Meyer, D. E. (2001). Virtually perfect timesharing in dual-task performance: Uncorking the central cognitive bottleneck. *Psychological Science, 12*, 101–108.

Schvaneveldt, R. W., & Chase, W. G. (1969). Sequential effects in choice–reaction time. *Journal of Experimental Psychology, 80*, 1–8.

Schweickert, R. (1983). Latent network theory: Scheduling of processes in sentence verification and in the Stroop effect. *Journal of Experimental Psychology: Learning, Memory, and Cognition, 9*, 353–383.

Schweizer, T. A., Jolicœur, P., Vogel-Sprott, M., & Dixon, M. J. (2004). Fast, but error-prone, responses during acute alcohol intoxication: Effects of stimulus–response mapping complexity. *Alcoholism: Clinical and Experimental Research, 28*, 643–649.

Seibel, R. (1963). Discrimination reaction time for a 1,023-alternative task. *Journal of Experimental Psychology, 66*, 215–226.

Seymour, P. H. K. (1974). Asymmetries in judgments of verticality. *Journal of Experimental Psychology, 102*, 447–455.

Shaffer, L. H. (1965). Choice reaction with variable S-R mapping. *Journal of Experimental Psychology, 70*, 284–288.

Shaffer, L. H. (1966). Some effects of partial advance information on choice reaction with fixed or variable S-R mapping. *Journal of Experimental Psychology, 72*, 541–545.

Shafir, E. (1995). Compatibility in cognition and decision. In J.R. Busemeyer, R. Hastie, & D.L. Medin (Eds.), *Decision making from the perspective of cognitive psychology: The psychology of learning and motivation* (Vol. 32, pp. 247—274) New York: Academic Press.

Shannon, C. E. (1948). A mathematical theory of communication. *The Bell System Technical Journal, 27*, 379–423, 623–656. (Available online at http://cm.bell-labs.com/cm/ms/what/shannonday/paper.html.)

Shin, Y. K., Cho, Y. S., Lien, M.-C., & Proctor, R. W. (2005). Is the psychological refractory period effect for ideomotor compatible tasks eliminated by speed-stress instructions? Manuscript submitted for publication.

Shinar, D., & Acton, M. B. (1978). Control–display relationships on the four-burner range: Population stereotypes versus standard. *Human Factors, 20*, 13–17.

Shiu, L.-P., & Kornblum, S. (1996). Negative priming and stimulus–response compatibility. *Psychonomic Bulletin & Review, 3*, 510–514.

Shulman, H. G., & McConkie, A. (1973). S-R compatibility, response discriminability, and response codes in choice–reaction time. *Journal of Experimental Psychology, 98*, 375–378.

Simon, C. W. (1954). The presences of a dual perceptual set for certain perceptual-motor tasks. USAF WADC Tech. Rep. No. 54–286.

Simon, J. R. (1967). Choice–reaction time as a function of auditory S-R correspondence, age, and sex. *Ergonomics, 10*, 659–664.

Simon, J. R. (1968). Effect of ear stimulated on reaction time and movement time. *Journal of Experimental Psychology, 78*, 344–346.

Simon, J. R. (1969). Reactions toward the source of stimulation. *Journal of Experimental Psychology, 81*, 174–176.

Simon, J. R. (1990). The effects of an irrelevant directional cue on human information processing. In R. W. Proctor & T. G. Reeve (Eds.), *Stimulus–response compatibility: An integrated perspective* (pp. 31–86). Amsterdam: North-Holland.

Simon, J. R., Acosta, E., & Mewaldt, S. P. (1975). Effect of locus of warning tone on auditory choice–reaction time. *Memory & Cognition, 3*, 167–170.

Simon, J. R., Acosta, E., Mewaldt, S. P., & Speidel, C. R. (1976). The effect of an irrelevant directional cue on choice–reaction time: Duration of the phenomenon and its relation to stages of processing. *Perception & Psychophysics, 19*, 16–22.

Simon, J. R., & Craft, J. L. (1970). Effects of an irrelevant auditory stimulus on visual choice–reaction time. *Journal of Experimental Psychology, 86*, 272–274.

Simon, J. R., Craft, J. L., & Small, A. M., Jr. (1970). Manipulating the strength of a stereotype: Interference effects in an auditory information-processing task. *Journal of Experimental Psychology, 86*, 63–68.

Simon, J. R., Hinrichs, J. V., & Craft, J. L. (1970). Auditory S-R compatibility: Reaction time as a function of ear–hand correspondence and ear-response-location correspondence. *Journal of Experimental Psychology, 86*, 97–102.

Simon, J. R., & Pouraghabagher, A. R. (1978). The effect of aging on the stages of processing in a choice–reaction time task. *Journal of Gerontology, 33*, 553–561.

Simon, J. R., & Rudell, A. P. (1967). Auditory S-R compatibility: The effect of an irrelevant cue on information processing. *Journal of Applied Psychology, 51*, 300–304.

Simon, J. R., Sly, P. E., & Vilapakkam, S. (1981). Effect of compatibility of S-R mapping on reaction toward the stimulus source. *Acta Psychologica, 47*, 63–81.

Simon, J. R., & Small, A. M., Jr. (1969). Processing auditory information: Interference from an irrelevant cue. *Journal of Applied Psychology, 53*, 433–435.

Simon, J. R., & Wolf, J. D. (1963). Choice–reaction time as a function of angular stimulus–response correspondence and age. *Ergonomics, 6*, 99–105.

Simpson, G. C., & Chan, W. L. (1988). The derivation of population stereotypes for mining machines and some reservations on the general applicability of published stereotypes. *Ergonomics, 31*, 327–335.

Sinclair, R. C., Mark, M. M., Moore, S. E., Lavis, C. A., & Soldat, A. S. (2000). Psychology: An electoral butterfly effect. *Nature, 408*, 665–666.

Slovic, P. (1975). Choice between equally valued alternatives. *Journal of Experimental Psychology: Human Perception and Performance, 1*, 280–287.

Small, A. M., Jr. (1990). Foreword. In R. W. Proctor & T. G. Reeve (Eds.), *Stimulus–response compatibility: An integrated perspective* (pp. v-vi). Amsterdam: North-Holland.

Smith, G. A. (1989). Strategies and procedures affecting the accuracy of reaction time parameters and their correlation with intelligence. *Personality and Individual Differences, 10,* 829–835.

Smith, S. L. (1981). Exploring compatibility with words and pictures. *Human Factors, 23,* 305–315.

Smith, W. J., & Cronin, D. T. (1993). Ergonomic test of the Kinesis keyboard. *Proceedings of the Human Factors and Ergonomics Society 37th Annual Meeting* (pp. 318–322). Santa Monica, CA: Human Factors and Ergonomics Society.

Smulders, F. T. Y., Kok, A., Kenemans, J. L., & Bashore, T. R. (1995). The temporal selectivity of additive factor effects on the reaction process revealed in ERP component latencies. *Acta Psychologica, 90,* 97–109.

Soetens, E. (1998). Localizing sequential effects in serial choice–reaction time with the information reduction procedure. *Journal of Experimental Psychology: Human Perception and Performance, 24,* 547–568.

Soetens, E., & Notebaert, W. (2005). Response monitoring and expectancy in random serial RT tasks. *Acta Psychologica, 119,* 189–216.

Sommer, W., Leuthold, H., & Hermanutz, M. (1993). Covert effects of alcohol revealed by event-related potentials. *Perception & Psychophysics, 54,* 127–135.

Stelmach, G. E., Goggin, N. L., & Amrhein, P. C. (1988). Aging and restructuring of precued movements. *Psychology and Aging, 3,* 151–157.

Stelmach, G. E., Goggin, N. L., & Garcia-Colera, A. (1987). Movement specification time with age. *Experimental Aging Research, 13,* 39–46.

Stevanovski, B., Oriet, C., & Joliecœur, P. (2002). Blinded by headlights. *Canadian Journal of Experimental Psychology, 56,* 65–74.

Stevanovski, B., Oriet, C., & Joliecœur, P. (2003). Can blindness to response-compatible stimuli be observed in the absence of a response? *Journal of Experimental Psychology: Human Perception and Performance, 29,* 431–440.

Stins, J. F., & Michaels, C. F. (1997). Stimulus–response compatibility is information-action compatibility. *Ecological Psychology, 9,* 25–45.

Stock, A., & Stock, C. (2004). A short history of ideomotor action. *Psychological Research, 68,* 176–188.

Stoffels, E. J. (1996a). On stage robustness and response selection routes: Further evidence. *Acta Psychologica, 91,* 67–88.

Stoffels, E. J. (1996b). Uncertainty and processing routes in the selection of a response: An S-R compatibility study. *Acta Psychologica, 94,* 227–252.

Stoffels, E.-J., Van der Molen, M. W., & Keuss, P. J. G. (1989). An additive factors analysis of the effect(s) of location cues associated with auditory stimuli on stages of information processing. *Acta Psychologica, 70,* 161–197.

Stoffer, T. H. (1991). Attentional focusing and spatial stimulus–response compatibility. *Psychological Research, 53,* 127–135.

Stoffer, T. H., & Umiltà, C. (1997). Spatial stimulus coding and the focus of attention in S-R compatibility and the Simon effect. In B. Hommel & W. Prinz (Eds.), *Theoretical issues in stimulus–response compatibility* (pp. 181–208). Amsterdam: North-Holland.

Stürmer, B., Aschersleben, G. & Prinz, W. (2000). Correspondence effects with manual gestures and postures: A study of imitation. *Journal of Experimental Psychology: Human Perception and Performance, 26,* 1746–1759.

Stürmer, B., Leuthold, H., Soetens, E., Schröter, H., & Sommer, W. (2002). Control over location-based response activation in the Simon task: Behavioral and electrophysiological evidence. *Journal of Experimental Psychology: Human Perception and Performance, 28,* 1345–1363.

Tagliabue, M., Zorzi, M., & Umiltà, C. (2002). Cross-modal remapping influences the Simon effect. *Memory & Cognition, 30*, 18–23.

Tagliabue, M., Zorzi, M., Umiltà, C., & Bassignani, F. (2000). The role of LTM links and STM links in the Simon effect. *Journal of Experimental Psychology: Human Perception and Performance, 26*, 648–670.

Teichner, W. H., & Krebs, M. J. (1974) Laws of visual-choice–reaction time. *Psychological Review, 81*, 75–98.

Telford, C. W. (1931). The refractory phase of voluntary and associative responses. *Journal of Experimental Psychology, 14*, 1–35.

ten Hoopen, G., Akerboom, S., & Raymakers, E. (1982). Vibrotactual choice–reaction time, tactile receptor systems, and ideomotor compatibility. *Acta Psychologica, 50*, 143–157.

Tipper, S. P., & Milliken, B. (1996). Distinguishing between inhibition-based and episodic retrieval-based accounts of negative priming. A. F. Kramer, M. G. H. Coles, & G. D. Logan (Eds.), *Converging operations in the study of visual selective attention* (pp. 337–363). Washington, DC: American Psychological Association.

Tlauka, M. (2004). Display–control compatibility: The relationship between performance and judgments of performance. *Ergonomics, 47*, 281–295.

Tlauka, M., & McKenna, F. P. (1998). Mental imagery yields stimulus–response compatibility. *Acta Psychologica, 98*, 67–79.

Toth, J. P., Levine, B., Stuss, D. T., Oh, A., Winocur, G., & Meiran, N. (1995). Dissociation of processes underlying spatial S-R compatibility: Evidence for the independent influence of what and where. *Consciousness and Cognition, 4*, 483–501.

Tucker, M., & Ellis, R. (1998). On the relations between seen objects and components of potential actions. *Journal of Experimental Psychology: Human Perception and Performance, 24*, 830–846.

Tucker, M., & Ellis, R. (2004). Action priming by briefly presented objects. *Acta Psychologica, 116*, 185–203.

Tversky, A., Slovic, P., & Kahneman, D. (1990). The causes of preference reversal. *American Economic Review, 80*, 204–217.

Umiltà, C. (1991). Problems of the salient-feature coding hypothesis: Comment on Weeks and Proctor (1990). *Journal of Experimental Psychology: General, 120*, 83–86.

Umiltà, C., & Liotti, M. (1987). Egocentric and relative spatial codes in S-R compatibility. *Psychological Research, 49*, 81–90.

Umiltà, C., & Nicoletti, R. (1985). Attention and coding effects in S-R compatibility due to irrelevant spatial cues. In M. I. Posner & O. S. M. Marin (Eds.), *Attention and performance XI* (pp. 457–471). Hillsdale, NJ: Erlbaum.

Umiltà, C., & Nicoletti, R. (1990). Spatial stimulus–response compatibility. In R. W. Proctor & T. G. Reeve (Eds.), *Stimulus–response compatibility: An integrated perspective* (pp. 89–143). Amsterdam: North-Holland.

Umiltà, C., & Nicoletti, R. (1992). An integrate model of the Simon effect. In J. Alegria, D. Holender, J. Junça de Morais, & M. Radeau (Eds.), *Analytic approaches to human cognition* (pp. 331–350). Amsterdam: Elsevier.

Urcuioli, P. J., Vu, K.-P. L., & Proctor, R. W. (2005). A Simon effect in pigeons. *Journal of Experimental Psychology: General, 134*, 93–107.

Usher, M., & McClelland, J. L. (2001). The time course of perceptual choice: The leaky competing accumulator model. *Psychological Review, 108*, 550–592.

Usher, M., Olami, Z., & McClelland, J. L. (2002). Hick's law in a stochastic race model with speed-accuracy tradeoff. *Journal of Mathematical Psychology, 46*, 704–715.

Valle-Inclán, F., de Labra, C., & Redondo, M. (2000). Psychophysiological studies of unattended information processing. *The Spanish Journal of Psychology, 3*, 76–85.

Valle-Inclán, F., Hackley, S. A., & de Labra, C. (2002). Attention and response activation in the Simon effect. In W. Prinz & B. Hommel (Eds.), *Attention and performance XIX: Common mechanisms in perception and action* (pp. 474–493). Oxford: Oxford University Press.

Valle-Inclán, F., Hackley, S. A., & de Labra, C. (2003). Stimulus–response compatibility between stimulated eye and response location: Implications for attentional accounts of the Simon effect. *Psychological Research/Psychologische Forschung, 67*, 240–243.

Valle-Inclán, F., & Redondo, M. (1998). On the automaticity of ipsilateral response activation in the Simon effect. *Psychophysiology, 35*, 366–371.

Vallesi, A., Mapelli, D., Schiff, S., Amodio, P., & Umiltà, C. (2005). Horizontal and vertical Simon effect: Different underlying mechanisms? *Cognition, 96*, B33–B43.

Van Duren, L., & Sanders, A. F. (1988). On the robustness of the additive factors stage structure in blocked and mixed choice–reaction designs. *Acta Psychologica, 69*, 83–94.

Van Selst, M., Ruthruff, E., & Johnston, J. C. (1999). Can practice eliminate the psychological refractory period effect? *Journal of Experimental Psychology: Human Perception and Performance, 25*, 1268–1283.

Van Zandt, T., Colonius, H., & Proctor, R. W. (2000). A comparison of two response-time models applied to perceptual matching. *Psychonomic Bulletin & Review, 7*, 208–256.

Verfaellie, M., Bowers, D., & Heilman, K. M. (1988). Attentional factors in the occurrence of stimulus–response compatibility effects. *Neuropsychologia, 26*, 435–444.

Vickrey, C., & Neuringer, A. (2000). Pigeon reaction time, Hick's law, and intelligence. *Psychonomic Bulletin & Review, 7*, 284–291.

Vidulich, M. A., & Wickens, C. D. (1985). Stimulus–central processing–response compatibility: Guidelines for the optimal use of speech technology. *Behavior Research Methods, Instruments, & Computers, 17*, 243–249.

Vince, M. A. (1945). Direction of movement of machine controls. I. Medical Research Council. F.P.R.C. 637.

Vince, M. A. (1947). The intermittency of control movements and the psychological refractory period. *British Journal of Psychology, 38*, 149–157.

Vince, M. A. (1950). Learning and retention of an "unexpected" control–display relationship. Medical Research Council. A.P.U. 125/50.

Vince, M. A., & Mitchell, M. J. H. (1946). Direction of movement of machine controls. II. Ministry of Supply Report No. SM 2861.

Virzi, R. A., & Egeth, H. E. (1985). Toward a translational model of Stroop interference. *Memory & Cognition, 13*, 304–319.

Vu, K.-P. L. (2005). Influences on the Simon effect of prior practice with spatially incompatible mappings: Transfer within and between spatial dimensions. Manuscript submitted for publication.

Vu, K.-P. L., Pellicano, A., & Proctor, R. W. (2005). No overall right–left prevalence for horizontal and vertical Simon effects. *Perception & Psychophysics, 67*, 929–938.

Vu, K.-P. L., & Proctor, R. W. (2001a). Determinants of right–left and top–bottom prevalence for two-dimensional spatial compatibility. *Journal of Experimental Psychology: Human Perception and Performance, 27*, 813–828.

Vu, K.-P. L., & Proctor, R. W. (2001b). Stimulus–response compatibility in interface design. In M. J. Smith, G. Salvendy, D. Harris, & R. J. Koubek (Eds.), *Usability evaluation and interface design: Cognitive engineering, intelligent agents, and virtual reality* (Vol. 1, pp. 1368–1372). Mahwah, NJ: Erlbaum.

Vu, K.-P. L., & Proctor, R. W. (2002). The prevalence effect in two-dimensional S-R compatibility is a function of the relative salience of the dimensions. *Perception & Psychophysics, 64*, 815–828.

Vu, K.-P. L, & Proctor, R. W. (2003). Naïve and experienced judgments of stimulus–response compatibility: Implications for interface design. *Ergonomics, 46*, 169–187.

Vu, K.-P. L, & Proctor, R. W. (2004). Mixing compatible and incompatible mappings: Elimination, reduction, and enhancement of spatial compatibility effects. *Quarterly Journal of Experimental Psychology, 57A*, 539–556.

Vu, K.-P. L., & Proctor, R. W. (2006). *Mixing compatible and incompatible mappings: Influence of bias on spatial compatibility effects.* Manuscript in preparation.

Vu, K.-P. L., & Proctor, R. W. (in press). Emergent perceptual features in the benefit of consistent stimulus–response mappings on dual-task performance. *Psychological Research.*

Vu, K.-P. L., Proctor, R. W., & Pick, D. F. (2000). Horizontal versus vertical compatibility: Right–left prevalence with bimanual responses. *Psychological Research, 64*, 25–40.

Vu, K.-P. L, Proctor, R. W., & Urcuioli, P. (2003). Transfer effects of incompatible location-relevant mappings on subsequent visual or auditory Simon tasks. *Memory & Cognition, 31*, 1146–1152.

Walker, B. N., & Ehrenstein, A. (2000). Pitch and pitch change interact in auditory displays. *Journal of Experimental Psychology: Applied, 6*, 15–30.

Wallace, R. J. (1971). S-R compatibility and the idea of a response code. *Journal of Experimental Psychology, 88*, 354–360.

Wallace, R. J. (1972). Spatial S-R compatibility effects involving kinesthetic cues. *Journal of Experimental Psychology, 93*, 163–168.

Wang, D.-Y. D., Proctor, R. W., & Pick, D. F. (2003). The Simon effect with wheel-rotation responses. *Journal of Motor Behavior, 35*, 261–273.

Wang, D.-Y. D., Proctor, R. W., & Pick, D. F. (2006). *Factors that determine the direction and size of the Simon effect with wheel-rotation responses.* Manuscript submitted for publication.

Wang, H., & Proctor, R. W. (1996). Stimulus–response compatibility as a function of stimulus code and response modality. *Journal of Experimental Psychology: Human Perception and Performance, 22*, 1201–1217.

Warrick, M. J. (1947). Direction of movement in the use of control knobs to position visual indicators. USAF Air Materiel Command Report No. 694–4C.

Warrick, M. J. (1948). Direction of motion stereotypes in positioning a visual indicator by use of a control knob. II. Results from a printed test. USAF Air Materiel Command Report MCREXD-694–19A.

Warrick, M. J. (1949). Effects of motion relationships on speed of positioning visual indicators by rotary control knobs. USAF Air Materiel Command Report No. 5182.

Wascher, E., Schatz, U., Kuder, T., & Verleger, R. (2001). Validity and boundary conditions of automatic response activation in the Simon task. *Journal of Experimental Psychology: Human Perception and Performance, 27*, 731–751.

Way, T. C., & Gottsdanker, R. (1968). Psychological refractoriness with varying differences between tasks. *Journal of Experimental Psychology, 78*, 38–45.

Weeks, D. J., Chua, R., & Hamblin, K. (1995). Attention shifts and the Simon effect: A failure to replicate Stoffer (1991). *Psychological Research, 58*, 246–253.

Weeks, D. J., & Proctor, R. W. (1990). Salient-features coding in the translation between orthogonal stimulus–response dimensions. *Journal of Experimental Psychology: General, 119*, 355–366.

Weeks, D. J., Proctor, R. W., & Beyak, B. (1995). Stimulus–response compatibility for vertically oriented stimuli and horizontally oriented responses: Evidence for spatial coding. *Quarterly Journal of Experimental Psychology, 48A*, 367–383.

Welford, A. T. (1952). The "psychological refractory period" and the timing of high-speed performance — A review and a theory. *British Journal of Psychology, 43*, 2–19.

Welford, A. T. (1968). *Fundamentals of skill.* New York: Methuen.

Welford, A. T. (1976). *Skilled performance: Perceptual and motor skills.* Glenview, IL: Scott, Foresman.

Welford, A. T. (1987). Comment on "Exceptions to Hick's law: Explorations with a response duration measure" (Longstreth, El-Zahhar, & Alcorn, 1985). *Journal of Experimental Psychology: General, 116*, 312–314.

Werheid, K., Ziessler, M., Nattkemper, D., & von Cramon, D. (2003). Sequence learning in Parkinson's disease: The effect of spatial stimulus–response compatibility. *Brain and Cognition, 52*, 239–249.

Wickens, C. D. (1984). *Engineering psychology and human performance.* Columbus, OH: Merrill.

Wickens, C. D., & Andre, A. D. (1990). Proximity compatibility and information display: Effects of color, space, and objectness on information integration. *Human Factors, 32*, 61–77.

Wickens, C. D., & Carswell, C. M. (1995). The proximity compatibility principle: Its psychological foundation and relevance to display design. *Human Factors, 37*, 473–494.

Wickens, C. D., Sandry, D. L., & Vidulich, M. (1983). Compatibility and resource competition between modalities of input, central processing, and output. *Human Factors, 25*, 227–248.

Wickens, C. D., Vidulich, M., & Sandry-Garza, D. L. (1984). Principles of S-C-R compatibility with spatial and verbal tasks: The role of display–control location and voice interactive display–control interfacing. *Human Factors, 26*, 533–543.

Wickens, D. D. (1938). The transference of conditioned excitation and conditioned inhibition from one muscle group to the antagonistic muscle group. *Journal of Experimental Psychology, 22*, 101–123.

Wickens, D. D. (1943). Studies of response generalization in conditioning: II. The comparative strength of the transferred and nontransferred responses. *Journal of Experimental Psychology. 33*, 330–332.

Wiegand, K., & Wascher, E. (2005). Dynamic aspects of S-R correspondence: Evidence for two mechanisms involved in the Simon effect. *Journal of Experimental Psychology: Human Perception and Performance, 31*, 453–464.

Wiegand, K., & Wascher, E. (in press). The Simon effect for vertical S-R relations: Changing the mechanism by randomly varying the S-R mapping rule? *Psychological Research.*

Wilkinson, L., & Shanks, D. R. (2004). Intentional control and implicit sequence learning. *Journal of Experimental Psychology: Learning, Memory, & Cognition, 30*, 354–369.

Wong, C. K., & Lyman, J. (1988). American and Japanese control–display stereotypes: Possible implications for design of space station systems. *Proceedings of the Human Factors Society 32nd Annual Meeting* (pp. 30–34). Santa Monica, CA: Human Factors Society.

Woodworth, R. S. (1938). *Experimental psychology.* New York: Holt.

Worringham, C. J., & Beringer, D. B. (1989). Operator orientation and compatibility in visual-motor task performance. *Ergonomics, 32*, 387–399.

Worringham, C. J., & Beringer, D. B. (1998). Directional stimulus–response compatibility: A test of three alternative principles. *Ergonomics, 41*, 864–880.

Wu, S. P. (1997). Further studies on the spatial compatibility of four control–display linkages. *International Journal of Industrial Ergonomics, 19*, 353–360.

Wühr, P. (2004). Sequential modulations of logical-recoding operations in the Simon task. *Experimental Psychology, 51*, 98–108.

Wühr, P. (2005). Evidence for gating of direct response activation in the Simon task. *Psychonomic Bulletin & Review, 12*, 282–288.

Wühr, P., & Ansorge, U. (2005). Exploring trial-by-trial modulations of the Simon effect. *Quarterly Journal of Experimental Psychology, 58A*, 705–731.

Wühr, P., & Müsseler, J. (2001). Time course of the blindness to response-compatible stimuli. *Journal of Experimental Psychology: Human Perception and Performance, 27*, 1260–1270.

Xun, X., Guo, S., & Zhang, K. (1998). The effect of inputting modality of secondary task on tracking performance and mental workload. *Acta Psychologica Sinica, 30*, 343–347.

Zhang, J., & Kornblum, S. (1997). Distributional analysis and De Jong, Liang, and Lauber's (1994) dual-process model of the Simon effect. *Journal of Experimental Psychology: Human Perception and Performance, 23*, 1543–1551.

Zhang, H., & Kornblum, S. (1998). The effects of stimulus–response mapping and irrelevant stimulus–response and stimulus–stimulus overlap in four-choice Stroop tasks with single-carrier stimuli. *Journal of Experimental Psychology: Human Perception and Performance, 24*, 3–19.

Zhang, H., Zhang, J., & Kornblum, S. (1999). A parallel distributed processing model of stimulus–stimulus and stimulus–response compatibility. *Cognitive Psychology, 38*, 386–432.

Zimba, L. D., & Brito, C. F. (1995). Attention precuing and Simon effects: A test of the attention-coding account of the Simon effect. *Psychological Research, 58*, 102–118.

Zorzi, M., & Umiltà, C. (1995). A computational model of the Simon effect. *Psychological Research, 58*, 193–205.

Author Index

A

Acosta, E., 83, 123
Acton, M. B., 299–301
Adam, J. J., 35–36, 38, 50, 174, 176–179,
    184–186, 193, 257, 259
Aglioti, S., 53
Akerboom, S., 65
Alcorn, M. B., 30
Allen, P. A., 242
Allport, A., 146, 164
Alluisi, E. A., 27–28
Amodio, P., 102
Amrhein, P. C., 33
Anderson, J. R., 248
Anderson, R. E., 85
Andre, A. D., 3, 171, 192–193, 281, 290, 303–304
Andrews, T. G., 199–200
Ansorge, U., 85, 112, 135
Anzola, G. P., 53, 105
Archer, W., 97–98
Arend, U., 122–123
Aschersleben, G., 21–22, 116, 269, 271–272
Atkins, P., 1
Atkins, S. G., 70
Azuma, R., 238

B

Bächtold, D., 68–69
Baeyens, F., 94
Bahrick, H. P., 15, 290
Bailey, R. W., 303
Baldo, J. V., 92
Barber, P., 20, 92, 129
Barnes, V. E., 3, 279
Bashore, T. R., 13
Bassignani, F., 139, 260
Bathhurst, K., 29
Bauer, D. W., 172–174, 179–180, 182
Baumüller, M., 68
Baylis, G., 39–40
Beckers, T., 143, 161
Beh, H. C., 29
Bekkering, H., 66, 93
Beller, H. K., 63
Bennett, K. B., 283, 285–286
Bergum, B. O., 195–196, 218–220
Bergum, J. E., 195–196, 218–220

C

Beringer, D. B., 279, 291
Berlucchi, G., 53
Berson, B. L., 303
Bertelson, P., 39, 171
Bertolini, G., 53
Beyak, B., 38, 180–181
Biederman, I., 47, 97–98
Biel, G. A., 192
Blanchard, H. E., 299
Boon, B., 176
Boroughs, J. M., 90
Bosbach, S., 93
Bossini, S., 191
Bourne, L. E., Jr., 63
Bowers, D., 89
Brass, M., 66
Brebner, J., 51, 61, 123, 204–207, 209, 211
Briggs, G. E., 15, 290
Brito, C. F., 89
Broadbent, D. E., 50, 196
Brown, S., 42
Brugger, P., 68
Buchtel, H. A., 53
Bullemer, P., 43
Burle, B., 35
Byrne, M. D., 248

C

Cairney, P., 51
Camacho, M. J., 303
Campbell, J. A., 81, 116
Campbell, K. C., 37, 40, 192
Canham, L., 90
Card, S. K., 252
Carnahan, H., 33
Carswell, C. M., 192, 281–284
Cataldo, G., 299
Cave, C., 75
Chan, A. H. S., 212, 217–219
Chan, W. L., 203–204
Chapanis, A., 299–302, 304
Chase, W. G., 39, 174–175, 190
Cho, Y. S., 175–178, 181, 186–191, 193, 243, 259,
    274
Chorev, Z., 108
Christensen, C. A., 24
Chua, R., 33, 38, 89, 275, 291

Subject Index